The Sciences' Media Connection –
Public Communication and its Repercussions

Sociology of the Sciences Yearbook

VOLUME XXVIII

For further volumes:
http//www.springer.com/series/6566

Simone Rödder · Martina Franzen · Peter Weingart
Editors

The Sciences' Media Connection – Public Communication and its Repercussions

Springer

Editors
Simone Rödder
Institute for Science and Technology
Studies (IWT)
Bielefeld University
Germany
sir@uni-bielefeld.de

Martina Franzen
Institute for Science and Technology
Studies (IWT)
Bielefeld University
Germany
martina.franzen@uni-bielefeld.de

Peter Weingart
Institute for Science and Technology
Studies (IWT)
Bielefeld University
Germany
weingart@uni-bielefeld.de

ISSN 0167-2320
ISBN 978-94-007-9248-7 ISBN 978-94-007-2085-5 (eBook)
DOI 10.1007/978-94-007-2085-5
Springer Dordrecht Heidelberg London New York

Printed on acid-free paper

Springer is part of Springer Science+Business Media (www.springer.com)

Preface

It is no longer 'breaking news' that science, scientists, research practices and results have become news items in the popular press. At the same time the researchers in universities and laboratories are confronted, either directly or indirectly, with the expectation to 'report about their activities to the public.' These expectations are raised in modern societies whose citizenry as a whole now claims the right to participate in decisions relevant for their own fate. There can be no doubt that the academic world which – at least in the public's mind – had been insulated from the 'real world,' residing in the ivory tower of self-referential intellectual pursuit and communication, has reacted to these expectations. Thus, the 'opening up' of science to the media is associated with the hope for its 'democratization.'

The interactions between science and the public assume many new institutional arrangements and involve different parties with diverging objectives. Inevitably, conflicts between norms, interests and styles of practice arise. Thus, the more cautious observers point to possible negative effects of the orientation of science to the media on the pursuit of knowledge. The present Yearbook addresses the overriding question: What are the effects of the interaction between these two very different worlds on science itself? Theoretical considerations and a host of empirical studies covering different configurations provide an in-depth analysis of the sciences' media connection and help to form a sound judgement on this recent development.

Preface

Acknowledgements

The Yearbook Sociology of the Sciences is a fully peer reviewed publication. The concept of the book is first reviewed by members of the editorial board. The articles are solicited by the editors. First drafts are reviewed first by the editors and, if accepted as pertinent to the topic, subsequently by two external reviewers. In this volume, the practitioners' perspectives are an exception in that they have only been editorially reviewed. In addition, a conference is held to allow the authors to interact and profit from further critique and suggestions.

We thank Everett Mendelsohn, Richard Whitley and Stephen Hilgartner for critical suggestions on the book plan at the early stage. We further thank the Center for Interdisciplinary Research (ZiF) at Bielefeld University for funding and hosting the conference which took place May 6–7, 2010. Sincere thanks are also extended to the anonymous reviewers who have a great share in shaping the ultimate product, to Marc Weingart for his meticulous language editing and to Lilo Jegerlehner for handling the manuscript professionally giving it its final appearance.

June 2011

Simone Rödder
Martina Franzen
Peter Weingart

Contents

Contributors

Franziska Badenschier, M.A., Institute and School of Journalism, TU Dortmund University, D-44221 Dortmund, Germany, franziska.badenschier@udo.edu

Martin W. Bauer, Prof., Institute of Social Psychology and Methodology Institute, London School of Economics, Houghton Street, London WC2A 2AE, UK, m.bauer@lse.ac.uk

Richard Elliott, MSc., The Institute for Science and Society, University of Nottingham, Law and Social Sciences Building, University Park, Nottingham NG7 2RD, UK, LBXRE@nottingham.ac.uk

Ulrike Felt, Univ.-Prof. Dr., Department of Social Studies of Science, University of Vienna, Sensengasse 8/10, A-1010 Wien, Austria, ulrike.felt@univie.ac.at

Maximilian Fochler, Dr., Department of Social Studies of Science, University of Vienna, Sensengasse 8/10, A-1010 Wien, Austria, maximilian.fochler@univie.ac.at

Fiona Fox, Director, Science Media Centre, 215 Euston Road, London, NW1 2BE, UK, fiona@sciencemediacentre.org

Martina Franzen, Dr., Institute for Science and Technology Studies (IWT), Bielefeld University, P.O. Box 100131, D-33501 Bielefeld, Germany, martina.franzen@uni-bielefeld.de

Liza Gross, Senior Science Writer/Editor, PLoS Biology, Public Library of Science, 1160 Battery Street, Suite 100, San Francisco, CA 94111, USA, lgross@plos.org

Joan Haran, Dr., ESRC Centre for Economic and Social Aspects of Genomics (Cesagen), Cardiff University, 6 Museum Place, Cardiff CF10 3BG, UK, HaranJ@Cardiff.ac.uk

Stephen Hilgartner, Assoc. Prof., Department of Science and Technology Studies, Cornell University, 306 Rockefeller Hall, Ithaca, NY 14853, USA, shh6@cornell.edu

Arlena Jung, Dr., Social Science Research Center Berlin (WZB), Berlin, Germany, arlena.jung@wzb.eu

Paul Nolte, Univ.-Prof. Dr., Friedrich-Meinecke-Institut, Free University Berlin, Koserstraße 20, D-14195 Berlin, Germany, paul.nolte@fu-berlin.de

Fran Osrecki, Dr., Fachbereich 1: Sozialwissenschaften, Osnabrück University, Seminarstraße 33, D-49074 Osnabrück, Germany, fran.osrecki@uni-osnabrueck.de

Hans Peter Peters, Prof. Dr., Institute of Neuroscience and Medicine: Ethics in the Neurosciences (INM-8), Research Center Jülich, D-52425 Jülich, Germany, h.p.peters@fz-juelich.de

Simone Rödder, Dr., Institute for Science and Technology Studies (IWT), Bielefeld University, P.O. Box 100131, D-33501 Bielefeld, Germany, sir@uni-bielefeld.de

Brian Trench, M.A., School of Communications, Dublin City University, Dublin 9, Ireland, brian.trench@dcu.ie

Susann Wagenknecht, M.A., Department of Science Studies, Aarhus University, C.F. Møllers Allé, build. 1110, DK-8000 Aarhus C, Denmark, su.wagen@ivs.au.dk

Peter Weingart, Univ.-Prof. em. Dr., Institute for Science and Technology Studies (IWT), Bielefeld University, P.O. Box 100131, D-33501 Bielefeld, Germany, weingart@uni-bielefeld.de

Holger Wormer, Univ.-Prof., Chair of Science Journalism, Institute and School of Journalism, TU Dortmund University, D-44221 Dortmund, Germany, holger.wormer@udo.edu

Part I
Introduction

Chapter 1
Exploring the Impact of Science Communication on Scientific Knowledge Production: An Introduction

Martina Franzen, Peter Weingart, and Simone Rödder

1.1 Science in the News

Dolly and Ida – the Edinburgh sheep and the primate fossil from the Messel Pit in Germany are only two examples for the phenomenon that scientific achievements nowadays come with names and faces, readily available to be pictured and portrayed in TV news and on print media pages. Claims of achievements even in the absence of a journal publication have taken center stage in a number of cases. To name but a few classics: Cold Fusion where scientists who wanted to replicate the results had to rely on media information (Lewenstein 1995; Bucchi 1998), and the Human Genome Project where a "PR version" of the genome sequence has been presented in a White House ceremony (Nerlich et al. 2002; Rödder 2009a). The latest "breakthrough" – hitting the headlines in spring 2009 – was an unusually complete Eocene primate fossil. In time with the publication of a paper in the open access journal *PLoS ONE* (Franzen et al. 2009), the presentation of the fossil was staged in the New York Museum of National History, featuring oversized pictures of the fossil.

The presentation made global news, but especially in print media and blogs, there was hardly any article without a critical note. In interviews, scientists repeatedly faced the question why they had staged an event "as glamorous as it is typical for the worlds of art and fashion" (*Süddeutsche Zeitung*, May 21, 2009) – implying that this is not the usual way of presenting a research result. The scientists involved pointed to the need for refinancing the fossil, which had been bought from a private collection for several hundred thousand Euros, and insisted that there was nothing wrong with promoting research that has been published in a scientific journal. Some colleagues, however, were quoted with critical comments: "It's not how I like to do science" (*The Australian*, May 21, 2009) and "The PR campaign on this fossil is I think more of a story than the fossil itself" (*LiveScience*, May 20, 2009). Interestingly,

M. Franzen (✉)
Bielefeld University
e-mail: martina.franzen@uni-bielefeld.de

S. Rödder et al. (eds.), *The Sciences' Media Connection – Public Communication and its Repercussions*, Sociology of the Sciences Yearbook 28,
DOI 10.1007/978-94-007-2085-5_1, © Springer Science+Business Media B.V. 2012

the media frenzy was also rather critically commented on in the high-profile journal *Nature*. An editorial pointed to "potential dangers of publicity machines" introduced by "conflicting incentives that all too easily undermine the process of the assessment and communication of science" (*Nature* 2009: 484); this all despite the fact that *Nature* as well as its counterpart *Science* promise their authors a maximum exposure of their published work by their professional press relation efforts and are often themselves involved in the craze for publicity (Franzen 2009).

Irrespective of the underlying motives, the intense reactions to media releases *prior to* or *accompanying a scientific publication* indicate a problem inherent in the process of science communication. That scientists are criticized by their colleagues for making scientific claims that have not (yet) been peer reviewed is characteristic, and even in time with a publication, highly sensational announcements are considered inappropriate (see Weingart and Pansegrau 1999; Rödder 2009b). To prevent the pre-publication of results in the public press, the scholarly journals have established a strict embargo policy.[1] These reactions point towards a systematic tension between the course of research communication within the peers' circles and a widening of the communication process towards the general public via media channels.

Starting point for our analysis are trends in the relationship between science and the mass media that are associated with the relevance of both scientific knowledge and media communication for modern societies. While sensational mass media coverage makes public dialogue unsatisfactory at times, this is better than no coverage at all – a point of view sometimes expressed by science policy-makers (e.g., House of Lords 2000), by press offices of research institutes (Peters et al. 2008) as well as by scientists (Rödder 2009b). In some cases, media attention to scientific communication appears welcome, as in the case of climate change where the mobilization of widespread attitude change among the public depends on it. In any case, the surge of media reporting about science is a phenomenon that deserves systematic study. Is the growing intensity of mass media coverage (e.g., Bauer 1995; Nelkin 1995; Elmer et al. 2008) an indication of science's orientation towards the standards and formats of the media? Or more generally: what are, if any, the consequences of intense media attention for science?

1.2 The Concept of Medialization

We use the concept of "medialization" to refer to the mutual relation between science and the mass media. It is based on the assumption that – due to the importance of the mass media in framing public opinion – there is an increasingly tighter coupling of science and the mass media. Our understanding of mass media includes all media which provide observations of society, i.e., which are processing such

[1] The embargo system, however, has been criticized for being a control instrument in science communication which serves to maximize the publicity that a journal gets from the popular media (Kiernan 1997).

observations in the framework of an organization (e.g., news departments) for a mass audience. The most obvious ones are newspapers, radio, television and web-based news media. Their organization and their addressing large unspecific audiences constitute particular selection criteria, the so-called news values, such as scandals, human interest, celebrity, etc. Thus, the term medialization has to be distinguished from the much broader concept of "mediatization." The notion of mediatization is often used to describe mediated communication processes and the role of technical media from mobile phones to online games in everyday life and society at large as an indicator for social change (e.g., Krotz 2001; Schulz 2004; and the contributions in Lundby 2009). While the interrelations of all these media with individual daily lives and institutional practices are relevant topics in media and communication studies, there is a point in distinguishing media in general (understood as technologies that allow for mediated communication) and mass media understood as a social system with its own logic based on specific selection processes of what becomes news; notwithstanding that mass media products differ in their underlying editorial programs.

As we will argue, these mass media form an important resource for the public legitimacy of science in modern mass democracies. Media attention for scientists, research institutes and scientific journals becomes crucial for public support. Under these circumstances it appears likely that the "media gain an indirect influence on or compete with the self-steering mechanisms of science" (Weingart 1998: 878). The staging of media events, the pre-publication of research results in the mass media prior to their scientific publication, the media orientation of scholarly journals and the appearance of visible scientists have been described as empirical indicators for an orientation of science towards the rationalities of the media system (Weingart 1998, 2001; Franzen 2009, 2011; Rödder 2009a). Medialization as an analytical concept refers to different degrees of orientation to the selection criteria of mass media organizations.

The central question is if this orientation towards the media by incorporating mass media-related criteria of relevance into communication strategies has an impact on the specific societal function of science, i.e., the production of new and reliable knowledge. Can an orientation of science to the mass media remain limited to activities on the front stage produced just for public view or does it extend to the backstage, thus affecting the criteria of relevance in knowledge production?

Before we put forward our argument in more detail, the broader context in which these developments take place needs some consideration. The urgency of looking at changes with regard to the institutional and epistemic characteristics of science is enhanced by the fact that political demands on scientists and scientific organizations to widen the scope of communication have recently been intensified.

1.3 The View from Science Policy

Public debates of science have been a matter of political attention for a number of years, albeit with oscillating interest in certain phases and in different countries. Felt et al. (1995: 245) have argued that resources for science communication programs are granted if, and only if, policy-makers see a need to regain public trust in times when the relationship of science and society is perceived to be hit by what is often called "crises of confidence" (House of Lords 2000). Between 2007 and 2013, the European Union spends 280 million Euros for the *Science in Society* part of its seventh Framework program while the US "Scientific Communication Act of 2007" foresees a National Science Foundation investment of 10 million US Dollars per year from 2008 to 2012 to train scientists in media communication (see Greco 2007).

Science communication in terms of public understanding of science was set up as a policy goal in 1985 when the British Royal Society published a groundbreaking report entitled "The Public Understanding of Science." The report encompassed the results of inquiries into the relationship of science and society and a set of recommendations, addressing the scientific community, the government, the industry, the education system and the mass media (The Royal Society 1985). The improvement of science communication was established as a way to enhance the quality of a nation as a location for industry in the so-called "knowledge society" (Bell 1973; Stehr 1994) and was seen as a major element "in raising the quality of public and private decision-making and in enriching the life of the individual" (The Royal Society 1985: 9). The underlying assumption was that there is a need for an overall awareness of the nature of science and the way science and technology pervade modern life. The report pictured science and technology as main factors for the prosperity of the British economy as well as playing a major role in most aspects of an individual's daily life, both at home and at work. It concludes that for personal decision-making, everybody needs some understanding of science, its accomplishments and its limitations.[2] While the report addresses a variety of audiences, its first audience were the scientists themselves: "Improving the general level of public understanding of science is now an urgent task for the well-being of the country, requiring concerted action from many sections of society including, most importantly, the scientific community itself" (ibid.: 6).

In Germany, science policy reacted somewhat later. A 1999 science policy document entitled "PUSH-Memorandum," signed by all major German research organizations, declares the legitimation of research funding, and securing public confidence as the main objectives of the campaign it initiated (Stifterverband für die Deutsche Wissenschaft 1999). Again, the scientific community was regarded as the memorandum's prime audience.

In 2000, the House of Lords followed the Royal Society report up with a review of the UK situation. The Select Committee on Science and Technology Third Report

[2] The term "science" includes mathematics, technology, engineering and medicine as well as the natural sciences. The non-scientific "public" is divided into five categories: (i) private individuals for their personal satisfaction and well-being; (ii) individual citizens for participation in civic responsibilities as members of a democratic society; (iii) people employed in skilled and semi-skilled occupations, the large majority of which now have some scientific content; (iv) people employed in the middle ranks of management and in professional and trades union associations; (v) people responsible for major decision-making in our society, particularly those in industry and government (The Royal Society 1985: 7).

stated a current lack of trust, induced by incidences such as the BSE crisis and GM food. Despite 15 years of public understanding of science efforts, the public was "uneasy about the rapid advance of areas such as biotechnology and IT" (House of Lords 2000: 2). Social science studies provided evidence that the assumption that informed most campaigns – namely, that enhancing understanding in terms of scientific literacy leads to more understanding in terms of acceptance – was empirically false (Evans and Durant 1995; Bucchi and Neresini 2002; for a critique of this so-called deficit model, see Hilgartner 1990; Irwin and Wynne 1996). In the report, the persistent mismatch between society's values and science's progress was seen as a major threat and produced a "call for increased and integrated dialogue" (House of Lords: 4). But the move from "Public Understanding of Science" to "Public Engagement with Science and Technology" – from "PUS to PEST," as the journal *Science* commented (*Science* 2002) – seemed to concern the means of communications rather than the policy goals, as it was intended "to *secure* science's 'licence to practice,' not to *restrict* it" (ibid., italics in original).

While the rhetoric of the political programs that address the relationship of science and society has evolved from public understanding of science to public understanding of research (see Nowotny 2005) to public engagement and public participation, the underlying policy goals seem to have hardly changed. The interest of science policy-makers in funding public understanding of science research and practice can be summarized as follows:

- To enhance *scientific literacy* as an essential requirement for everyday life and as a prerequisite for tax payers willingness to fund scientific research,
- To equip all with *science for citizenship*, i.e., *public participation* for improved public and private decision-making,
- To legitimate *public research funding*,
- To provide *legitimation to (the mainstream) views* in science- and technology-intensive policy controversies.

1.4 Roots: Studies of Science and the Mass Media

The relationship between science and the mass media has attracted the attention of sociologists of science and communication scholars alike. The intellectual interest in this particular coupling is motivated by the difference between scientific knowledge as an esoteric, elitist and partly inaccessible knowledge system, on the one hand, and everyday knowledge shared by the public at large, on the other. This difference is seen – and evaluated – from different perspectives at different times. Some scholars see it as an analytical problem: e.g., how can the efficacy of scientific knowledge production be maintained if the highly specialized and condensed communication is broadened to the general public? How is the scientific communication process affected by interventions from the public that does not share the same level of competence (Bucchi 1996)? Other scholars are concerned with the political implications of the perceived hierarchy of specialized and popular knowledge in democratic societies (Wynne 1991; Gibbons et al. 1994): e.g., how can expert and lay competencies be accommodated? Are there viable forms of the general public's participation in the production of scientific knowledge (Jasanoff 2004)? Some go as far as to deny the specificity of scientific knowledge and want to demonstrate that it is basically *the same* as everyday knowledge, i.e., that it cannot claim epistemic privilege (Knorr-Cetina 1995). But doubts about the status of scientific knowledge cannot ignore that there is an empirically observable difference between scientific and other kinds of knowledge, both in terms of epistemic characteristics and social organization. No other kind of knowledge can claim the same rigor of validation and certification, professionalized training, reliability and degree of institutionalization as scientific knowledge. Not least the "third wave" of science studies returns to acknowledging expert knowledge and, on these grounds, rephrases the question of participation as the problem of selecting the appropriate expertise for decision-making in science and society (Collins and Evans 2002).

It is noteworthy that different approaches to the study of the relationship between science and the media differ not so much in the perception of the phenomenon as such, i.e., what characterizes this relationship, but rather in its interpretation. Obviously, differences of interpretation stem from different theoretical perspectives. The dominant one, in sociological jargon, is action theory. The minority view, favored by the editors, is the perspective of differentiation theory based on the analysis of communications rather than individual actions. To shift from using "knowledge" as a basic concept to "communication" appears to us to provide a superior analytical distinction between science and other social spheres not least because focusing on communication brings different publics as its reference into the picture.

Before tracing the theoretical differences and their genealogy, some terminological definitions are warranted. When we speak of "science," any closer look at empirical phenomena reveals differences between "sciences," i.e., disciplines, research fields, etc. which can become crucial in concrete cases. Thus, the generic term is just a shortcut. Likewise, when we speak of the "public," things are getting even more complex because the public is also empirically elusive. Sociologically, the public, unless socially organized as in voluntary associations or semi-organized as in temporary demonstrations or town-hall meetings, only exists as an abstract "referent" of actions or communications. Publics in this sense are constructed only virtually by the media or by (scientific) authors when they communicate. These "constructed" publics are loosely institutionalized in the audiences addressed, say by newspapers and television broadcasts in their targeted reader- or viewership. This relation between an intangible and unorganized "public" and the "public" constructed by the mass media justifies regarding the mass media as an analytical proxy for the public.

The "dominant view" (Hilgartner 1990) of the science/media relationship is based on two questionable assumptions. First, there exists a clear hierarchy between a "high" scientific and a "low" popular (media generated) knowledge based on the criterion of "truth"; second, the communication of scientific knowledge to the public (media) unavoidably entails the "distortion" of this knowledge (Kepplinger 2001; Kohring 2005). Both assumptions inform among other sources public understanding of science campaigns. But a long line of observers have pointed out that the reality is more complex.

1.5 Science and Its Publics

Ludwik Fleck, in characterizing his notion of "thought collective" (Denkkollektiv), distinguished between the "specialized expert" (spezieller Fachmann) and general experts (allgemeine Fachmänner) as working on a particular problem, where the former is in the center of what he called the esoteric circle and the latter are disciplinary contributors (Fleck [1935] 1980). In the exoteric circle one would find the "educated dilettantes" (gebildete Dilettanten). This structure is reflected in the distinction between "expert" (fachmännisch) and popular knowledge. Even within the esoteric circle one can, according to Fleck, differentiate between "journal science" (Zeitschriftwissenschaft) and "handbook science" (Handbuchwissenschaft). Popular science (populäre Wissenschaft) as the scientific knowledge shared by generally educated dilettantes, to Fleck, is made up of a broad base of knowledge, simplified, stripped of the disputes among experts, apodictic in nature, with only gradual transitions to "Weltanschauung" (ibid.: 147ff). Fleck's differentiation of "internal" publics of science which correspond to different types or representations of knowledge was taken up by Whitley in his distinction between knowledge producers and knowledge acquirers. Whitley, correctly in our view, draws a sharp distinction between the two spheres which constitutes popularization as the marker of the boundary. "Popularization thus implies the separation, however minimal, of knowledge constitution from other social processes and activities" (Whitley 1985: 12).

Several important conclusions follow from this. Popularization is any communication of knowledge, including intra-scientific communication across organizational boundaries (ibid.: 13). The more removed the reception of knowledge is from the production site, the more authoritative, certain and apodictic it may be presented by the scientists. Simplification and apodictic certitude in the presentation of scientific knowledge are not features of science reporting in the media alone, but already occur with gradual variations in textbooks designed for students or for a broader academic audience (see also Stichweh 2003). Hilgartner also represents the range from knowledge production to the communication to a broader public as a continuum of contexts of communication and concludes that "'popularization' is a matter of degree" (Hilgartner 1990: 528, Fig. 2). Likewise, Bucchi, on the basis of Whitley and Hilgartner, represents scientific communication as passing through different stages, from intraspecialistic to interspecialistic to a pedagogical and finally to a popular stage and terms it an "expository continuum" (Bucchi 1996: 381). This picture of popularization differs significantly from the simple hierarchical view in that it perceives the publics of the knowledge producers as the crucial reference. Communications are directed to these publics according to their assumed competence, interests and functionality. Of course, those audiences that are perceived to be influential receive special attention. As Whitley writes:

> Clearly, the more important an audience is to a group of researchers, the more they have to demonstrate the validity and importance of their results to that audience in terms which follow its standards and goals. Thus popularisation has to fit in to the audience's framework and concerns rather than simply expressing the researchers' priorities and approaches (ibid.: 19).

This is, in essence, the mechanism underlying the science/media coupling.

Fleck's original differentiation of publics and the kind of communication based on it has thus become the "dominant view" among students of science while the traditional view is still dominant among scientific experts, science administrators and science policy-makers. The traditional view has its particular functions, especially in reiterating the authority of scientific knowledge and the experts representing it, its distinctive features that set it apart from laymen's knowledge (Hilgartner 1990: 530). Just as this view is informed by the interests of scientific experts and policy-makers and at least supports them by implication, the perception of the science/public continuum can be translated into a normative view justifying public participation in the production of scientific knowledge (e.g., the concept of "extended peer review" as promulgated by Funtowicz and Ravetz 1993; see also Gibbons et al. 1994). An opening of science to a wider public is then associated either with hopes for democratization (Nowotny et al. 2001; Jasanoff 2003) or with the risk of a dilution of quality criteria and, thus, the erosion of the credibility of scientific knowledge.

1.6 Change of Perspective: A Differentiation Theory Approach

Against this background we want to argue that a communication-based instead of an actor-based perspective provides the more precise and productive questions. Do the interactions between science and the media really change epistemic practices and criteria of validation of knowledge? Or are they limited to staging? In other words, is the autonomy of science affected by new demands of the media and their increasing importance for public legitimacy of science?

Autonomy in this context refers to the specificity of scientific (in distinction from political, economic or media-related) communication. The concept is thus not normative but rather an analytical abstraction of the empirically observable characteristics of scientific communication. For example, the validity of a scientific claim cannot be substantiated by the size of the research budget. It is the mode of communication that allows distinguishing science from economics (and other social systems), and it is in scientific communication that the autonomy of science becomes manifest.

From the perspective of action theory, it is more difficult to come to a concrete conclusion. Scientists perform different roles. Individual scientists can personally be visible in the public sphere, engage in politics and be involved in private companies. Does that mean that science as a whole is merged into mass media, politics or the economy? The often cited argument that the boundaries between these different social spheres are "blurred" is merely based on the observations of actors' interactions and/or their own statements but does not offer any further explanation. Instead, the claim that previously differentiated spheres are merged and their boundaries blurred cannot account for the fact that, nevertheless, identifiable structures that define science in distinction from politics, the economy and others, still persist. They become apparent not least in the recurring troubles and tensions associated with scientific policy advice and encounters of scientists and journalists.

In contrast to the constructivist positions in the social studies of science (Latour and Woolgar 1979; Knorr-Cetina 1981) and SSK (the sociology of scientific knowledge) (see Barnes et al. 1996 for an overview), we insist that one cannot ignore that the scientific endeavor is oriented to the "attainment of truth" (rather than power or profit) no matter how successful. Nor can it be ignored that the attribution of reputation functions as a reward mechanism specific to the institution of science (and constitutes its social structure) in contrast to the attribution of prominence of celebrities by the mass media.

We thus argue for a differentiation theoretical approach which is not referring to actors but is based on the concept of communication (Luhmann 1984, 1997). The term "communication" in this theoretical context denotes ways in which different social systems (like science, politics or the mass media) describe themselves and are perceived from outside. In a very general sense, science can be characterized by its communications oriented to the distinction of "true" and "not true." This pertains to the general symbolic content of the communication, not to the question if a particular communication is actually true or false. This perspective allows sharpening the question which kinds of mutual relations between different systems really lead to the disappearance of one or the other, i.e., to a blurring of boundaries between science and the media. It enables the analyst to distinguish between any adaptations to "external" expectations and to locate them in organizations, roles or interactions. On this basis, the empirical investigation can distinguish differentiation and de-differentiation. As the latter is rather unlikely in terms of social evolution, one avoids a premature diagnosis that it happens. Thus, it is an open question, for example, if and how prominence in the media does affect reputation in science. Only if the former would replace the latter could it be said that science has become medialized in the sense of a blurring of the boundaries between them.

With this level of detail in trying to locate structural changes, the consequences of the science-media coupling have rarely been systematically addressed. We propose to study the medialization of science on three different levels that also structure the book.

On a first level, the individual responses to media attention ought to be discussed. One phenomenon of a media orientation is the visibility of scientists. What kind of attitudes and adaptation strategies to media logic can be observed in interactions of scientists and their publics? From a sociology of science perspective, it is of particular interest to evaluate whether expectations of visibility alter the professional role of scientists.

A second analytical approach aims at the organizational responses. In what ways do media routines affect the self-governance of scientific institutions? Do the press relation efforts of research institutions and scientific journals show repercussions on scientific practise?

On a third level, the scientific publication system comes into focus. Since science reproduces itself by way of publications – the final product of scientific activity – a crucial point is whether mass media related criteria are anticipated in the presentation of scientific knowledge. Currently, the traditional publishing system is shifting from the print era to new electronic modes of communication. New forms of media might change scholarly communication and can widen the scope of participants in communication processes.

Each part presents papers devoted to the sciences' media connection from different perspectives, ranging from practitioners' views to empirical investigations and theoretical contributions. To set the stage for the study of repercussions of media attention on science on the three levels, the volume starts off with a theoretical elaboration of the medialization concept (Chapter 2) and then assembles studies on media coverage of science as a reference for the subsequent analyses that point to the dimensions of science news reporting as well as the audiences to which it is addressed.

References

Barnes, B., D. Bloor, and J. Henry (1996). *Scientific knowledge. A sociological analysis*. London: Ahtlone.

Bauer, M. W. (1995). *Science and technology in the British Press, 1946–1990*. London: Science Museum and Wellcome Trust.

Bell, D. (1973). *The coming of post-industrial society: A venture in social forecasting*. New York: Basic Books.

Bucchi, M. (1996). When scientists turn to the public: Alternative routes in science communication. *Public Understanding of Science*, 5(4), 375–394.

Bucchi, M. (1998). *Science and the media. Alternative routes in scientific communication*. London: Routledge.

Bucchi, M. and F. Neresini (2002). Biotech remains unloved by the more informed. *Nature*, 416(6878), 261.

Collins, H. M. and R. Evans (2002). The third wave of science studies: Studies of expertise and experience. *Social Studies of Science*, 32(2), 235–296.

Elmer, C., F. Badenschier, and H. Wormer (2008). Science for everybody? How the coverage of research issues in German newspapers has increased dramatically. *Journalism & Mass Communication Quarterly*, 85, 878–893.

Evans, G. and J. Durant (1995). The relationship between knowledge and attitudes in the public understanding of science in Britain. *Public Understanding of Science*, 4(1), 57–74.

Felt, U., H. Nowotny, and K. Taschwer (1995). *Wissenschaftsforschung: Eine Einführung*. Frankfurt/M., New York: Campus.

Fleck, L. ([1935] 1980). *Entstehung und Entwicklung einer wissenschaftlichen Tatsache*. Frankfurt/M.: Suhrkamp.

Franzen, J. L., P. D. Gingerich, J. Habersetzer, et al. (2009). Complete primate skeleton from the Middle Eocene of Messel in Germany: Morphology and paleobiology. *PLoS ONE*, 4(5), e5723.

Franzen, M. (2009). Torwächter der Wissenschaft oder Einfallstor für die Massenmedien? Zur Rolle von *Science* und *Nature* an der Schnittstelle zwischen Wissenschaft und medialer Öffentlichkeit. In S. Stöckel et al. (eds.), *Das Medium Wissenschaftszeitschrift seit dem 19. Jahrhundert. Verwissenschaftlichung der Gesellschaft – Vergesellschaftung von Wissenschaft*. Stuttgart: Franz Steiner Verlag, pp. 229–252.

Franzen, M. (2011). *Breaking News: Wissenschaftliche Zeitschriften im Kampf um Aufmerksamkeit*. Baden-Baden: Nomos.

Funtowicz, S. O. and J. R. Ravetz (1993). The Emergence of Post-Normal Science. In R. V. Schomberg (ed.), *Science, politics and morality: Scientific uncertainty and decision making*. Dordrecht: Kluwer, pp. 85–123.

Gibbons, M., C. Limoges, H. Nowotny, et al. (1994). *The new production of knowledge. The dynamics of science and research in contemporary societies*. London, Thousand Oaks, New Delhi: Sage.

Greco, P. (2007). University in the 21st century. *Journal of Science Communication*, 6(2), np.

Hilgartner, S. (1990). The dominant view of popularization: Conceptual problems, political uses. *Social Studies of Science*, 20(3), 519–539.

House of Lords Select Committee on Science and Technology. (2000). Third Report. London.

Irwin, A. and B. Wynne (1996). *Misunderstanding science? The public reconstruction of science and technology*. Cambridge: Cambridge University Press.

Jasanoff, S. (2003). Technologies of humility: Citizen participation in governing science. *Minerva*, 41, 223–244.

Jasanoff, S. (2004). *States of knowledge: The co-production of science and social order*. London: Routledge.

Kepplinger, H. M. (2001). *Die Kunst der Skandalisierung und die Illusion der Wahrheit*. München: Olzog.

Kiernan, V. (1997). Ingelfinger, embargoes, and other controls on the dissemination of science news. *Science Communication*, 18, 297–319.

Knorr-Cetina, K. (1981). *The manufacture of knowledge*. Oxford: Pergamon Press.

Knorr-Cetina, K. (1995). Laboratory studies. The cultural approach to the study of science. In S. Jasanoff et al. (eds.), *Handbook of science and technology studies*. Thousand Oaks: Sage, pp. 140–166.

Kohring, M. (2005). *Wissenschaftsjournalismus. Forschungsüberblick und Theorieentwurf*. Konstanz: UVK.

Krotz, F. (2001). *Die Mediatisierung kommunikativen Handelns. Der Wandel von Alltag und sozialen Beziehungen, Kultur und Gesellschaft durch die Medien*. Wiesbaden: Westdeutscher Verlag.

Latour, B. and S. Woolgar (1979). *Laboratory life*. Beverly Hills: Sage.

Lewenstein, B. V. (1995). From fax to facts: Communication in the cold fusion saga. *Social Studies of Science*, 25(3), 403–436.

Livescience. (May 20, 2009). C. Moskowitz: Amid media circus, scientists doubt 'Ida' is your ancestor, available at: http://www.livescience.com/animals/090520-fossil-reactions.html (last accessed on November 9, 2010).

Luhmann, N. (1984). *Soziale Systeme*. Frankfurt/M.: Suhrkamp. [English translation: Social Systems (1995), Stanford: Stanford University Press].

Luhmann, N. (1997). *Die Gesellschaft der Gesellschaft*. Frankfurt/M.: Suhrkamp.

Lundby, K. (ed.). (2009). *Mediatization: Concept, changes, consequences*. New York: Peter Lang.

Nature (2009). Media frenzy. *Nature*, 459(7246), 484.

Nelkin, D. (1995). *Selling science. How the press covers science and technology*. New York: Freeman.

Nerlich, B., R. Dingwall, and D. D. Clarke (2002). The book of life: How the completion of the Human Genome Project was revealed to the public. *Health*, 6(4), 445–469.

Nowotny, H. (2005). Science and society: High- and low-cost realities for science and society. *Science*, 308(5725), 1117–1118.

Nowotny, H., P. Scott, and M. Gibbons (2001). *Re-thinking science: Knowledge and the public in an age of uncertainty*. Cambridge: Polity Press.

Peters, H. P., H. Heinrichs, A. Jung, et al. (2008). Medialization of science as a prerequisite of its legitimation and political relevance. In D. Cheng, et al. (eds.), *Communicating science in social contexts: New models, new practices*. Dordrecht: Springer, pp. 71–92.

Rödder, S. (2009a). Reassessing the concept of a medialization of science: A story from the "book of life." *Public Understanding of Science*, 18(4), 452–463.

Rödder, S. (2009b). *Wahrhaft sichtbar. Humangenomforscher in der Öffentlichkeit*. Baden-Baden: Nomos.

Schulz, W. (2004). Reconstructing mediatization as an analytical concept. *European Journal of Communication*, 19(1), 87–101.

Science (2002). From PUS to PEST. *Science*, 298(5591), 49.

Stehr, N. (1994). *Knowledge societies*. London: Sage.

Stichweh, R. (2003). The multiple publics of science: Inclusion and popularization, available at: http://www.uni-bielefeld.de/(de)/soz/iw/pdf/stw_science_popular.pdf (last accessed on July 7, 2010).

Stifterverband für die Deutsche Wissenschaft (1999). Dialog Wissenschaft und Gesellschaft. *PUSH-Symposium*, Essen.

Süddeutsche Zeitung (May 21, 2009). Ressort Wissen. P. Illinger: Viel Rummel um ein Urzeit-Äffchen, available at: http://www.sueddeutsche.de/wissen/grube-messel-viel-rummel-um-ein-urzeit-aeffchen-1.441196 (last accessed on November 9, 2010).

The Australian (May 21, 2009). L. Dayton: Scientists divided on Ida as the missing link, available at: http://www.theaustralian.news.com.au/story/0,25197,25515021-2702,00.html (last accessed on November 9, 2010).

The Royal Society (1985). *The public understanding of science*. London: The Royal Society.

Weingart, P. (1998). Science and the media. *Research Policy*, 27(9), 869–879.

Weingart, P. (2001). *Die Stunde der Wahrheit? Zum Verhältnis der Wissenschaft zu Politik, Wirtschaft und Medien in der Wissensgesellschaft*. Weilerswist: Velbrück.

Weingart, P. and P. Pansegrau (1999). Reputation in science and prominence in the media: The Goldhagen Debate. *Public Understanding of Science*, 8(1), 1–16.

Whitley, R. (1985). Knowledge producers and knowledge acquirers. In T. Shinn and R. Whitley (eds.), *Expository science. Forms and functions of popularisation*. Sociology of the Sciences Yearbook IX. Dordrecht: Reidel, pp. 3–28.

Wynne, B. (1991). Knowledges in context. *Science, Technology & Human Values*, 16(1), 111–121.

Part II
Medialization of Science – Theoretical Considerations

Chapter 2
The Lure of the Mass Media and Its Repercussions on Science

Peter Weingart

2.1 The Issue

The thesis of the 'medialization of science' stipulates that the relationship between science and the media has changed substantially over the few last decades due to a growing dominance of the mass media[1] in public communication. It has become problematic because science (like politics) is adapting to the criteria of media communication which seems to imply that the criteria of scientific knowledge generation are losing their orienting function. In fact, the issue has become more pressing exactly because the opening up of science to the public has become the accepted expectation by policymakers everywhere. The demands for accountability are motivated by the public legitimation of science needed in democratic societies and have seeped into management and evaluation schemes governing academic institutions.

Assessments about the nature of the new relationship between science and the media and its effects on science diverge. One established view is that communication of scientific knowledge to and in the mass media 'distorts' that knowledge because of its esoteric nature. There is also the opposite stance, that scientific knowledge *should* be communicated to the public since they have a right to that knowledge either because they have paid for it and/or because, in democratic societies, no single group has a right to monopolize superior and specialized knowledge. Finally, there is a middle ground position claiming that scientific knowledge is already being communicated to different kinds of publics that may be differentiated and ordered along a continuum ranging from the specialists' communities to the broad public and therefore the issue is not new. The diversity of the discussion has, by now,

---

[1] The term 'media' in this context relates to mass media only, i.e., any media which are produced by editorial staffs and are addressed to an unspecific public (regardless of its special profile of interests targeted by the media, see below). The obvious ones are newspapers, television, radio and web-based news media (see Chapter 1).

P. Weingart (✉)
Bielefeld University
e-mail: weingart@uni-bielefeld.de

S. Rödder et al. (eds.), *The Sciences' Media Connection – Public Communication and its Repercussions*, Sociology of the Sciences Yearbook 28,
DOI 10.1007/978-94-007-2085-5_2, © Springer Science+Business Media B.V. 2012

initiated an entire research field which is not only characterized by different theoretical approaches, but also by a mix of explicit and implicit normative assumptions and objectives, both of which are often conflated.

In the following, I want to develop the argument that science is being medialized and give reasons why and to what extent this is problematic for science (and indirectly also for society). First, to set the stage I will briefly define different kinds of knowledge and different kinds of publics among which they are communicated in order to show why, in spite of many similarities, the communication between science and the general public via the mass media is not without problems. Then I will develop the concept of medialization itself from an analytical perspective that differs from those either claiming the distortion, if not disappearance, of science or denying a boundary between its publics. In that context, I will differentiate different levels (interaction, organization, system/program) on which medialization takes place and the respective repercussions on science.

2.2 Similarities and Differences Between Communication in Science and by the Mass Media: Types of Knowledge and Publics Addressed

There are interesting similarities between scientific and mass media communication which allow for a certain degree of ambivalence when trying to distinguish between them. Both rely on technologies of diffusion such as printing or digitalization. Both attempt to catch the attention of the recipients, and both do so by trying to establish credibility. For both 'novelty' is a primary value. Earlier in history, the publics of science and the media were not quite as sharply separated as they are now. That is part of the issue, however, insofar as both science and the media have differentiated from one another (see Section 2.3).

But then, the differences appear. Knowledge production and communication within science, although also searching for attention, basically rely on the recognition of 'relevance' and 'novelty' with reference to knowledge that is already known and questions that are posed. Relevance and novelty are ideally determined by the entire community of specialists involved in the process of knowledge production and communication. The production of scientific knowledge is guided by more or less explicit research agendas, based on the application of an elaborate set of methods. There are no comparable mechanisms in mass communication. Mass media communicate 'new events' and – like science and all other social systems – 'create' their own 'reality' by selecting and shaping them according to so-called 'news values' – interpreted and applied by editors and journalists – which steer the attention of the media. (Even if they repeat known facts they have to give them the appearance of newness). The media *observe* society.

Although even the media know a certain amount of self-referentiality (i.e., they cite each other), this is not systematic, leading to cumulative knowledge production like in science. Note that this criterion distinguishes more clearly between the media and the natural sciences than between the media and the social sciences and humanities. Knowledge production in the latter is not cumulative or only so to a much lesser degree. Thus, scholars of literature publish both in their specialized journals and in high-brow mass media. Controversies within disciplines of the social sciences and humanities such as that over Daniel Goldhagen's book *Hitler's Willing Executioners*, or between the German archaeologists Korfmann and Kolb over the interpretation of archeological finds to determine the location and importance of Troy, evidently find their way more easily into the media (Weingart and Pansegrau 1999; Chapter 15). The content is more accessible and the knowledge involved is part of a discourse among the broader educated public.

An important distinction between the types of knowledge communicated by science and the mass media emanates from their differences in credibility. In the communication among scientists (intra-specialist), credibility is created by an elaborate system of 'checks' that establish the objectivity and validity of the knowledge produced. These checks take the form of critique by specialists before publication, of 'peer review' controlling publication, the concealment of authors and reviewers to avoid conflicts of interest, and strict rules governing the originality and authorship of publications to prevent fraudulent communication. Most indicative of the sensitivity of these mechanisms with respect to their function of quality assurance are the reactions to their violations. Incidents of fraud, fabrication of data or just plain plagiarism are not only sanctioned by the scientific community but are formally examined by institutional review boards, and they are reported widely in the media reflecting their perception that the trustworthiness of science is of high public interest. Although many newspapers and public TV channels claim impartiality vis-à-vis governments, political parties and their ideologies, and quality journalism tries to establish credibility the quality assurance even in good newspapers or television does not match that in science. Boulevard journalism operates completely outside these concerns. As a result, trust in science as an institution is invariably higher than trust in the media. This pattern is remarkably stable across different societies and over time (see Europäische Kommission 2010).

The *differences in knowledge communication between science and the media* are inextricably connected with the social organization of the knowledge producers and with the publics addressed by their communications. Communication in science relies on a clearly defined set of criteria of exclusion (or inclusion resp.) delineating the public to be addressed. These criteria may be summarized implicitly as *competence*. Competence is constituted, first of all and formally, by certificates indicating the successful conclusion of an education and, secondly, by past participation in contributions to the creation of knowledge (research) and their communication (publications) which have been certified by 'competent' peers. Communication in science involves *communities* of scholars who are bound by work on problems which they perceive as being of mutual interest. Thus, such communities are constituted by a network of topics, problems, and answers. Individual members providing inputs (research results), and who often know each other personally, attribute reputation to each other as a reflection of the perceived value of their contributions to the common endeavor. This is a self-reinforcing mechanism since the hierarchy of reputation in each field also provides the orientation of the work agenda and thus the consensually perceived relevance of knowledge. Individuals hardly matter, they may enter and leave the respective community without affecting its continuity which demonstrates the degree of institutionalization of research fields. The mass media, on the other hand, do not address communities delineated by criteria of inclusion and exclusion, let alone constituted by criteria of competence. Most commonly they address unspecific publics, as large as possible because that usually translates into profits, the chief objective of the media as commercial enterprises. As publics have become differentiated according to their interests, the media try to capture their attention by, first of all, researching their specific profiles: demographic, social structural, life style, etc. Although they may be very successful in 'profiling' their prospective audiences, the public they address remains unknown to them in principle. (Mass) Media and their publics are connected through contingent expectations rather than direct reciprocal communication.

Even if the media cater to quite specific publics, e.g., bikers, hobby gardeners, or teenagers, they address a particular interest profile of potential readers/viewers. Neither does the reception of and contribution to the communication require proof of competence, nor do they constitute an institutionalized community with rules of inclusion and exclusion. 'Letters to the editor,' although known to both communication cultures, do not fulfil the same function in each of them. Some mass media mimic the reciprocity of scientific communication by staging debates or controversies over a certain topic, but they are not sustained, they do not constitute communities and they are not subject to similar processes of quality control.

One could try to play down the differences by pointing out that quality controls do not always work well in scientific communication while some journalists meet very high quality standards. Empirical variations are not relevant, however. What count are the systematic, institutionalized differences.

The systematic differences between communication in and by science and by the mass media explain (1) why scientific communication enjoys trust both among the scientists and the public; and (2) why a conflation of media used and publics addressed in scientific communication is perceived as problematic by scientists. The suspicion or even downright contempt of scientists towards any trespassing of the demarcation line between the 'internal' and the 'external' communication channels is relevant insofar as it indicates a well based fear that the proper operation of the mechanism of the attribution of reputation may be disrupted. The consequences of such disruption could be grave, indeed. If criteria other than those deemed relevant by the competent members of the community would be applied in the evaluation of contributions to knowledge, the basis for trust in the communications of other peers would erode, and with that the orientation which questions to pursue and which to leave unanswered, which colleagues to pay attention to and which to ignore would be lost. In other words: The evolutionary advantage of differentiation that the system of knowledge production as we know it represents would be reversed.

This does not mean that scientists do not address other communities than their own. They do in many ways. Most common is their communication with the broader public through various forms of popularization ranging from books and articles to exhibitions and image campaigns. But authors who become recognized as popularizers do not normally compete with scientists for reputation within science.[2] If scientists themselves popularize their research findings or write textbooks, they distinguish this activity from research proper. This institutional separation of social roles as well as activities and their products was gradually achieved in the course of the nineteenth century and completed at its end. In other words, popularization and research are clearly distinguishable activities involving different publics.

Today's image campaigns for science, 'science exhibitions,' and various types of science events are typically commissioned by governments and/or science administrations such as research councils and foundations. They are designed and executed by professional PR firms which enlist scientists' help to provide substance. In contrast to the earlier forms of popularization, they are targeted to large unspecific audiences. Their success is usually measured in attendance, not in sustained behavioral change although raising interest in the sciences among children and acceptance in the general population is their purported objective.

Fleck as well as Cloître and Shinn have argued that the communication from science to the public is a continuum rather than dichotomous (Fleck 1980; Cloître and Shinn 1985; see Chapter 1). Here the point is that even though communication from one discipline to another (inter-specialist) may involve popularization and therefore be seen on a continuum, it is still considered 'internal' to science, in contrast to popularization to the outside public. The evaluation of competence is still the crucial mechanism in upholding that distinction. The difficulties to institutionalize interdisciplinary research illustrate this succinctly: Specialists from different disciplines have fewer problems understanding each other and, consequently, less need to 'popularize' for their colleagues if their subject matters and specialized languages are 'close together' like physics and mathematics. In such cases, they are also able to mutually judge each others' competence. On the other hand, if their disciplines are far apart like atmospheric physics and economics (as in climate research), the inability to understand each others' terminology implies the inability to judge each others' competence and thus the need to popularize. Experts on both sides then have to rely on the established reputation as it 'is known in the field.' However, there is still a borderline between this 'internal' popularization – to the extent that it happens – and that which is directed to the so-called lay public.

Massimiano Bucchi has convincingly argued that – beyond the continuity model that adequately describes the information flow from science to the public – there are exceptional cases. These are cases when scientists turn directly to the public such as the formation of new disciplines or research fields ('constitutive boundary work,' as Bucchi calls it and cites environmental and information sciences (1996: 382)). These cases are seen as 'deviations' from the normal pattern of popularization, and although they count as a specific resource for the scientists, they are at the same time looked upon with great apprehension. There is a

---

[2] Historically, the role of popularizer has become quasi professionalized at the end of the nineteenth century. Till today borderline examples come to mind: Gould, Dawkins, Sagan, etc. However, they do not falsify the claim.

> tension [...] within the scientific community between absorbing [...] deviation into ordinary expository practice (popularization) to avoid its 'uncontrolled abuse,' and at the same time preserving it as a sort of 'emergency exit' for specific situations as well as a potential source of scientific change (ibid.: 387).

This tension reflects the borderline between the 'internal' and the 'external' publics and the conflicting expectations (and opportunities) to communicate to them.

A final argument could be to claim that the lay public may, in certain constellations, gain competence in a particular field of scientific research and that such a development would level the distinction between the intra- and inter-specialist publics, on the one hand, and the lay public on the other. There are examples where this claim has some justification: so-called 'round tables,' citizens fora, and similar arrangements in which representatives of the lay public are brought together with scientists to negotiate questions of hazard and risk with respect to particular experiments (e.g., genetically modified corn) or technologies (nuclear power). Perhaps even more compelling are those cases where activist groups negotiated with experts over changing research priorities as the incidents reported by Epstein about AIDS groups (Epstein 1995). AIDS treatment activists were able to establish themselves as credible participants in the process of knowledge construction and actually affected changes in the epistemic practices of biomedical research. But in order to achieve this role, activists and citizens had to obtain considerable competence in the respective research fields to meet with the scientists on the same level. In fact, their estrangement from their fellow citizens due to their acquired competence is a topic of research on its own.

It must be concluded, then, that in spite of many similarities between types of knowledge and publics addressed in the communication of science and the mass media, crucial differences remain. They explain the uneasiness among scientists vis à vis the direct communication with the mass media. This uneasiness or rather the institutionalized assessment is an indication of the differences between scientific and mass media communication. To conclude this argument: The more cumulative the knowledge, the more firmly institutionalized the certification of this knowledge and of its producers' competence, *the more pronounced are the differences to the mass media*. A confounding of science's publics can create conflicting expectations for science communication. The question, then, is what dimensions this conflict has and which repercussions for science it implies.

2.3 Democratization and the Emergence of Mass Media

Before pursuing the analytical argument, it must be briefly explained which rather recent developments have led to the present state of what is being described as the medialization of science. Historically, *science* has communicated to a broader public. But in this process both science and the publics have changed fundamentally and the same must be said for the media. Modern science emerged first in the mid seventeenth century but although communication among scientists (the very term 'scientist' was coined by Whewell only in 1833) began in correspondences and specialized journals once the Royal Society had been founded, it also extended to different publics: curious members of the aristocracy at the courts of Europe, onlookers on markets watching experiments executed in public, later on members of the educated bourgeoisie and working class in the lecture halls in Paris, London, Berlin and Vienna where scientists themselves or, gradually taking over from them, popularizers presented the newest wonders of science. By then the communication among scientists had withdrawn into their more and more numerous specialized journals, experiments were no longer carried out in public but in laboratories and were reported about later. Circa 1830 disciplines in the modern sense had emerged replacing natural philosophy. A century later, scientific research especially in the natural sciences had become so specialized that communication was de facto incomprehensible to the lay public and even across disciplinary lines (Bensaude-Vincent 2001).

The crucial point in the development of the *media* is their shifting nature to become true mass media. This has a technical, a political and an economic aspect. Technically, the major advances are the invention of high pressure rotation printing, of the radio in the 1890s and of television in the 1930s. By the mid 1930s radio broadcasting had become a mass medium, i.e., counting millions of listeners. Economically, newspapers that had been published by political parties or by small publishers as 'opinion press' began to address a mass public, beginning in the 1870s. By the turn of the century newspapers had become economic enterprises rather than instruments of political organizations, even though their publishers used them for political purposes like Hearst in the US and Ullstein and (in the 1920s) Hugenberg in Germany. This commercialization of the print media (later also of radio and TV) constituted or reinforced 'news values' as operational criteria which from now on governed them. Politically, these developments were framed by the emergence of a mass society, first triggered by capitalism and the creation of a large industrial workforce organizing in the labour movement in the nineteenth century. This movement brought about the eventual democratization of the Western industrializing nations after World War II. Democratization and the development of mass media go hand in hand as the general public, i.e., the entire population, becomes, in principle, the target audience of the media. From now on the legitimacy of political organizations, individual politicians and governments but also of societal institutions like science is largely determined by the mass media as they assume the central communicating function in mass democracies. The mass media articulate expectations *as if* representing the general public.

Here an important difference between science and politics in mass democracies becomes apparent: Politics derives its legitimacy from the general public only. For science two publics are relevant: The general public is the source of legitimacy with respect to the funds and the institutional support it provides, whereas the specialized scholarly public is the source of legitimacy for judgments of the quality of truth claims and the attribution of reputation to scientists and/or their institutions.

Democratization has led to a recent shift in weight to the former in the sense that accountability to the general public is considered more important than before.

Given their ubiquity the mass media have dramatically changed the *attention economy*. Space and time to make oneself heard and to actually be listened to are expanded continuously but, parallel to that, the inflationary use of communication creates an ever growing, self-accellerating pressure to participate in the competition for attention. Science is no exception.

2.4 Conceptualizing the 'Medialization' of Science

The thesis of the 'medialization' of science (see Chapter 1) has initiated a number of studies that have attempted to determine the actual effects of the orientation to the media (see Weingart 1998; Peters et al. 2008; Rödder 2009; Franzen 2011). There is general agreement on at least three observations: (1) The orientation of science to the mass media has grown more intense; (2) This may create tensions of different degrees of severity within science because the orientation to the media is in conflict with rules and values prevailing in science; (3) These tensions are expressed in the dilemma in which scientists find themselves because the demand to communicate with the public has become part of their legitimating exercises in the context of mass democracies whose publics and political leaderships no longer recognize and accept the professional elites' privilege of virtual unaccountability. Differences exist over the reach and impact of media orientation on science. They range from discarding them as 'just show' which has no effect whatsoever, to a doomsday scenario in which science ultimately dissolves into the media. The differences of interpretation are due to different theoretical frameworks in which they are developed, and sometimes also due to different normative positions.

In order to come to a common interpretation, one has to choose a theoretical framework that allows to take account of the distinctions between communication of science and of the mass media developed above. At the same time it should make it possible to identify the impacts of the orientation of science communication to the mass media.

The framework chosen here is that of systems theory for the simple reason that it proceeds from the very fundamental distinction between science and the media as the result of the (historical) functional differentiation of modern societies. Without this differentiation the fairly recent developments in the relation between science and the mass media would not even be a problem, just as the communication of scientific knowledge to the public was not a problem before that differentiation occurred in the seventeenth to mid nineteenth centuries. This theoretical framework has the advantage over others that it maintains distinctions that allow descriptions in a seemingly confused real world. These other approaches are based on vague concepts (e.g., 'Mode-2') which require considerable descriptive and interpretive input without providing criteria that allow to delineate change. Even though they often postulate convergence or a 'blurring of boundaries,' they still have to rely on the very distinctions which are supposedly disappearing. Not only is it highly questionable to claim the complete diffusion of science into the mass media in the face of counter evidence and, likewise, to promote 'dialogues,' 'engagement' and 'participation' between science and the public as if there were no barriers of specialized knowledge and training. It can neither be justified nor explained theoretically in these frameworks.

2.4.1 Science as a Social System and the Science-Media Coupling

Sociological systems theory suggests that modern societies are differentiated into a set of functional systems: politics, the economy, law, religion, science and (with some qualification) the media (Luhmann 1995). These systems are autonomous in the sense that they are defined by their respective operating codes, in the case of science it is *truth* (see also Chapter 1), for the political system it is *power*, for the economic system it is *profit*, etc. In reality, this means that in any pertinent communication it is possible to identify the distinction between 'truth' and 'non-truth,' no matter how convoluted it may appear, and that it cannot be merged with others. A simple expression of this is the famous dictum that 'truth cannot be established by majority vote, nor bought with money, nor mandated by edict.' As long as this distinction is present in communications, no matter if it may be violated from time to time, functional differentiation is a social reality.

A further consequence of this categorization is that systems cannot merge, nor can one dominate the other. If that were to happen, it would be tantamount to a reversal of the epochal evolution of human societies. Systems can only 'irritate' each other. They can act upon each other creating resonance, and they can be coupled on the basis of mutual expectations. *Differentiation of codes implies that systems have their own 'frequencies' with which they react to external irritations*. This means that there cannot be a one-dimensional and uni-linear communication of meaning between systems. What appears relevant in one system is not equally relevant in another. Typically, the fault between systems becomes apparent in scientific advice to policymakers when what appears to be certified and objective knowledge is transformed into decisions in a highly selective and often seemingly irrational manner. Thus, irritations from one system can only be processed in another system in its specific mode of communication. Assumptions about causality as in theories of steering are too simple in view of a more complex reality.

Luhmann used two parallel concepts to describe inter-systems relations: *coupling* and *resonance*. Coupling as a metaphor is taken from biology and refers to the mutual dependencies between systems and their environments. Applied to the social realm this mutual dependency exists in the form of expectations and services. In our case: Science provides a steady stream of information to the media. Not all of it is interesting to them but some is: The discovery of a new star, the spread of a virus, the extinction of a particular species are all information communicated by science on which the media rely for their news reporting. The mass media, on the other hand, are coupled with science because science relies on the media's focusing of public attention on important discoveries and, indirectly, demonstrating its utility and legitimating its costs. The political expectation that science be accountable involves, to a large degree, the media which have to spread the message to the general public. The coupling of systems, therefore, constitutes the connections and dependencies between them without which mutual impacts, i.e., resonance, would not be possible.

2.4.2 Resonance Between Science and the Media

Resonance is a metaphor borrowed from physics where the term denotes the irritation or agitation of a system capable of oscillation. The crucial variable is resonance frequency which means that oscillations caused by energy input from outside the system may cumulate and, in the extreme case, lead to catastrophic destructions (like in the case of bridges which can collapse under the impact of marching troops). The opposite is that irritations from outside have no effect at all, i.e., the system does not resonate. The medialization thesis focuses interest on the science side of resonance effects. To cite an example: In the mid 1970s, climate scientists postulated anthropogenic global warming, sometimes with alarming predictions in order to capture media (and ultimately political) attention. They used exaggerated claims, thereby adapting to the media's criteria of relevance. The issue enjoyed a relatively high level of interest from the media for years. At least some climate researchers use this attention to advance their political message. But they do not anticipate the cyclical nature of this attention. When the attraction of climate change as an issue seems to wear out the media suddenly begin to question the credibility of the climatologists' thesis of imminent anthropogenic global warming. Controversies among climate researchers are interpreted as conflicts, suppression of opposition, and manipulation of information. A simple error in an authoritative report on the state of research is reason for the media to accuse the scientists of forging their data. Now the scientists struggle to regain their credibility (see Weingart et al. 2007).

To exploit the metaphor a little more: If the 'irritation' meets the particular 'resonance frequency' of the system it has an amplifying effect. In the example, the crucial irritation is not the reporting of disputes between global warming advocates and sceptics but the questioning of the formers' motives as being politically interested and, thus, untrustworthy. The allegation of untrustworthiness on account of manipulation, fraud or politicization is the strongest that can be raised against science. One reaction is for scientists to stage public relations campaigns to gain public acceptance. This has already happened in several instances, e.g., nuclear power, genetic engineering, stem cell research to name the most conspicuous.

The next question concerns the actual effects, i.e., the amplitudes or kinds of resonance. The interest in gauging the possible effects on science points to different levels on which such effects can occur or not.

To gauge the effect of medialization, systems theory suggests three levels on which these effects may occur. These levels incorporate various empirical observations and at the same time allow to distinguish between them according to their range and impact on science. These are the interactional, the organizational and the program level.

(1) The strongest resonance would be on the systems level, i.e., if the orientation of science to the media would effectively lead to the displacement of epistemic criteria of novelty, relevance and robustness (all with reference to the scientifically certified body of knowledge) by the media's criteria of news values. In this case, the production of knowledge would clearly be affected, one could not speak of scientific communication anymore. This is *not very likely to happen.*

    Very well imaginable, however, is that the orientation to the media may have effects on what in systems theory is called the 'program level.' Programs are the variable aspect of codes (which are constant). 'Theories' and 'methods' are programs of the code of 'truth.' If one method is replaced by another, this does not affect the code (see Luhmann 1990: 401ff).

(2) A probably lesser impact would be one on the interactional level. On the *level of interaction,* scientists do not 'normally' communicate with journalists but only with their peers. This is exemplified by the fact that *reputation* within the scientific community, which is the basis of the social structure of science, cannot be gained by communicating with the media.[3] In fact, as described above, there is a clearly defined demarcation line based on the criterion of 'competence' that separates science (or rather research fields) from other systems (i.e., the environment of science). In the media, the analogon to reputation in science is *prominence.* It shares with reputation that it denotes visibility based on achievement, but the crucial systemic difference is what counts as achievement. In science, in accordance with its code, it is the generation of new knowledge. In the media system, in accordance with its code, it is attraction of public interest based on news values. Prominence may be gained by movie stars, politicians, bank robbers and murderers alike. Scientists can also acquire media prominence but prominence cannot *normally* be translated into scientific reputation and vice versa (see Weingart and Pansegrau 1999). In the logic of systems theory this implies: If media prominence were to be transformable into reputation, the mechanism of allocation of reputation would not operate any more.

(3) The least far reaching effects may be expected on the level of organization. Organizations exist to achieve certain decisions. They can cut across social systems and have references to each of them. Universities are a case in point. They are linked to the science system through research. At the same time, the other of their chief functions is teaching. Increasingly, they are involved in technology transfer which requires them to operate in terms of the economic system. The same holds for their administrations – if they are financially at least partially independent – which have to calculate like commercial companies.

---

[3] As argued above, this differs among disciplines. In some fields in the humanities (e.g., history, literary sciences), the educated public of highbrow newspapers is a legitimate source of reputation. But even in these fields the ultimate criterion is acclaim from peers, as the Goldhagen case has demonstrated.

The effects of medialization on science can, thus, be defined more precisely by differentiating between these three levels giving guidance to the interpretation of the variety of empirical observations usually subsumed under the term *medialization*.

Ad 1) Orientation to the media may have an impact on the *program level* in different ways. For example, if a positivist epistemology is replaced by a constructivist epistemology or if linear causality loses out against chaos theory, this may be due to amplifying effects of media reporting and is, in fact, criticized in the respective scientific communities as such. It does not put the code of science itself into question but it implies that the self-direction of science is weakened and no longer functions in the traditional manner of disciplinary self-referentiality. Instead, value preferences of the public as communicated by the media serve as *additional* references, when, for example, new kinds of research fields emerge such as 'gender studies' or 'environmental studies' (Maasen and Weingart 2000; Chapter 5; Luhmann 1990: 401ff). These fields reflect the resonance of the science system to public discourse staged by the mass media. They are not products of the self-referential development of the landscape of disciplines.

Ad 2) There are rare cases which may be taken to illustrate that media prominence has helped scientists to gain reputation in science but upon closer examination they do not quite prove the point. Historian Daniel Goldhagen's media prominence has not survived scrutiny by peers when he was up for an academic appointment. (As a humanities scholar whose audiences are also outside his discipline he represents a borderline case anyway). Pons and Fleischman, the two scientists who claimed to have achieved 'cold fusion,' only temporarily succeeded in communicating exclusively through the mass media (TV and newspapers) until the peer review mechanism prevailed (Lewenstein 1995).

Rödder (2009) has shown that among geneticists, even though this is a community that is more exposed to the media's limelight, there is still a hierarchy of acceptance of different motives to search media attention. Lobbying for science in general or for one's own discipline is accepted while advertising for personal gain and prominence is not. In contrast, in the social science and humanities, crossing the line between 'internal' communication and communication to and in the mass media is much easier and generally more accepted.

Ad 3) The resonance on the organizational level, i.e., what effects it has on science when scientific organizations deal with expectations of communicating with the public, is less obvious. Universities and research organizations have set up their own PR offices which produce reports directed to the broader public and the media in order to attract students and sponsors. They also engage in the formulation of 'mission statements' that are directed to their boards of trustees, their community and to policymakers legitimating their actions and performance. These are ways of representing the university to the public, and the means to do this are media oriented. (The growth and increased influence of media trained staff in science organizations is a clear indicator). None of these developments affects the research process itself, at least not immediately. They are independent from it. In some cases, universities have even established rules that shield scientists from communicating with the media by controlling media access. The multitude of systems references and their apparent combinations, therefore, is not yet an indication of medialization in the sense of 'blurring boundaries.' Yet, it can still be asked, for example, if the costly PR activities of universities are actually fulfilling their purpose of reaching a target group whose decisions are supposed to be influenced in favor of the university in question, or if they are simply an activity reacting to a general trend of media orientation with no specific public addressed. There is no doubt that these activities drain the resources, both in terms of time and money, of any organization.

But it is common knowledge that symbolic actions and rituals have an impact on organizations, as do bureaucratic rules. It is by no means clear what kind of representation remains just that on the front stage and how much effort on the level of representation will ultimately have repercussions on the backstage, i.e., on the production process itself. It can be assumed that PR campaigns on behalf of particular research fields or the popularization of its discoveries will not affect the research process and the epistemic criteria involved.[4] However, if reference to the public (via the media) is perceived to be so important for the legitimacy of research that scientists exaggerate their truth claims as in the case of climate change, stem cell research or nano-technology, this may have repercussions on the validity of theories and methods. In such cases, media orientation on the program level may imply that the respective communication can no longer be considered scientific.

There are also indirect effects. The introduction of performance measures and their focus on quantitative indicators that can be expressed in simple numbers has in some cases been introduced by the media and rankings are continuously reported in widely visible print media such as THES. They play ideally to the news value of competition and picking winners. Their effects, both intended and unintended, on the organizations which they purport to measure are profound (Weingart 2005; Espeland and Sauder 2007).

Another resonance of 'medialization' has been identified in a particular sector of scientific communication (see Chapter 17). The link of top journals such as *Science* and *Nature* to the mass media by way of pre-publication press releases and related promotional activities that play to the news values of novelty and sensation has an impact on the communication process. The *acceleration* of the publication process together with the ambitious search for high-impact papers comes at the price of increased incidents of exaggerated claims that had to be taken back by the respective journals in retractions. It has to be noted that the communication of results in scientific journals (!) is a part of knowledge production considering that the communication process never ends and through peer review contributes to the certification of knowledge.

The top journals' orientation to the mass media also seems to result in a selection in favor of spectacular, surprising research results while reporting on 'normal science' is relegated to less visible journals (see Franzen 2011). This selectivity tends

[4] It is interesting, though not a counter argument, that nano scientists, after having had success with their public propaganda in capturing media and political attention, have shrugged back from it, presumably for fear of becoming its victims (Lit. in Kaiser et al. 2010; for genome researchers see Rödder 2009).

to contribute to the creation of topical cycles which steer attention in the selection of research topics. Focusing attention in this way usually also implies the extension of the public of this communication from the narrower scientific community to the broad political public.

But when it becomes more important to publish an article containing 'sensational news' rather than to make sure that the news is 'true' and recognized as such by the scientific community, obviously the mass media's public is considered more important than the 'public' of peers. At the same time, this obstructs the internal communication process and, thus, indirectly the production of knowledge. Strictly speaking, orientation to the media has then replaced orientation to 'truth.'[5]

2.5 Conclusion: Medialization as Coupling of Systems

From the preceding analysis a number of conclusions can be drawn: (1) It is very useful to conceptualize 'medialization' of science distinguishing effects that can be identified on the organizational, interaction and program levels. (2) Rather than limiting the term 'medialization' to effects with regard to the code, i.e., to de-differentiation, only to conclude that it does not happen it appears more productive to subsume a broad variety of phenomena of science's media orientation under the concept and to use the previous distinction to specify the degree of conflict. There is, furthermore, a theoretical and a substantive conclusion.

The theoretical conclusion leads back to the concept of coupling. Coupling describes dependencies between systems due to mutual services (performance). In one sense, this description is quasi definitional: The science system is coupled to the system of politics because it provides knowledge for decision making and legitimacy in exchange for public funds. In this sense, couplings either exist or they do not. In another sense, however, they can differ in degree of tightness: Public funds may be replaced by private funds. Scientific knowledge may compete or even be replaced by knowledge produced by religious institutions which also provide legitimacy for governments. Just as much as couplings between systems can therefore change in intensity, so can medialization. In fact, medialization is a special case of the coupling between science and the media. By conceptualizing their relationship in this way, the danger of dramatizing phenomena or confounding them is avoided while, at the same time, allowing for the differentiation of various kinds of configurations that surprise the observer time and again. Only in this way can the conceptual apparatus cope with ever new developments among which the complete disappearance of science is the most unlikely one.

The substantive conclusion is that the theoretical approach allows for a differentiated view of medialization. The medialization of science is not good or bad as

---

[5] This illustrates, by the way, why the orientation to other relevant references is analytically akin to scientific misconduct. In many cases (e.g., Hwang, Schön) media hype triggered by *Science* and/or *Nature* played a role.

such but must be seen as a new kind of coupling of science with the media and – through them – with other social systems. This will not leave science unaffected.

Thus, the focusing of topics through the media is part of their attention management and concerns the agenda of science. Likewise, extending the public to include economics and politics may be functional for science if it mobilizes research funds, for example. However, if political or economic considerations enter into the evaluation of research results or the recruiting of staff, that becomes problematic because it affects the validity and credibility of scientific knowledge. Even the acceleration of the publication process may be positive in that it promotes the development of science as long as it does not compromise the quality assessment mechanisms. The exaggeration of results is seen as an undesirable side-effect of the medialization of science (see Chapter 17). It can be traced to the combination of media orientation and its implementation in incentive schemes that reward publications in top journals with the prospect of career advancement.

Cases like these are (still) isolated incidents. They occur in particular research areas which enjoy unusual public attention already and which have considerable economic implications for the media and the scientists involved. But even the cases which are supposedly limited to effects on the organizational level may be significant. A university that is forced to compete with others for public funds and students and puts an emphasis on communication oriented to the mass media may, as a consequence, decide to prioritize research fields which promise greater acceptance among policymakers. A scientist who is remunerated at least in part on the basis of his/her publications in high impact journals and the number of citations they receive may decide to work on topics that are more likely to be 'news worthy' and have a higher probability of being published in *Science* or *Nature*. The one overall conclusion is that the orientation to the media weakens science's self-direction. The self-referentiality of the disciplines that characterized the development of science through much of the twentieth century is gradually complemented by reference to the mass public. This should not come as a surprise as it indicates a new place for science in a changed social and political environment, i.e., mass democratic societies: The interests and values of their publics communicated by the media become an important referent for science. While the boundaries of science may become more permeable with respect to problems addressed, this does not necessarily imply that the production of certified knowledge will cease to happen. Science will be more responsive to society at large but the welfare of society will continue to depend on reliable knowledge.

References

Bensaude-Vincent, B. (2001). A genealogy of the increasing gap between science and the public, *Public Understanding of Science*, 10(1), 99–113.

Bucchi, M. (1996). When scientists turn to the public: Alternative routes in science communication. *Public Understanding of Science*, 5(4), 375–394.

Cloître, M. and T. Shinn (1985). Expository practice: Social, cognitive and epistemological linkages. In M. Cloître and T. Shinn (eds.), *Expository science: Forms and functions of popularization*. Dordrecht: D. Reidel, pp. 31–60.
Epstein, S. (1995). The construction of lay expertise: AIDS activism and the forging of credibility in the reform of clinical trials. *Science, Technology & Human Values*, 20(4), 408–437.
Espeland, W. and M. Sauder (2007). Rankings and reactivity: How public measures recreate social worlds. *American Journal of Sociology*, 113(1), 1–40.
Europäische Kommission (ed.) (2010). Eurobarometer Science and Technology Report.
Fleck, L. (1980). *Entstehung und Entwicklung einer wissenschaftlichen Tatsache*. Frankfurt/M.: Suhrkamp.
Franzen, M. (2011). *Breaking news. Wissenschaftliche Zeitschriften im Kampf um Aufmerksamkeit*. Baden-Baden: Nomos.
Lewenstein, B. V. (1995). From fax to facts: Communication in the cold fusion saga. *Social Studies of Science*, 25(3), 403–436.
Luhmann, N. (1990). *Die Wissenschaft der Gesellschaft*. Frankfurt/M.: Suhrkamp.
Luhmann, N. (1995). *Social systems*, Stanford: Stanford University Press.
Kaiser, M., et al. (eds.) (2010). *Governing future technologies. Nanotechnology and the rise of an assessment regime, Yearbook Sociology of the Sciences 27*. Dordrecht: Springer.
Maasen, S. and P. Weingart (2000). *Metaphors and the dynamics of knowledge*. London/New York: Routledge.
Peters, H. P., et al. (2008). Medialization of science as a prerequisite of its legitimation and political relevance. In D. Cheng et al. (eds.), *Communicating science in social contexts: New models, new practices*. Dordrecht: Springer, pp. 71–92.
Rödder, S. (2009). *Wahrhaft sichtbar. Humangenomforscher in der Öffentlichkeit*. Baden-Baden: Nomos.
Weingart, P. (1998). Science and the media. *Research Policy*, 27(9), 869–879.
Weingart, P. (2005). Impact of bibliometrics upon the science system: Inadvertent consequences? *Scientometrics*, 62, 117–131.
Weingart, P. and P. Pansegrau (1999). Reputation in science and prominence in the media – The Goldhagen Debate. *Public Understanding of Science*, 8(1), 1–16.
Weingart, P., A. Engels, and P. Pansegrau (2007). *Von der Hypothese zur Katastrophe. Der anthropogene Klimawandel im Diskurs zwischen Wissenschaft, Politik und Massenmedien. With cooperation of Tillmann Hornschuh* (2nd ed.). Leverkusen Opladen: Barbara Budrich.

Part III
Media Coverage of Science

Chapter 3
Public Attention to Science 1820–2010 – A 'Longue Durée' Picture

Martin W. Bauer

The longitudinal observation of science in the modern mass media is not a major research enterprise. It compares in no way with the efforts that go into the tracking of climate change, economic performance of national economies, stock markets, or single corporations for that matter. But the long-term climate of public opinion about science deserves more of our attention than it has hitherto received.

Social history enquiries often start with the construction of long time-series, be that of birth and mortality rates, prices of basic goods, household income, or some other indicator of social activity. These series often demonstrate rhythms of activities that neither coincide with the reigns of great men or women (e.g., Queen Victoria), the annual cycle of the sun (January to December), nor with the convenience of the Gregorian calendar (e.g., 1800–1899). Researchers thus discovered the late medieval expansion of the thirteenth century, the inflation of the long sixteenths century and of the late eighteenth century (see Braudel 1958; Fischer 1996), and the long-waves of the world economy (see Van Duijn 1983; Maddison 1995). These secular trends often mark the shifting mentalities within which faster moving events and actions take place.

In the following I would like to mimic these 'longue durée' approaches for the public attention to science. A long-term perspective on public attention to science should stimulate new insights into the history of public and popular science by offering a criterion for structuring its chronology, and by putting the present level of attention into perspective. Science reportage in the modern circulation media of print and broadcasting is an integral part of the history of science. We easily recognise many different genres of public discourse of science produced outside the laboratory and government policy making. Depending on the context, this discourse can be informative, celebratory or critical of science. The attribute 'public' shall mark a difference to both the discourses of science and to those of government.

M.W. Bauer (✉)
London School of Economics
e-mail: m.bauer@lse.ac.uk

S. Rödder et al. (eds.), *The Sciences' Media Connection – Public Communication and its Repercussions*, Sociology of the Sciences Yearbook 28,
DOI 10.1007/978-94-007-2085-5_3,

The present yearbook explores the consequences of a historical increase in science reportage in the mass media. This expansion of news space involves professional news management, e.g., events and stunts that celebrate the mapping of the human genome, raise the alarm over asteroids or climate change, or bring science to public scrutiny over such issues as nuclear fission, genetically modified crops or embryonic stem cells research. The thesis suggests a transfer from quantitative change in public attention to a qualitative shift in the conduct of science: Science risks a closer coupling with public opinion by incorporating media logic and attention seeking into knowledge production. Prominence increasingly competes with reputations in certifying knowledge (see Weingart 1991; Chapter 1).

The present long-term perspective contributes to this debate. First, I will be able to ascertain whether the recent increase in media coverage is indeed historical. Second, I compare the present with previous periods during which public attention to science was higher, and thus sharpen the boundary conditions of the 'medialization' thesis.

This chapter will offer three things. First, I will clarify my take on the relationship between public spheres, public attention to and mass media coverage of science. Second, I review a number of studies of long-term trends in mass media coverage of science and 'scavenge' their data to construct an *integrated index of fluctuations in science news* over the past 200 years. I will demonstrate the viability and validity of such an index. News references shall be our indicator of public attention to science. Clearly the time period of two centuries is ambitious. The challenge of mapping public discourse of science over that period would involve writing a history of public and popular science. This, however, is not my present objective. Others like Jacques and Raichvarg (1991) or recently Knight (2006) offer narrative histories of changing actors, institutional arrangements, genres and agendas in communicating science to the wider public and how this practice consolidated the role of the 'scientist' in the first place, not a term in common use before 1850. My database has an Anglo-Saxon bias. But as this part of the world has been an important location of science, this might offer a useful prototype against which other evidence can be compared. Finally, I will explore several hypotheses to explain the ebbs and flows of public attention to science. The latter will be presented in the form of an apophasis, i.e., confronting different explanations with the evidence and ordering them according to plausibility. I hope this exercise will hone the problem for research. Here, I am building on earlier efforts to map the changing intensity and motives of science communication (Bauer 1995, 1998a).[1]

[1] I worked on the problem of long-term trends in media coverage of science in the early 1990s, but with little resonance, except maybe among Chinese colleagues who keenly discussed my presentation in Beijing in 1995, and at a colloquium at Paris VII in 1996, where the late Dorothy Nelkin very much encouraged this seemingly irrelevant post-doctoral concern. Subsequently, the historical literature and the data assembled at the time moved into an office box and collected dust. I was rather delighted when the Bielefeld group invited me to revisit this idea for the present project, maybe an indication that its time has come.

3.1 Public Spheres, Public Attention and News Intensity

Public attention received by societal actors or institutions, be that of government, a political party, the family, religion and churches, social movements or business corporations, is commonly assessed on three basic variables: intensity, framing and positioning. *News intensity* is measured by the dedicated news space, *issue framing* qualifies the structure of the discourse, and by *positioning*, the actors are evaluated and the attitude of the media outlet is presented. I consider 'science' to be such a societal activity that receives attention from print media and their audiences, which can be characterised by intensity, framing and positioning, and this varies over time.

The following explores mainly the intensity of science news over nearly 200 years, while questions of issue framing and positioning are dealt with in passing. Public attention to science will be our key concept. But is public attention identical with mass media coverage? And is public attention a passive or an active phenomenon?

For the researcher working at present, any definition of public attention must consider different arenas of the public sphere where attention might focus differently. The public sphere itself is a modern form of mediation between government and citizens, a space of display, an arena of inspiration, and a forum of scrutiny. This sphere of public debate moderates private interests towards common concerns and developed since the seventeenth century in Europe and North America (see Taylor 2007: 185ff). The public sphere constitutes a common space where attention focuses and opinions can converge if needed. Its key feature is freedom from censorship and interference from governments: the hard won statutory (Human) Right to Free Expression, in particular of a free press. The public sphere has a utopian core centred on procedural ideals of non-exclusion, arguments only, good arguments shall prevail and absence of violence, undue pressure and self-delusion; elements that can decay (see Habermas [1962] 1989). This sphere is central to modern secular society and differentiated into mutually resonating arenas: formal and informal deliberations, everyday conversations in cafes, canteens, public houses and on the streets, and mass media of all kinds including books, newspapers, magazines, radio, television and recently the internet.

The public sphere includes the mass media among others; and each medium being selective on its own terms creates a system of distributed attention (Neidhardt 1993). While mass media are not identical with the public sphere, they are nonetheless a valid indicator of its operations, particularly in the long run. Currently we might observe that newspaper coverage of science is very different from that of TV and radio, and neither of these is identical to people's everyday concerns and conversations. To reassure ourselves we compare news analysis and opinion polls of people's concerns, and might conclude that indeed Middle America reads and worries about little else than the 3Gs: Gays, Guns and God. While on a daily basis it remains open whether mass media drive public concerns or are driven by them, in aggregate over the year this distinction is obsolete because news and public attention are closely coupled. Any lag between attention and news will be levelled by convergence over time. There is also good evidence of 'news herding', i.e., convergence between different news outlets on certain topics and convergence between news and public attention, particularly in moments of crisis when distributed attention focuses on the same few issues (McCombs 2004). For the purposes of a long-term indicator, we would therefore conclude that a measure based on newspaper and magazine coverage is a valid index of the trends in the level of public attention to science.

Attention is selective and has an active and a passive modality. Attention is pathetic, raised involuntarily by unusual events. We are passive in relation to an attention seeking stimulus. On the other hand, we focus our attention voluntarily, we make an effort, follow an event, and we are vigilant because we expect something to happen. A recent theory postulates a healthy equilibrium of intention and inclination; and the system might be taxed by active attention having to compensate for too many passive distractions (Kaplan and Berman 2010). In science communication, the emerging imbalance between active news-making, i.e., investigated news, towards attention seeking through news-feeding and press-releases is indeed a crisis index of science journalism (see Goepfert 2008).

Public attention works as a set of spotlights, and what they highlight is stereotypical, conforms to expectations, news values and perceptual schemata (see Lippmann 1922; Waldenfels 2004) which elaborate certain features of science and downplay others. Here, notions of framing and social representations of science and technology have their analytic importance (Farr 1993; Wagner and Hayes, 2005; Jovchelovitch 2006; Bauer and Gaskell, 2009).

All considered, it seems clear that the flow of science news is no historical constant, but driven by factors internal and external to science. Science passively gets and actively seeks public attention. Science stories compete for the attention of news editors and broadcast producers who second guess their audiences. Over time, we can expect resonance between media coverage and public attention. Hence, mass media such as books, periodicals, and newspapers are covariates of public attention, either given or sought, and we can take the frequency of mass media references as a valid index of public attention to science.

3.2 Studies of Long-Term Science News Intensity

There are few longitudinal studies of science coverage that applied systematic methodology such as content analysis and included continuous estimates of news intensity. My observations will be based on these few studies that jointly cover the past 200 years.

The topic ‘science in the mass media’ has recently become a cottage research industry, not least enabled by the availability of keyword search facilities in online press archives (for recent reviews, see Lewenstein 1995 or Schäfer 2011). A keyword such as ‘human genome’ in a database such as *Lexis/Nexis* creates with little effort an index of news intensity on this topic over the past 20 years. There is no limit to the specificity of the topic that can be investigated. Similarly, *Google Trends* offers instantaneous time-lines for keywords and time period; though remains opaque as to the exact metric. Very recently Google's *Books Ngram Viewer* has increased the stakes in this game of the instant intensity index. These facilities make the construction of news intensity measures very easy indeed. However, in a frenzy of specificity, the overall picture of 'science and technology' can easily get lost; one cannot see the wood for the trees. It is unclear which keyword combination to use to study 'science and technology' in the news, and one cannot just sum up the specific studies because of the semantic overlap of many keywords. Simple summation would result in massive double counting.

To create an overall picture of science coverage, one best resorts to estimates based on systematic sampling of relevant material using a broad definition of science news and reportage. However, only a handful of studies use such a broad criterion. On these we will focus our attention. We only consider the variability they report by normalising each study to the same scale. Figure 3.1 shows the estimates of news intensity as z-values for the entire study. Z-values have a mean of 0 and a standard deviation of 1: z (SD = 1; $M = 0$).

Starting with the early nineteenth century, Cooter (1984) explores the meaning of popular science in a genre which for present readers seems unusual but that was a popular way of engaging science at the time. The focus is on periodicals publishing in and around phrenology, the idea that human character manifests itself in cranial cavities. The study lists periodical publications making reference to phrenology, ethology, spiritualism, etc. that were in press between 1823 and 1914. By creating a year-on-year count of how many of these outlets were in press, assuming a similar number of articles for each edition, I arrive at an index of changing news intensity,

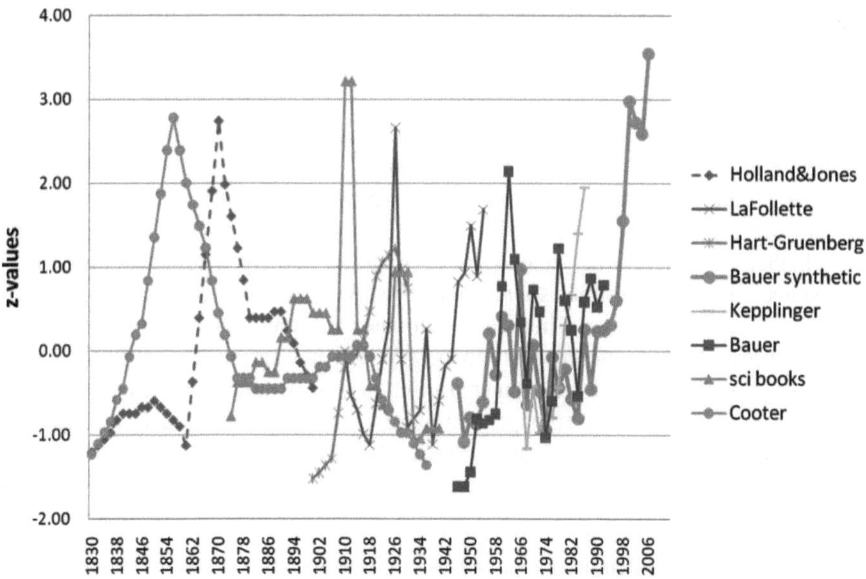

Fig. 3.1 The longitudinal studies presented as deviations from their mean

in Fig. 3.1 marked as 'Cooter'. This shows the peak in the 1840s, a decline and a modest revival of this genre of public science in the late nineteenth century.

Holland and Jones' (1994) offer an index of science items in the *Athenaeum*, an important English Victorian periodical serving an elite audience. Their effort, covering the period 1830–1900, is incomplete but gives estimates of the ebb and flow of science reportage in nineteenth century Britain in 10-year intervals. Science coverage in the *Athenaeum* is higher in the 1860s and 1870s compared to the periods before and after.

Popular science book publishing, itself subject to waves of boom and bust (see Turney 2008), equally provides an indication of public attention to science. Based on mainly French and British sources, I counted popular book publications year-on-year which are listed (see Ring 1988; Beguet 1990). 'Sci books' in Fig. 3.1 show increased numbers in the 1870s, then a decline, a spike in the early years of the twentieth century, a decline during WWI and an increase again in the 1920s.

The third study is LaFollette's (1977 and 1991) investigation of science in US periodicals from 1905 to 1955. This study showed the decline of science coverage in the early years of the twentieth century, an increase in the mid 1920s, a short spike in the mid 1930s, and a continuous rise after WWII.

Gruenberg (1935) reports on a study which maps the decline in religious and increase in scientific themes in US periodicals. During the first decades of the 20th century, periodical circulation increased from 5 million to 33 million, and six popular science magazines expanded their reach from 57,000 to 1.2 million readers in the USA. The share of science periodicals in the US increased from 1 percent in 1900 to over 4 percent, peaking in the mid 1920s; at the same time, the share of Protestant religious periodicals declined from 4 percent to about 1 percent (ibid.: 135ff). This trend appears as 'Hart-Gruenberg' in Fig. 3.1. Over the same period, the uptake of science in the school curriculum seems to decline within an expanding curriculum, pointing to a paradox of less attention to science in schools and more public attention.

Our own map of science reportage in the British Press between 1946 and 1994 (Bauer et al. 1995) is based on a random sample of news. Every second year, we randomly selected 10 days of several British newspapers from Left to Right, popular and quality, and within them all relevant materials on microfiche.[2] Judgement on whether an article was 'science news' was guided by a broad definition of 'relevance', including crime, sports and life-style news if it contained reference to science and technology. This study estimates the ebb and flow of all science reportage and its thematic contents in post-WWII newsprint. The overall picture was surprising to us, shown as 'Bauer' in Fig. 3.1. In Britain, science news, after a period of limited space due to paper rationing, expanded continuously through the 1950s, reaching a peak in the early 1960s. The news flow then declined into the 1970s, to recover during the 1980s and into the 1990s.

[2] We conducted our research in the early 1990s, when on-line resources were not yet at hand. Systematic data collection was a rather laborious undertaking involving many visits to the British Newspaper Library making photocopies from microfiche. Today's bottleneck is no longer data collection, but data analysis.

The decline of the late 1960s into the 1970s and the subsequent expansion of science news is also documented by Kepplinger (1989) in Germany between 1965 and 1985. His indicator, 'Kepplinger' in Fig. 3.1, shows an increase in coverage during the late 1970s, mainly in science-critical and politicised terms. A new elite 'on the march through the establishment' changed science news into an 'artificial horizon'. This interpretation of the expansion and framing is controversial, not least as it assumes a quasi-Leninist role of science journalism as an extension service, which is an unrealistic expectation of the role of mass media in modern society (see Ruhrmann 1991, specifically on Kepplinger).

To update the picture after 1990, I combined the results of a number of keyword searches on *Lexis/Nexis* and the *UK Guardian Archives,* including nuclear power, environment, internet, biotechnology and nanotechnology. Our own studies on biotechnology and genetic engineering showed rapidly expanding news coverage peaking in the early 2000s (Bauer et al. 1995; Bauer, 2007). Environmental news went through various cycles of public attention from the 1960s, reaching high intensity with the debate over global warming in the 1990s and beyond. Nuclear power news regained prominence with the apparent 'renaissance' of nuclear energy in the early 2000s. Various waves of news have accompanied the 'computer revolutions' since the late 1940s, from mainframe in the 1950s, to microcomputer in the 1960s and 1970s, to home and personal computers in the 1980s and beyond, and the internet since 1990. The news of computing, information technology and the WWW-internet reached unprecedented levels in the 1990s with the millennium hype. My index of computer news requires a logarithmic scale; the news intensity dwarfs previous coverage into insignificance. Taking all these topics – nuclear, environment, biotech and computing – together, each standardised to its own trends, I reach an estimate of how science news has evolved after 1990. Figure 3.1 includes this index as 'Bauer synthetic', showing the massive expansion of news space into the new millennium.

3.3 Constructing an Index of Fluctuations and Ignoring Volume

This cliometric approach to public science might be desirable for several reasons. First, we can demonstrate the ups and downs of public attention. Second, we gain a criterion to periodize the history of popular science on the intensity of news flow. This could foreground correlated changes in genres, format, mediums, themes and framing of science. Third, the emerging picture might inform communicators, historians and sociologists of science.

The above studies constitute a series of overlapping 'probes' taken from the communication system of science: Victorian England, late nineteenth century France, early twentieth century USA, post-war Britain and elsewhere. They can provide a synthetic picture, if we are prepared, at least temporarily, to ignore the specificity of each context. We must be aware that these studies do *not* provide an index of the volume of public science. Focus on a few print outlets cannot deliver such an estimate anyway. The genres of science communication proliferate over the past 200 years from public lecturing and debating in defined places, to itinerant lecturing and education, to high and low brow periodicals, to expositions, to the mass circulation newsprint in the late nineteenth and early twentieth century, including the popular science magazines, to the arrival of broadcasting radio and television in the early and mid twentieth century, and recently the internet at the turn of the twenty-first century (see Whalen and Tobin 1980; Kuritz 1981; Sheets-Pyenson 1985; Raichwarg 1990; Cooter and Pumphrey 1994; Knight 2006). With all this, we must assume a massive expansion of volume through different formats along with the expansion of news space, media genres and formats in general. Price (1963) and Wagner (1985) documented exponential growth in matters of science communication since the mid eighteenth century.

Unable to represent the volume of science news, nor to differentiate the UK, France, the US or Germany, our index offers at best an overall sense of fluctuation, above and below the trend for a global system of science news: At times the news stays below, at other times above the expected trend.

How might it be possible to link these studies into a continuous time-series? Some statistical 'trickery' is required. In order to achieve a coherent picture, I chain the data presented in Fig. 3.1 and produce Fig. 3.2. This 'chaining' is achieved in several steps which are outlined in the Appendix. The consolidated long-term picture suggests secular waves of science news. It is prudent to talk of waves rather than cycles to stress the irregularity of the dynamic. Let us focus our interpretation on the 10-year moving averages that are presented in Fig. 3.2 with a band of uncertainty.

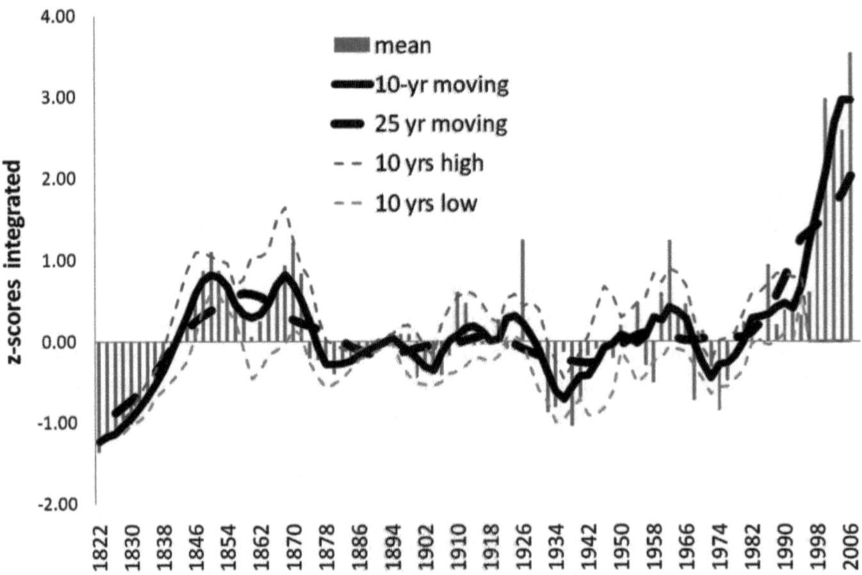

Fig. 3.2 An index of science news fluctuations between 1820 and 2006, above and below the long-term trend. The graphic shows estimated z-scores and 10-year and 25-year moving averages within a band of moving +/– 1 SD of estimates for 10 years averages

After an assumed retraction from an earlier expansion in the late eighteenth century (see Shapin 1974; Mortureux 1978; Moss 1983; Niderst 1991; Wardhaugh 2009), public attention to science stays above the trend for 36 years between 1842 and 1878, stalling slightly in the late 1850s. The late nineteenth century brings to an end what one might call the 'double wave' of the mid century.

A second wave peaks after WWI and declines in the 1940s. Science news stays above the trend for 20 years between 1908 and 1928, stalling slightly during WWI.

A third wave shows steady rise of news intensity into the mid 1960s, staying above the trend for 16 years from *1950 to 1966*, declining until the mid 1970s below the trend. Since 1982 we see a renewed rise of science news above the trend. This expansion is particularly accentuated in the first decade of the twenty-first century with two standard deviations above the expected level.

Hereby, a long-term picture of public attention to science is emerging: four irregular waves of expansion and contraction around an unknown trend between 1820 and 2010. This picture allows to draw two broad conclusions.

First, the fluctuation in public attention to science is stronger in the nineteenth than in the twentieth century. The double wave of the mid nineteenth century is more accentuated than those of the early and mid twentieth century. A the time, science news was a real signal above and beyond the background noise.

Second, the expansion of science news since the 1990s and into the new millennium to over two standard deviations from the trend is without precedent over the past 200 years. This result seems to corroborate an assumption of this yearbook edition: The recent increases in science news are historically unique and point to qualitative change in the system. A new technical term such as 'medialisation' might thus be justified to mark this fact.

3.4 Additional Evidence of Ups and Peaks, Downs and Troughs

One might opine that the above observations are based on slim evidence, which is undeniably the case. So let us take these propositions as hypothetical and confront them with further evidence. Additional evidence is already scattered in PhD theses and in scholarly research of different periods, or it might only be forthcoming. Here are some examples.

Turner (1980) talks of three phases of science in public in the nineteenth century that involve shifts in the framing of science and changing actors. From 1800–1851, science is established as an enterprise to make a living. From the mid 1840s to the 1870s, scientists mobilise an attack against religion in an attempt to gain access to the public education system and financial resources and to create a 'scientific priesthood' as an alternative (Galton's term, see Turner 1988). Finally, from 1880 to 1919, science is reframed in public as a tool of national competition against Germany, culminating in the 'science guild' and its concern with industrial efficiency. Turner does not offer quantitative evidence, but hints at rising expectations, i.e., increasing public attention, until the 1870s. When these are not met by government resources and curriculum reform, decline sets in that only recovers on the new efficiency agenda.

This story is consistent with a secular increase of public attention into the 1870s and a subsequent decline.

By contrast, Hinton (1979) documents an early Victorian decline in public attention between the 1820s and 1860s. An analysis of the contents of public book collections in Birmingham, Newcastle, Aston and other places of Victorian England showed that, while total book numbers increase between 1825 and the 1860s, though by no means matching the collections of the twentieth century public libraries, the relative number of science and engineering publications held declined from one third to a fifth and less (see Hinton 1979: 233). Assuming that these collections responded to popular demand, this suggests a declining public interest in science topics relative to others between 1825 and 1860. A similar picture arises from the contents of popular nineteenth century life improvement magazines. Titles such as *Chamber's Edinburgh Journal, The Penny Magazine* and *The Saturday Magazine* carried between 20 percent and 35 percent science content in 1835. This declined to between 11 percent and 16 percent in similar titles like the *Leisure Hour*, *People's Magazine* and *Good Words* by 1869. Consistent with such decline is the crisis in the Royal Society, its competition with the Royal Institution (since 1799), and the foundation of the British Association (in 1831), all reflecting dissatisfaction with the existing arenas of science communication (see Knight 2006). In 1831, the British Association was formed by leading men of science who were in contact with local artisans. For Sheets-Pyenson (1985) the decline and resurgence in the mid nineteenth century marks the change in the constitution of the scientific field by artisans, cultivated amateurs and naturalist, self-made men and country parsons, into a more professional activity. These observations do not clearly match my calculations. There seem to have been a decline and a recovery of public attention in the early nineteenth century; however, the timing of the reversal remains unclear on this evidence.

I also retain doubts over the validity of a sharp decline after 1870. By that time, 15 in 10,000 Englishmen and women were members of one of 1,200 scientific societies across the country (Sheets-Pyenson 1985: 562). More likely we have to assume a plateau of some length. Ellegard (1990) gauges British public opinion on Darwin's evolutionary theory between 1859 and 1972 and shows a stable interest in this topic, though mainly in high brow periodicals; in low brow outlets it is much lower; and in the middle ground interest is declining (ibid.: 346ff). The carrying capacity for science reportage expanded in the 1870s and 1880s. The steep decline in the *Athenaeum*, my indicator, might point to increased specialisation as science contents move from general societal to specialist science periodicals like *Nature,* founded in 1869.[3] Victorian periodicals served different segments of an increasingly literate British public (see Lancashire, 1988: 46). Similar impressions appear in Broks (1996: 25ff) who estimated between 4 percent and 10 percent science news

[3] In the nineteenth century *Nature* was a journal specializing in science, while the *Athenaeum* was a general read of the educated classes with an interest in science among other things; the proliferation of specialist science journals which nowadays makes *Nature* a very general basic research journal only happened in the early twentieth century.

in the popular press, more in the monthly than in the weekly periodicals, and more so in the upmarket ones, much of it natural history and health matters. The exact timing of these figures is unclear, but they clearly compare favourably to the 1 to 4 percent science in the news 100 years later (see below).

In the US, Ehrhardt (1993: 49ff) observed a decline in public science after a peak in the 1870s, which only recovers in the early 1900s. This is consistent with my observations. The educated elite had been infused with the ideals of science, and the growth of the 'yellow press' was not expanding science reportage, though they included science stories. Ironically, several studies that provide quantitative estimates of science news are motivated by moral panics over the yellow press and a moral decline that manifested itself in the displacement of religious by other interests such as science in everyday life (see Krippendorff 2004: 3ff).

In the late nineteenth century, science had morphed from a marginal and amateurish pastime of gentlemen, and of the occasional artisan with a sense for curiosities, into a professional activity, aspiring to take a central role in modern society with claims to promote social progress, overall prosperity and public health (see Knight 2006).

LaFollette's mapping of a boom in science news in the 1920s is well corroborated. Gruenberg's (1935) data for the period 1900–1930 overlaps with LaFollette's for 25 years and confirms the increase between 1900 and 1925, declining thereafter. Caudill's (1987) study of US news suggests that the Darwin debate centred in the famous 'monkey trial' of 1925 contributed to this picture. Before WWI this expansion of popular science was in part a socialist effort to broaden workers' education (Cotkin 1984).

Other research on the first half of the twentieth century confirms a rapid expansion and correlates it with an emerging new actor: the division of labour between scientists doing research and specialist communicators brings to the fore the 'popularisers', employed and encouraged by professional societies (Ehrhardt 1993). This period saw campaigns of the US Medical Association against quacks and dubious remedies and the beginning of large scale public health campaigns sponsored by life insurance companies. By 1928, it was reported that science book publishing had expanded much faster than that of other books; though book sales stalled with the depression of the 1930s (ibid.: 213ff). The 1930s saw a radical movement fusing the concerns of science with political change. Both in the US and in Britain, prominent scientists-cum-activists formed 'visible colleges' (Werskey 1988), though a boom in public attention to science does not seem to have come from this mobilisation effort; more likely it set the scene for things to come.

The post-WWII expansion of science news is confirmed by the overlap between LaFollette's and our own data between 1945 and 1955. This is a period of expansion of public attention to science, not least on nuclear issues, both civil and military. 1953 saw the launch of the 'Atom for Peace' programme, designed to control proliferation of nuclear weapons and to give the atom a peaceful image as the energy provider for the post-war reconstruction efforts (see Weart 1988). This expansion of science news continued until the mid 1960s. The period 1950–1966 is clearly significant in public attention to science: It saw the emergence of concern over science literacy and the public understanding of science in the US and elsewhere after Sputnik of 1957 (see Lewenstein 1992).

The reversal of the low point of the 1970s is confirmed by an Australian study for the period 1980–1990 documenting a fourfold increase of science and technology news (ADITC 1991). A Japanese study for the period 1985–1992 shows an increase of science news from 1 percent to about 1.5 percent of the total newshole, which is a 50 percent increase over 7 years (Niva and Kobayashi 1994). A US study covering 1966–1990 observes a science news increase in leading outlets from 0.42 percent to 2.04 percent over the period, a five-fold increase often in specialist science sections. In 1978, only the *NY Times* had a science section. By 1989, 95 US papers had one (Pellechia 1997), many of which have probably gone again with the recent US crisis of journalism, though the overall science newshole seems to reach 10 percent including science, technology, environment and health and medicine by 2010 (PEW 2011).

Bucchi and Mazzolini (2007) report on Italy, 1946–1997. Science coverage increased to a peak in the late 1960s and then declined, only to recover since the 1980s. This overall secular picture of an up-swing into the 1960s, a decline into the 1970s and a subsequent recovery of science news after 1980 is documented even across the Iron Curtain. Petkova and colleagues report on Bulgaria, 1946–1995, and confirm this long swing even for the official Communist Party newspaper (see Bauer et al. 2006).

Analysis of the social representation of science, i.e., the rhetorical framing and positioning of scientific actors in public would add an additional dimension to this long-term story, and one could do another round of scanning the scholarly sources such as Hayes (1994). Clearly the elaboration of science news in terms of issues, framing and positioning of science in public discourse is no constant either. For example, LaFollette (1991) showed that biology provided most story lines pre-WWII, while post-WWII was dominated by physics news, not least by reports from the nuclear frontier. Our own analysis confirms this for the UK: Until the 1970s science news is predominantly physics and engineering news, while ever since, bio-medical news slowly but steadily retakes the limelight; this 'medicalisation of science news' is more in evidence in the popular than in the quality press (Bauer 1998b). Also, post-WWII, we observe that science news is predominantly positive and celebratory when reporting on computers and space technology, ambivalent on genetics and biotechnology, but predominantly negative over nuclear power and the environment, leaving aside trends and phases for the moment (Bauer et al. 1995).

3.5 Why Is Public Attention to Science Not Constant?

All this being said, the evidence is fairly clear: Public attention to science is no constant, and over the past 200 years, several periods of higher and lower than expected public attention can be identified. How to explain this is rather a different matter. Let us explore a number of ad hoc ideas that might bring us closer to understanding these fluctuations. There might be *societal factors* that bring science into the limelight and *endogenous factors* that increase the need for science to actively seek public attention. The following is an attempt to explore some correlates of public attention and gage their explanatory power.

3.5.1 Changing Societal Contexts

The ups and downs in public attention might be a response to challenges and opportunities which society put at the door of science. Increased attention seeking might be a response to a crisis of legitimacy; changes in mass media technologies make news-making easier, and long-term economic trends encourage or discourage science news to surf the waves.

- *Hypothesis 1: Response to a crisis of legitimation.* Each expansion of public attention implies a legitimation crisis of science in society. When the social value of science is in doubt, efforts to mobilise public support result in increased public attention.

A crisis of legitimation of science is often invoked in discussions of science literacy or public understanding of science. Scientists fear that their licence to operate is under threat. Since the mid 1970s the post-war arrangements of unquestioned public support for pure science can no longer be taken for granted: 'Science is too important to leave to the scientists' became the rallying cry to public accountability. This call had different motives: an economic policy worrying about cost-effectiveness and a civil society worrying about science preparing for nuclear warfare and polluting the environment. Is science sustainable, delivering the goods, value for money, translating ideas into marketable products, bringing societal progress? Public attention seeking by science is thus a form of continuation of politics by other means. These efforts are born out of necessity to regain lost privileges, and lead to the mobilisation of scientists since the 1980s (see Bauer and Jensen 2011), manifested in the influential reports of the Royal Society (1985) and the House of Lords (2000).

However, a crisis of legitimation does not seem to cut much ice. It is unlikely that all waves of expanding public attention are a response to crisis, nor is it specific to science, because the crisis of legitimation affects all societal institutions. Earlier periods of public attention are less a crisis response than the building up of a legitimate position in society in the first place. The nineteenth century sees the formation of the scientific profession, and its activities repositioned from marginal pursuits to the centre of state policy by WWI (see above and Knight 2006). The crisis response does not seem to be a fruitful explanation for long-term developments, though maybe throws some light on what happened in the 1980s, kick-starting the current wave of public attention. More evidence to the point might emerge soon, as the Royal Society reflects back on 25 years of public understanding of science.

- *Hypothesis 2: New mass media affordances.* The formats and media of public science change, each wave expresses the enthusiasm created by a new technology of mass media which creates an overdrive of news activity

Science communication depends on the current state of mass circulation technology. The ups and downs of news intensity might reflect the simple fact of new opportunities to communicate which creates its own dynamics. New technology creates affordances. In that sense the expansion of the mid nineteenth century reflects the potentials of mass periodical printing in its high and low brow versions. The surge at the turn of the century reflects the new popular science magazines, complementing the specialist outlets for the professionals. The expansion into the 1920s comes with radio broadcasting, popular book publishing for an increasingly literate public, and the arrival of the 'yellow press' with focus on sensation rather than robust information. The expansion into the 1960s, though here measured on news print, rides on the back of television and the very latest, since the 1990s, on the back of the PC, internet and social media. New means of communication offer opportunities. This affordance is also welcomed by scientists who seek public attention in order to sway the opinions of peers in controversies over scientific matters such as 'big bang', 'gaia', evolutionary psychology or cold fusion (see Gregory 2005; Cassidy 2005).

- *Hypothesis 3: New actors make new waves.* Each wave is defined by novel actors of science communicators; as one type of communicators outlives the other, so does one wave follow the other.

The expansion of the mid nineteenth century might also reflect changes among the speakers of science: The new 'players' of science replaced the 'gentlemen of leisure' and country parson naturalists. Men like Humphry Davy and John Dalton sought to make a living from research and further their interests by pressure groups like the Royal Institution (RI founded 1799) and the British Association for the Advancement of Science (BAAS founded 1831; see MacLeod and Collins 1981). Both the RI and the BAAS competed with the Royal Society and pressured an old guard who closed shop in 1847 with a new admission system. Later in the 1870s, this competition between old and new actors manifests itself also in the polemic against religion. Here, we see the scientist-politicians getting to work.

The wave of public attention in the early twentieth century is carried by scientists-cum-popularisers writing popular books and for magazines. These publicly visible researchers address a wider audience to mobilise for social change, including Eugenics and efficient production. In Britain initially to save the Empire (MacLeod 1994), later in the 1930s, this activism includes radical socialism (Werskey 1988).

The post-WWII wave of public attention to science is carried by a new breed of journalists specialising in science reportage. These science writers organise in the American Association of Science Writers (AASW) and its British or French equivalent (BASW or AESF). The active scientists continue to lend their authoritative voices to radio and their faces to the glamour of TV, only to withdraw from either in the 1970s and to refocus their attention on lobbying and policymaking, leaving the field of public attention to science writers (see Gingras 1994, for Canada).

Critical science journalism marks the start of the latest up-swing in science news since the 1980s. Science journalists are on record in referring to Three Mile Island (1979) as a turning point in their ethos: The nuclear power issue and environmental degradation turned hitherto advocates of science into henceforth critical observers (Carsons [1962] 2000; Franklin 2007). Kepplinger (1989) observed the subversive influence of an activists baby-boom generation. It appears that the intensification of news since the 1990s arises from much professional Public Relations marketing genomics, climate change, stem cells, synthetic biology or nanoscience like any other commodity (Nelkin 1995). Every lab nowadays has its own press function that creates a backlash in form of knowledge consumer protection (Goldacre 2009).

The idea that each wave of public attention to science correlates with new actors who carry the effort seems to have much to go for. Science communication moves from the science-entrepreneur and science-politician (mid nineteenth century), to the science-popularisers (1920s), to the science communicator (1950s), to critical science journalism (1980s), and finally to the Public Relation professionals of the present.

Economic historians observe the long-term trends in the economy and identify longer periods of expansion, recession and depression. For some this rhythm is regular and endogenous to 'capitalism' with a cycle of around 50 years; for others this rhythm is irregular and due to external shocks to the world market. Table 3.1 lists the different chronologies proposed by these observations, the idea of endogenous Kondratjeff cycles (see Trebilcock 2002; Van Duijn 1983),[4] each upturn defined by a new set of technologies, and observations of phases of exogenously varying growth rates (Maddison 1991, 1995).

- *Hypothesis 4: long-waves of economic growth.* Public attention to science reflects economic activity. Hence, the phases of public attention to science co-vary with periods of economic growth.

[4] The name 'Kondratjeff cycle' honours its 'inventor', the Russian economist Nikolai Dimitriyevich Kondratjeff (1892–1928), an agricultural economist who founded in 1920 the Russian Conjuncture Institute and died in a Siberian labour camp. He developed the hypothesis that the world economy would expand and contract in cycles of about 50 years, and this was endogenous to the capitalist economy; what brought him into conflict with his masters was the possibility that capitalist economies might recover with as much regularity as they get into crisis. Later economists such as Schumpeter and others developed this hypothesis and linked it to inventions and innovations. Inventions are swarming in depression periods, and each new upswing is characterised by a new scientific technological impulse, hence the naming of the cycles as 'cotton-iron-steam' or 'railroadization' or 'electricity and automobile', etc. (see Van Duijn 1983). Many historians such as Maddison are critical of the very existence of these 'cycles' and see more exogenous shocks such as wars and policy ideas as the source of long-term changes in growth rates.

Table 3.1 Public attention to science, long waves, and phases of growth in world economy; comparing different chronologies of long waves of development

Public attention (10-year moving averages)	Long waves (by Kondratjeff)	Key technologies of the time	Phases of growth (by Maddison)
Decline >1800 Low: 1820s	I. 1790–1847 Up 1790–1825 Down: 1825–1847	Water, Textile, Iron	1820–1870 [1 percent]
Above 1842–1878 Peak: 1850 Peak: 1868 Low: 1880 and 1904	II. 1847–1893 Up: 1847–1873 Down: 1874–1893	Steam, Railways, Steel	1870–1913 [2.1 percent]
Above: 1908–1928 Peak: 1924 Low: 1936	III. 1893–1945 Up: 1893–1929 Down: 1929–1945	Electricity, chemicals, Automobile	1913–1950 [1.9 percent]
Above: 1950–1966 Peak: 1962 Low: 1972	IV. 1945–1990 Up: 1945–1973 Down: 1973–1990	Petrochemicals, Electronics, Aviation	1950–1973 [4.9 percent]
Above: 1982–present Peak: present	V. 1990–present Up: 1990–2008 Down: 2009–?	Digital networks, Biotechnology, Nanotech	1973–2008 [2.2 percent to 1992]

Source: For Kondratjeff cycles, see Trebilcock (2002) and Van Duijn (1983); for phases of economic growth, see Maddison (1991, 1995)

Recent ideas on science fiction suggest that public attention to science might have a role to play in anticipating innovation and economic development. The notion of 'diegetic prototyping' points out that fictional imagination creates public expectations for new developments, showing the benevolence and viability, normalises the innovation before the fact, and thus prepares the ground for the facts to follow (Kirby 2009). Such imagination cultivated in public attention to science is a growth factor. It creates political imperatives, moves young people into science careers, mobilises human resources, and stimulates consumer awareness and demand.

Table 3.1 brings into a synopsis the three chronologies of public attention to science, Kondratjeff cycles, and growth phases of the world economy. The comparison of these three historical rhythms shows no clear coincidence of public attention to science, neither with the long waves, nor with the phases of growth. The striking synchronicity between all three chronologies occurs in 1870 and again 1970. Around 1870, public attention to science is at its peak, and so is the second Kondratjeff cycle followed by a long depression. For Maddison, a new phase of higher economic growth starts around this time. One hundred years later, around 1970, public attention to science is at a low point, when the fourth Kondratjeff cycle reaches its peak, and a new phase of low economic growth begins with the 'oil crisis'.

The periods when public attention is above expectations – 1842–1874, 1908–1928, 1950–1966 and after 1982 – science news seems to overlap with expanding Kondratjeff cycles. Though not entirely synchronised, sometimes public attention anticipates the economic cycle as in the 1840s and in the 1980s; in the 1920s attention falls into the later part of the expansion, and in the early phase in the 1950s. Because the dating of Kondratjeff cycles is not entirely consensual among authors, this matching depends on setting the boundaries.

3.5.2 Endogenous Factors in the Operations of Science

There could be other factors bringing about the instability in public attention to science. Expanding scientific research creates a crisis of self-confidence. Seeking public attention might be a way of self-reassuring science of its epistemic authority. Second, if public attention to science involves myth making, but debunking and disenchantment is the core ethos of science, then public attention would create contradictions at the very core of science communication.

- *Hypothesis 5: Epistemic self-assurance.* Scientific knowledge expands, and with expansion comes an identity crisis and a crisis of self-confidence. Public attention gained or sought serves an epistemic self-reassurance.

Ludwik Fleck ([1935] 1979) suggested in a study of the consolidation of a scientific fact, that such facts are supported by thinking communities and thinking styles which enable research in the first place and validate the claims made. In addition, Fleck observed that knowledge that arises from thinking communities is 'esoteric' and uncertain without public attention and recognition. Only textbooks and popularised renditions of facts and theories turn 'esoteric uncertainty' into 'exoteric certainty' which reassures not only consumers, but also producers of knowledge. This attitude of certainty arises from the rhetoric of simplification, concrete examples and visualisation. Jurdant (1993) restated this function of outreach as the 'autobiography of science': Like autobiography, the attention seeking of science combines fact and fiction and self-serving apology. This hypothesis suggests that public attention lags behind waves of expanding knowledge. For a test we need a measure of knowledge production. Dicke (1975) traced the production of PhD dissertations since 1880 and observed an average annual growth rate of 6.8 percent and periods of high production above the trend in 1880–1900, the 1930s, and the 1950s. These periods are low in public attention. However, they anticipate the expansions that follow in the 1910s and 1920s and again in the 1950s and 1965. This correlation between spurts of knowledge production and subsequent public attention deserves further attention.

Jurdant (1969) and others have pointed to another feature of public attention to science, its inherent paradoxical nature as ideology and myth making. In order to bridge the gap between science and common sense, communicators resort to visual images, analogy, drama and story telling to accommodate to existing commonsense understanding of the world. This includes emotional appeals and references to authorities. However, commonsense, visualisation, emotional appeal and cult of authority are historically contrary to the iconoclastic ethos of science; take for example the Royal Society's motto *nullius in verba* (i.e., take nobody's word for it). The difficult relationship between scientists and journalists, and the perennial complaints of scientists about news that is sensational, hyperbole, emotional and biased by ideological prejudice testifies to this tension and that not only since 'Frankenfood' of the 1990s.

- *Hypothesis 6: The science myth making paradox.* Public communication of science is paradoxical, and therefore cyclical. Public attention for science involves drama, enchantment and myth making, which is contrary to the scientific ethos of debunking, disenchantment and demystification. As public attention to science expands, myth making becomes subject to iconoclastic polemic and science communication will decline until the need for it becomes again stronger than the concerns about it.

This contradiction between a rationalist ethos of science and a myth-making practice of public communication might explain why the expansion of public attention to science is unstable, comes in waves, and turns when it turns from expansion to contraction. When the polemic against 'sensational' and 'ideological' science news leads to a crisis of science communication, the wave of attention turns. To test this hypothesis, we need to trace the polemics against sensationalism and attention seeking in science communication, and we would expect this to be most accentuated around the peaks, hence around 1850 and 1870, the mid 1920s and mid 1960s, and again at present.

The analysis of the ideological work in public attention to science is a feature of the late 1960s and 1970s. At this time we see the proliferation of studies on the 'accuracy of science reportage' (see Tichenor et al. 1970) addressing scientists' discomfort with news values. Inconsistent with our expectation is Kepplinger's (1989) polemic of against 'science-made-politics' in the 1970s, and equally so is Burnham's (1987) diagnosis of the replacement of science reportage with superstitions in the disguise of life-style advice. However, these voices coincide with the beginning of a new phase of mobilisation of public attention, rather than with its peak years. Little is known about previous expansions in that respect, though the present medialization hypothesis ponders the risks of public attention right at the peak of expansion, and could therefore be a case in point. It seems that the paradox of communication of science deserves further investigations.

3.6 Conclusion

In this chapter, I painted a picture of long-term fluctuation in the public attention to science. Combining the quantitative evidence from several studies and my own data, I constructed an integrated index of public attention to science based on deviations from the long-term trend. My sources recorded references to science in periodicals, newspapers, and books covering the years between 1820 and 2006. Biased towards the English speaking world, the emerging chronology needs to be tested in other contexts.

The picture that emerges is one of long-term fluctuations in four waves where public attention to science is above expectations: 1842–1878, 1908–1928, 1950–1966, and 1982 to the present. This conclusion is reached on the basis of 10-year moving averages for bi-annual estimates of public attention.

This purely quantitative picture suggests that the most recent period of public attention to science, which began in the 1980s, is indeed historical. It is longer than any other wave of expansion in the past 200 years, and the amplitude is without precedent. This assumption of the medialization thesis, explored in this volume, seems to rest on firm ground.

A number of ad hoc hypotheses were explored to explain these instabilities in public attention to science: a response to a crisis of the legitimacy, affordances of new mass media technologies, new actors coming on stage, surfing the waves of economic growth, epistemic self-reassurance, and the paradox of science communication. The evidence for these schematic explanations is inconclusive, but some are more promising for further research than others. Furthermore, narrative history must illuminate the details of these waves of public attention. This is clearly a different challenge. However, the waves 'discovered' in the present effort might define the chronology for such a history of public attention to science over the past 200 years.

Appendix: How to Calculate the Index of Public Attention?

The data for the continuous 'index of science news fluctuations between 1820 and 2006' as shown in Fig. 3.2 (above) is constructed by way of the following steps involving the transformation and standardisation of data from different sources. This allows us to chain up several data series for different periods and reach a continuous estimate of long-term fluctuations around an unknown, but most likely upward trend in overall volume.

1. The individual data series are transformed into standardised z-values, irrespective of what is being counted: $z = (X - M_{\text{series}})/\text{SD}_{\text{series}}$. With this procedure any trend is taken out of the data. The missing years are interpolated. The result is a bi-annual data series for each study for which the variability becomes comparable on the same scale with an average, $M = 0$, and standard variation, $\text{SD} = 1$. One SD contains 68 percent of the variability assuming a normal distribution. The result of this transformation is presented in Fig. 3.1.
2. In an Excel file, these data series are then ordered on a continuous time-line from 1820 to 2010. However, z-values do not guarantee that the end of one series neatly matches with the beginning of the next series; therefore, step three:
3. Several of the data series overlap. This allows to average the overlapping years and to chain the discontinuous data series on these estimates, e.g., LaFollette's

and my own studies overlap for 10 years between 1946 and 1955; for these years, I average the z-values of step 1 and create a combined series from 1905 to 1992. For some periods where there are several studies, I average z-values for that period with $z_t = (z_1 + z_2 + z_k)/k$ and $SD_z\ (z_1,\ z_k)$. k is the number of overlapping studies in each year. Working through all the data series results in what is shown as bars in Fig. 3.2: bi-annual estimates of deviations from an overall, most likely exponential trend of volume.

4. Because each bi-annual z-value remains an uncertain estimate, moving averages give a more reliable picture by smoothing over short-term fluctuations. Thus in addition to bi-annual averages (bars), I report 10-year moving average, the continuous thick line in Fig. 3.2, and a 25-year moving average, the dashed thick line in Fig. 3.2. A moving average is a mean of means for a fixed number of years: e.g., $Z_t = (z_{t-5} + \ldots + z_{t-1} \ldots + z_{t+1} + \ldots + z_{t+5})/10$.
5. To document the uncertainties in the overall picture, I plot the moving average of the standard deviations (SD) that arise from combining the overlapping data series in step 3. Figure 3.2 shows the band of confidence for 10-year averages, considering one moving standard deviation on either side: $BC = Z_{10year} +/-1\ SD_{10year}$. Based on my database, with 68 percent certainty the 'true' value of public attention should fall within that band.

References

ADITC (1991). Australian Science and Innovation impact brief, Section 5: Science and Technology News in the Media, 35–43.

Bauer, M. W. (1995). The opening and closing of the window of science, 1820–1990. Paper presented to the Chinese Association for Science and Technology. Public Understanding of Science Conference, 15–19 October 1995, Beijing.

Bauer, M. W. (1998a). 'La longue durée' of popular science, 1830-present. In D. Deveze-Berthet (ed.), *La promotion de la culture scientifique et technique*: *Ses acteur et leurs logic*. Actes du colloque des 12 et 13 decembre 1996. Paris: Publications de Universite Paris 7 – Denis Diderot, pp. 75–92.

Bauer, M. W. (1998b). The medicalisation of science news: From the 'rocket-scalpel' to the 'gene-meteorite' complex. *Social Science Information*, 37, 731–751.

Bauer, M. W. (2005). The mass media and the biotechnology controversy. *International Journal of Public Opinion Research*, 17(1), 5–22 [special issue].

Bauer, M. W. (2007). The public career of 'genes' – trends in public sentiment from 1946 to 2002. *New Genetics and Society*, 26(1), 29–45.

Bauer M. W. and G. Gaskell (2009). Social representations theory: A progressive research programme in social psychology, *Journal for the Theory of Social Behaviour*, 38(4), 335–354.

Bauer, M. W. and P. Jensen (2011). The mobilisation of scientists for public engagement. *Public Understanding of Science*, 20(1), 3–11.

Bauer, M. W., J. Durant, A. Ragnarsdottir, and A. Rudolfsdottir (1995). *Science and technology in the British Press, 1946–1990* (Vol. 1–4). Report to the Wellcome Trust for the History of Medicine, July, London: The Science Museum.

Bauer, M. W., K. Petkova, P. Boyadjieva, and G. Gornev (2006). Long-term trends in the representations of science across the iron curtain: Britain and Bulgaria, 1946–95. *Social Studies of Science*, 36(1), 97–129.

Beguet, B. (ed.) (1990). *La science pour tous: Sur la vulgarisation scientifique en France de 1850 a 1914*. Paris: Bibliothèque de CNAM.

Braudel, F. (1958). La longue durée. *Annales E.S.C.* no 4, Oct–Nov, 725–53 (reprinted in 'Ecrits sur l'Histoire', 1969. Paris: Flammarion, 41–83).

Broks, P. (1996). *Media science before the Great War*. London: MacMillan Press.

Bucchi, M. and R. G. Mazzolini (2007). Big science, little news. Science coverage in the Italian daily press, 1946–1997. In M. W. Bauer and M. Bucchi (eds.), *Journalism, science and society. Science communication between news and public relations*. London: Routledge, pp. 53–70.

Burnham, J. (1987). *How superstition won and science lost: Popularizing science and health in the US*. New Brunswick: Rutgers University Press.

Carsons, R. ([1962] 2000). *Silent spring*. London: Penguin Classics.

Cassidy, A. (2005) Popular evolutionary psychology in the UK: An unusual case of science in the media. *Public Understanding of Science*, 14, 115–141.

Caudill, E. (1987). A content analysis of press views of Darwin's evolution theory 1860–1925. *Journalism Quarterly*, 4, 782–786.

Cooter, R. (1984). *The cultural meaning of popular science, Phrenology and the organisation of consent in 19th century Britain*. Cambridge, UK: Cambridge University Press.

Cooter, R. and S. Pumphrey (1994). Science in popular culture. Special issue: Science Popularisation. *History of Science*, 32(3), 237–267.

Cotkin, G. (1984). The socialist popularization of science in America, 1901 to the First World War. *History of Education Quarterly*, 24, 101–214.

Dicke, R. H. (1975). Are there inherent limits to the natural sciences? *Journal of the Franklin Institute*, 300, 93–98.

Ehrhardt, G. R. (1993). *Descendents of Prometheus: Popular science writing in the US, 1915–1948*. PhD Thesis, Department of History. Durham, NC: Duke University.

Ellegard, A. (1990). *Darwin and the general reader*. Chicago: Chicago University Press.

Farr, R. M. (1993). Common sense, science and social representations. *Public Understanding of Science*, 2, 189–204.

Fischer, D. H. (1996). *The great wave: Price revolutions and the rhythm of history*. New York: Oxford University Press.

Fleck, L. ([1935] 1979). *Genesis and development of a scientific fact* (Original in German). Chicago: University of Chicago Press.

Franklin, J. (2007). The end of science journalism. In M. W. Bauer and M. Bucchi (eds.), *Journalism, science and society – Science communication between news and public relations*. New York: Routledge, pp. 143–156.

Gingras, Y. (1994). *Pour l'avancement des science. Histoire de l'ACFAS 1923–1993*. Montreal: Boreal.

Goepfert, W. (2008). The strength of PR and the weakness of science journalism. In M. W. Bauer and M. Bucchi (eds.), *Journalism, science and society – Science communication between news and public relations*. New York: Routledge, pp. 215–226.

Goldacre, B. (2009). *Bad science*. London: Harper Perennial.

Gregory, J. (2005). *Fred Hoyle's Universe*. Oxford, UK: Oxford University Press.

Gruenberg G. C. (1935). *Science and the public mind*. New York: McGraw-Hill.

Habermas, J. ([1962] 1989). *The structural transformation of the public sphere. An inquiry into a category of bourgeois society*. (Original in German). Cambridge, MA: MIT Press.

Hayes, R. (1994). *From Faust to Strangelove. Representations of scientists in Western Literature*. Baltimore, MD: Johns Hopkins University Press.

Hinton, D. A. (1979). *Popular science in England, 1830–1870*. Unpublished PhD Dissertation, University of Bath.

Holland, S. and S. Jones (1994). *The ATHENAEUM scientific database*. Manuscript. Department of Information Science, City University, London.

House of Lords Select Committee on Science and Technology. (2000). *Science and society*, 3rd report. London: HMSO.

Jacques, J. and D. Raichvarg (1991). *Savants et ignorants: Une histoire de la vulgarisation des science*. Paris: Seuil, pp. 290.
Jovchelovitch, S. (2006). *Knowledge in context: Representations, community and culture*. London: Routledge.
Jurdant, B. (1969). Vulgarisation scientifique et ideologie. *Communication*, 14, 150–161.
Jurdant, B. (1993). Popularisation as the autobiography of science. *Public Understanding of Science*, 2, 365–374.
Kaplan, S. and M. G. Berman (2010). Directed attention as a common resource for executive function and self-regulation. *Perspectives on Psychological Science*, 5(1), 43–57.
Kepplinger, M. (1989). *Künstliche Horizonte, Folgen, Darstellung und Akzeptanz von Technik in der Bundesrepublik*. Frankfurt/M.: Campus.
Kirby, D. (2009). Diegetic prototypes and the role of popular films in generating real-world technological developments. *Social Studies of Science*, 40(1), 30.
Knight, D. (2006). *Public understanding of science – A history of communicating scientific ideas*. Oxon, UK: Routledge.
Krippendorff, K. (2004). *Content analysis – An introduction to its methodology*. London: Sage.
Kuritz, H. (1981). The popularization of science in 19th century America. *History of Education Quarterly*, 21, 259–274.
LaFollette, M. C. (1977). *Authority, promise, expectations: The image of science and scientists in American popular magazines, 1910–1955*. PhD dissertation Indiana.
LaFollette, M. C. (1991). *Making science our own. Public images of science, 1910–1955*. Chicago: University of Chicago Press.
Lancashire, J. A. (1988). *An historical study of the popularisation of science in general science periodicals in Britain, c. 1890–c. 1939*. PhD dissertation, University of Kent at Canterbury.
Lewenstein, B. (1992). The meaning of 'public understanding of science' in the US after World War II. *Public Understanding of Science*, 1, 45–68.
Lewenstein, B. (1995). Science in the media. In S. Jasanoff, G. E. Markle, J. C. Petersen, and T. Pinch (eds.), *Handbook of science and technology studies*. Thousand Oaks: Sage, pp. 343–360.
Lippmann, W. (1922). *Public opinion*. New York: HBC (new edition 2004, Dover publications).
MacLeod, R. (1994). Science for imperial efficiency and social change: Reflections on the British Science Guild, 1905–1936. *Public Understanding of Science*, 3(2), 155–194.
MacLeod, R. and P. Collins (eds.) (1981), *The parliament of science. The BA for the advancement of science, 1831–1981*. Northwood, Middlesex, UK: Science Reviews.
Maddison, A. (1991). *Dynamic forces in capitalist development. A long-run comparative view*. Oxford: Oxford University Press.
Maddison, A. (1995). *Monitoring the world economy, 1820–1992*. Paris: OECD.
McCombs, M. (2004). *Setting the agenda: The mass media and public opinion*. Cambridge: Polity Press.
Mortureux, M. F. (1978). *La formation et le fonctionnement d'un discours de la vulgarisation scientifique au XVIIIeme siècle a travers l'oeuvre de Fontenelle*. Thèse, Paris VIII, Juin (copy).
Moss, H. (1983). Scientists and sans-culotte: The spread of scientific literacy in the revolutionary year II. *Fundamentia scientiae*, 4, 101–115.
Neidhardt, F. (1993). The public as a communication system. *Public Understanding of Science*, 2, 339–350.
Nelkin, D. (1995). *Selling science: How the press covers science and technology*. New York: W. H. Freeman.
Niderst, A. (ed.) (1991). La diffusion des sciences au XVIIIe siecle. *Revue d'histoire des sciences*, 44(3/4), Paris: PUF.
Niva, F. and S. Kobayashi (1994). *A study of science and technology information dissemination by Japan's mass media*. Paper presented to the Conference 'Education for Science Literacy', 6–9 November 1994. London: Science Museum.

Pellechia, M. G. (1997). Trends in science coverage: A content analysis of three US newspapers. *Public Understanding of Science*, 6, 49–68.

PEW. (2011). Pew research center's project for excellence in Journalism – Year in the news. http://features.journalism.org/year-in-the-news. Last accessed 25 February 2011.

Price, D. J. De S. (1963). *Little science, big science*. New York: Columbia University Press.

Raichwarg, D. (1990). *400 années de diffusion de la science par le spectacle, formes, objectifs, moyens*. Thèse de doctorat, Paris VII.

Ring, K. (1988). *The popularisation of elementary science through popular science books, 1870–1939*. PhD Thesis, University of Canterbury.

Royal Society of London. (1985). *Public understanding of science*. London: The Royal Society.

Ruhrmann, G. (1991). Analyse von Technik- und Risiko-Berichterstattung. Defizite und Forschungsperspectiven. Kommentar zu Kepplinger. In J. Kruger and S. Russ-Mohl (eds.), *Risikokommunikation*. Berlin: Edition Sigma, pp. 145–174.

Schäfer, M. S. (2011). Taking stock: A meta-analysis of studies on the media's coverage of science. *Public Understanding of Science*. (online first published on December 1, 2010 as doi:10.1177/0963662510387559).

Shapin, S. (1974). The audience for science in 18th century Edinburgh. *History of Science*, June, 12, 95–121.

Sheets-Pyenson, S. (1985). Popular science periodicals in Paris and London: The emergence of low scientific culture 1820–1875. *Annals of Science*, 42, 549–572.

Taylor, C. (2007). *A secular age*. Cambridge, MA: Belknap Press.

Tichenor, P. J., C. N. Olien, A. Harrison, and G. Donohue (1970). Mass communication systems and communication accuracy in science reporting. *Journalism Quarterly*, 47, 673–683.

Trebilcock, C. (2002). Surfing the wave: The long cycle in the industrial centuries. In P. Martland (ed.), *The future of the past. Big questions in history*. London: Pimlico, pp. 66–88.

Turner, F. (1980). Public science in Britain 1880–1919. *Isis*, 21, 589–608.

Turner, F. (1988). The Victorian conflict between science and religion: A professional dimension. In G. Parsons (ed.), *Religion in Victorian Britain, Vol. IV Interpretations*. Manchester: Manchester University Press, pp. 170–197.

Turney, J. (2008). The latest boom in popular science books. In M. W. Bauer and M. Bucchi (eds.), *Journalism, science and society – Science communication between news and public relations*. New York: Routledge, pp. 81–92.

Van Duijn, J. J. (1983). *The long wave in economic life*. London: George Allen and Unwin.

Wagner, S. (1985). *Die Entwicklung der exakten Naturwissenschaften von der Antike bis zur Gegenwart. Eine Quantifizierung ihrer Geschichte*. Two volumes. Bielefeld: Science Studies Report no 27.

Wagner, W. and N. Hayes (2005). *Everyday discourse and common sense – The theory of social representations*. London: Palgrave MacMillan.

Waldenfels, B. (2004). *Phenomenologie der Aufmerksamkeit*. Frankfurt/M.: Suhrkamp.

Wardhaugh, B. (2009). Mathematics in English printed books, 1473–1800: A bibliometric analysis. *Notes and Records of the Royal Society*, 63, 325–338.

Weart, S. (1988). *Nuclear fear. A history of images*. Cambridge, MA: Harvard University Press.

Weingart, P. (1991). The loss of distance: Science in transition. In G. E. Allen and R. M. MacLeod (eds.), *Science, history and social activism: A tribute to Everett Mendelsohn*. Amsterdam: Kluver, pp. 167–184.

Werskey, G. (1988). *The visible college – A collective biography of British scientists and socialists of the 1930s*. London: Free Association Books.

Whalen, M. D. and M. F. Tobin (1980). Periodicals and the popularization of science in America, 1860–1910. *Journal of American Culture*, 3, 195–205.

Chapter 4
Issue Selection in Science Journalism: Towards a Special Theory of News Values for Science News?

Franziska Badenschier and Holger Wormer

4.1 Introduction

"Why is our breakthrough not in the news?" This is perhaps one of the most frequently asked questions concerning the media among scientists and their public relation officers: The research results were perfect, the competitors in Europe, the US and Japan are shocked, the rest of the scientific community is impressed; and what do the media do? They ignore it! – However, instead of an alleged "ignorance" of the media, the decision to ignore the "breakthrough" may be the result of a news selection process perfectly reasonable from the point of view of journalism.

To answer the question what news is, journalism schools and editorial offices worldwide provide simple answers which are common sense among journalists: "News is what's different" or the "man-bites-dog-formula" are probably among the best known. However, originally these simple formulas have been developed mainly for the classical sections of the media. Also most of the journalism theories are based on empirical research in general journalism (especially in political journalism). In contrast, the development of science sections and science journalism was widely separated. And for a long time, science journalism was more influenced by the world of science itself than by general journalism (Rensberger 2009). Therefore, it seems to be reasonable to validate general journalism theories for the special field of science journalism from case to case. Concerning especially the news selection processes in science journalism, several authors have pointed to the need of an empirically grounded reconstruction of these processes (Hömberg 1996; Schäfer 2007). Science journalists also report about special news selection processes in their sections as did Illinger:

> Different from the political section where the daily agenda is often determined by the actual events the science editor has to dig in a rather disordered box of news if he does not like

H. Wormer (✉)
TU Dortmund University
e-mail: holger.wormer@udo.edu

S. Rödder et al. (eds.), *The Sciences' Media Connection – Public Communication and its Repercussions*, Sociology of the Sciences Yearbook 28,
DOI 10.1007/978-94-007-2085-5_4,

to rely completely on the agenda of the big scientific magazines; and when there is not the hurricane of the century whirling at the American coast [...] (Illinger 2005).

But why should a theory of news values matter for a practitioner in the media? Isn't it true (as one of the reviewers of this article supposed) that "reporters continue doing what they do with or without us (scholars, editor's note) looking over their shoulders"? First of all, we have the strong feeling that any debate about the own way of working is a good starting point for self-reflection. This is true for medical doctors doing surgery as well as for pilots and many other professions. So why should we renounce such attempts of reflection as an element of quality management in journalism which plays a fundamental role in democracy? If such a theoretical reflection and observation gets systematically connected with journalistic practice and education, there should be a good chance to produce a next generation of journalists being used to looking over their own shoulders. And with regard to well established media and their journalists in charge, there may be some good news in the bad news of the recent "media crisis": In an era of newspaper dying and media fragmentation, a notable amount of journalists cannot afford to do business as usual. In a conference with about 20 editors-in-chief from German mass media,[1] we have noted more demand for scientific advice and answers in a dramatically changing media world than probably ever before. This is especially true for questions on how (that is, by which means) to attract readers, listeners and spectators. Such a question is closely related to a theory of selection processes.

A deeper reflection of selection processes is also interesting for science communicators and scientists in all fields in order to give them a better understanding of how the media work and to improve their communication. The question why different scientific disciplines achieve a different degree of media attention is not the basic topic of this article. However, it is a frequently asked question by scientists and science communicators, for example, when they are confronted with rankings of the most popular disciplines in the media (see Table 4.1). We believe that internal factors in journalism are part of the answer.

In this work, we try to shed light on the selection processes in science journalism from three different angles. First, we describe observations from the journalistic practice as well as from science journalism teaching and propose a simple

Table 4.1 Top three science issues covered in leading nationwide German newspapers in two periods of 13 weeks each (rounded to full percent; complete ranking in Elmer et al. 2008)

Scientific issue	Articles (sample 2003/2004)	Scientific issue	Articles (sample 2006/2007)
Medicine	455(28%)	Medicine	703(29%)
Biology	209(13%)	Environment	366(15%)
Technology	187(11%)	Biology	333(14%)

[1] "Journalistische Qualität in der Krise," Dortmund University, Jan. 2010; http://idw-online.de/en/news352415.

heuristic model for the selection of science topics.[2] Afterwards, we describe the existing theory of news values for *general* journalism. In the last part, we try to transfer the general theory to science journalism and test the applicability of an adapted "science news value theory" empirically by combining guided interviews and a content analysis. Finally, we propose further steps to examine the proposed model as well as the adapted catalogues of news factors for science coverage.

4.2 Favourite Topics in Media Coverage of Science

Regarding the top fields of science covered by the media, there are typical patterns which seem to be internationally consistent. Altogether, medicine/health and biology dominate science coverage worldwide (Bauer 2000; van Rooyen 2002; Bucchi and Mazzolini 2003). In a long-term study of the *New York Times*, health, medicine, and behavioural science are constant among the best-selling topics, with maximum values of some 58% (Clark and Illman 2006). According to our own recent data for German broadsheets, their top list is as given in Table 4.1.

Although the detailed rank order is strongly dependent on the definition of every field, such rankings raise questions concerning explanations for these differences in science coverage. One of the few studies that compare the coverage of different scientific fields concludes that, so far, there is no convincing explanation for the different degrees of medialization (Schäfer 2007). While Schäfer focuses on differences between the epistemic cultures of the scientific fields themselves,[3] we argue that this perspective needs to be complemented by studying the journalistic perspective and the decision making processes of journalists: Are there certain factors especially dedicated to medical or biological issues which make them – on average – more attractive for journalists (and their readers) than other fields of science? Are some topics, regardless of their detailed content, less attractive because already their "price tag" causes negative associations (e.g., the "complicated" chemistry that everybody hated at school)? Are journalists on average more familiar with certain

[2] The term "science journalism" is used in the sense of "reporting on (natural) sciences, medicine and technology" (see Wormer 2008) regardless whether the reporting is done by specialised (science) journalists or others.

[3] Aside from the concept of different epistemic cultures of science, we are convinced that some rather simple parameters should also have an influence on the amount of science coverage in the media: For example, the total scientific output in biomedical research is higher than in, let's say, archaeology (in terms of number of scientists and scientific publications). Therefore, in this case, the amount of coverage is in line with the expected situation because larger fields get more awareness in the media than smaller ones. Concerning the journalists themselves, we could also confirm (at least for Germany) a dominance of science journalists with a background in biology who may have a tendency to prefer biological and (bio-)medical issues. Among other explanations, such aspects are at least one part of the story on why these disciplines are nearly always at the top list of the most covered topics.

issues because of their educational background making these issues more attractive for them? Schäfer (2007) puts the matter in a nutshell: "A reliable empirical reconstruction of the news factors in science coverage is still missing."

4.3 Inside the Science Section: The Practitioner's Perspective

4.3.1 Time Dependent Selection Factors

Editors can "even decide differently in August than they did in July". This quotation of the former director of the Henri Nannen Journalism School and one of the most influential journalism teachers in Germany (Schneider 1986) illustrates that the work in the media has a highly dynamic component. Schneider refers mainly to the changing comprehensibility of different terms with which the reader is getting more and more familiar during a repeated reporting on a certain issue. However, the observation of a highly dynamic process is also true for the issue selection process: A topic interesting on one day can become unimportant the next day – and the other way around. From the practical perspective, it should be discussed if time dependent factors influencing the selection process of editors for science stories may be slightly different than for other fields of coverage (such as politics or sports). For this influence, at least two different time dependent effects can be distinguished.

4.3.1.1 Passive Background Effect (Crowding Out)

> Science rules our life, but politics (or what is considered to be politics) rules the first pages of the newspaper. [...] Therefore, science is taking place mostly on the first pages or in the top news only when a politician talks about science (Schütze 1996).

This observation of a journalist specialised in science politics may have changed only a little bit over time. But it is interesting to put it the other way around: Political news has the power to crowd out even the most interesting science news. In the internal hierarchy of the mass media, the political editors are usually the most powerful ones and, by far, most of the editors-in-chief are socialised as political journalists. This situation can kill the original front page position even of an exclusive science story.

On the other hand, a calm day without any notable news in politics increases the chance for science topics to appear at the front end of the paper. During such a day, the *news* editor may even come up by himself with a proposal to write about the lack of donors for organ transplantations. In such cases, the argument of a *science* editor that there is nothing new in this "news" would not help: The news (!) editor demands the story following the claim: "It's not new but interesting." If there is a lack of interesting events on the news market (that is, a background effect around zero), nearly any science story will be regarded as a timeless beauty. And in contrast to articles about politics or sports, the news editor can hope that only a few readers will realise that the chosen science "news" was actually not new.

Fig. 4.1 And politics takes it all: Political news can be the front-page-death even for the reporting on real scientific breakthroughs (picture by H. Becker; taken from Wormer 2006)

Interestingly, this, let's say, "passive background effect" on the selection process is widely ignored in studies dealing with science coverage. Indeed, the science section was clearly identified as a "delayed media section" (Hömberg 1989) installed only at the bottom of the classical sections. But hardly any communication researcher seems to consider that, in such a position, there are different rules for the selection of news than for the well established sections. However, at least in some media, the standing of the science section has improved during the last 15 years (Wormer 2006), becoming more than a "nice-to-have-section" (Illinger 2005). Concerning communicational studies, it should still be recommended to check any data for confounding news background. To put it in other words: "What other things happened on the news market the same day?" should become a more important question for studying the selection process for science news (Fig. 4.1).

4.3.1.2 Active Background Effect (Pulling In)

When the science editor from a nationwide newspaper presented the topics for the next science page in the daily editorial conference, he not only had to explain the meaning of the word "Tsunami"; he also had to justify that new scientific results of Tsunami researchers were newsworthy enough to select this "strange" topic for the next edition.[4] One month later, after Christmas 2004, when one of the biggest

[4] The corresponding article was published November 25, 2004.

Tsunamis in recent history had occurred in Asia, the editorial demand of articles about Tsunami research reached the limit of capacities of some science editors.[5]

The Tsunami example shows some of the patterns for the issue selection process in science journalism: The time dependent background factors of the news market can trigger a shift from classical *science* journalistic selection criteria ("real scientific news" from a conference as a trigger for reporting) to *general* news selection criteria. In the first case, the general actuality seems to be less important for science journalism; in the second case, the general news market creates a classical situation of actuality (despite of hardly any real *scientific* news). As our data indicate, even for some daily broadsheets, the distribution between "scientific" and "non-scientific" triggers for the reporting on science topics is not far from a 50:50 situation (Elmer et al. 2008).

4.3.2 Time Independent Selection Factors

Although the factors mentioned above clearly have a time dependent (i.e., a news market dependent) component, they also have timeless aspects that could improve the chance of a science topic to be selected in the media: Science news with any political aspect should always increase its importance for general journalists. The same is true for a scientific conference or a scientific publication (see Ten Eyck et al. 2001; Ten Eyck and Williment 2003; Elmer et al. 2008). However, especially in science journalism, topics can be selected even without any daily actuality.

A special situation that may help to identify rather time independent factors for the selection process is given in an editorial office which runs two different publications with two different time scales. One example is mentioned by Illinger (2006). Having been responsible for both the science section of a daily newspaper and a magazine, Illinger also had to answer the question which topics should be saved for the monthly magazine and which should be selected for the next day(s) in the newspaper. A question that he answered as follows:

> What we can illustrate opulently, is rather published in the magazine. The newspaper can react [...] much more on actuality. For the magazine [...] we are dealing rather with latent actuality. In principle most of the topics in the magazine could also be used for the newspaper; the other way round this is not true to the same extent. Articles for the magazine mostly need more human touch and a more narrative component than many newspaper reports that are more aligned to facts (Illinger 2006).

A method suitable for empirical research to figure out which rather time independent factors are important for the selection process is a situation where the factor "actuality" is constant for all topics. To a certain extent this is true for the weekly

[5] The fact that in the same year a best-selling author had published a novel also dealing with a Tsunami delivered an extra (cultural) angle for the reporting after the disaster (e.g., Wellershof 2005). Interview with Frank Schätzing, *Der Spiegel*, 1, 3.1.2005, pp. 114–115. www.spiegel.de/spiegel/print/d-38785544.html).

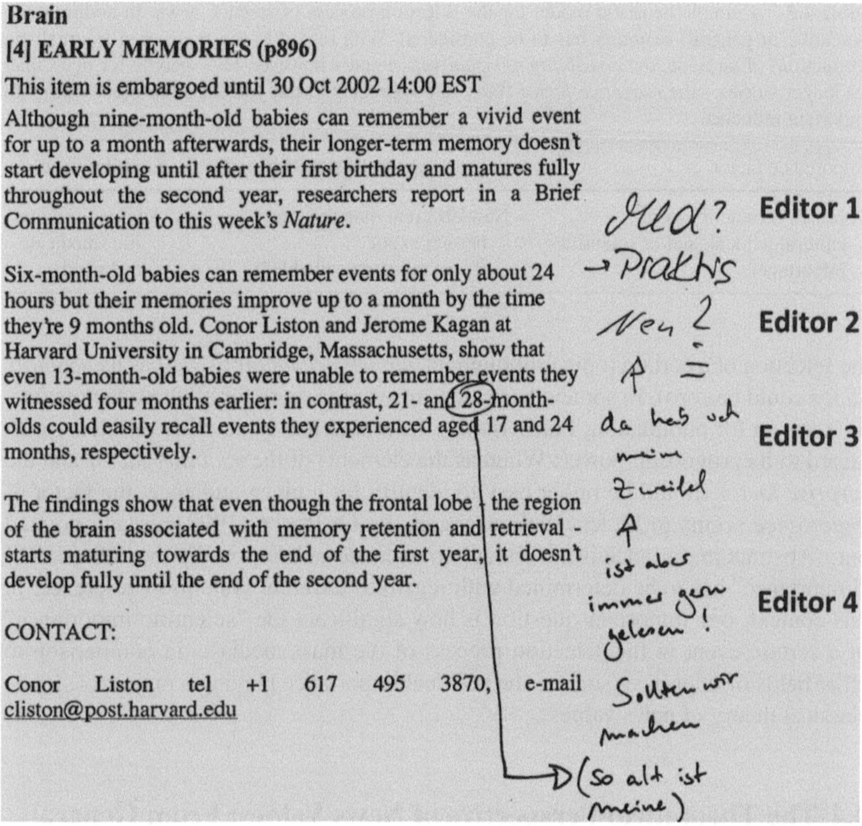

Brain

[4] EARLY MEMORIES (p896)

This item is embargoed until 30 Oct 2002 14:00 EST

Although nine-month-old babies can remember a vivid event for up to a month afterwards, their longer-term memory doesn't start developing until after their first birthday and matures fully throughout the second year, researchers report in a Brief Communication to this week's *Nature*.

Six-month-old babies can remember events for only about 24 hours but their memories improve up to a month by the time they're 9 months old. Conor Liston and Jerome Kagan at Harvard University in Cambridge, Massachusetts, show that even 13-month-old babies were unable to remember events they witnessed four months earlier: in contrast, 21- and 28-month-olds could easily recall events they experienced aged 17 and 24 months, respectively.

The findings show that even though the frontal lobe - the region of the brain associated with memory retention and retrieval - starts maturing towards the end of the first year, it doesn't develop fully until the end of the second year.

CONTACT:

Conor Liston tel +1 617 495 3870, e-mail cliston@post.harvard.edu

Fig. 4.2 Press releases from scientific magazines with a fixed embargo time would be an interesting research object to study the selection processes in the media. In this case we present an insight into some of the personal criteria noted by four different science editors from a nationwide German newspaper who are selecting news from a *Nature* press release. (For anonymisation the initials of the editors in the original document were replaced by "Editor 1–4"; Editor 1: "Short news?" Editor 2: "Is it really new?" Editor 3: "I have some doubts here." Editor 4 (with a daughter in the same age as mentioned in the press release): "But it is always loved by the readers! It should be done."). (Picture taken by H.W.)

press releases from magazines such as *Nature* or *Science*: Most of the offered news has the same embargo time (Fig. 4.2).

Our practical experience and observation of many hundreds of such standard selection processes as well as the discussions in editorial conferences should allow some hypotheses on the main factors used by journalists. Furthermore, such factors are also recommended to journalists in journalism teaching (for interns in the editorial office, at journalism schools and universities as well as in handbooks for journalistic practice). Regarding all this, the following simple model for the selection of science topics in the media is proposed (Table 4.2).

However, the weighting of each factor in the proposed model is not quantified. For example, in some cases, a very high *surprise-factor* alone may be sufficient for

Table 4.2 A simple heuristic model for the selection process of science news. In addition, the (scientific or general) *actuality* has to be considered. With regard to the realisation (journalistic production) of an issue, the *possibility of visualisation* and – however, less specific for news than for longer stories – the *narrative factor* (fairy tale approach: "I tell you the story of...") have an important influence

Importance factor	Surprise factor ("Astonishment")	Usability factor
Political, economic, social, cultural, ethical, and/or scientific importance	New/different than thought before; exotic ("cocktail-party-suitable")	Advice for daily life (medical, technical...)

the selection of a certain topic (although its *importance factor* as well as its *usability factor* could be zero). In some other cases, an average score for all three factors may be a reason for publication. Furthermore, the model has some other problems with regard to its prognostic power: Whereas the elements of the *usability factor* and the *surprise factor* should be rather easy to identify for a given audience, the factor of *importance* seems to be less well-arranged. As La Roche (1999) already pointed out: "Abstract importance is not enough to make news." Especially the definition of "importance" has to be determined with regard to different systems of reference. In this context, one important question is how significant the "scientific importance" of a certain event in the selection process of the mass media is in comparison to other fields of reference (such as the political importance playing a major role in the classical theory of news values).

4.4 The Theoretical Perspective of News Values: From General Journalism to Science Journalism

In journalism theory, four concepts of news selection have achieved a broader acceptance: gatekeeping, framing, news bias, and news values.[6] In this chapter, we can only focus on the theory of news values. Starting with a brief introduction of the well established concept, we will discuss former attempts to apply this general approach especially on science coverage.

4.4.1 The Theory of News Values

The theory of news values, also called the *concept of news factors*, describes why a topic is newsworthy and therefore published by the media. An event has several characteristics ("news factors") such as *geographical proximity*, *unexpectedness*, or

[6] Others such as cognitive and socio-psychologist concepts are also interesting but rare in the literature. For a short introduction to the different concepts, see the International Encyclopaedia of Communication (Donsbach 2008).

prominence (*reference to elite persons*). The total news value increases with the number of news factors that are accumulated by an event and also with the intensity of these factors.[7] If the resulting total news value of an event is higher, its chance to be selected for publication will be better. In this "model of two components" (number and intensity), journalists price the impact of each of these news factors more or less conscious (Kepplinger and Weißbecker 1991). It is proven that topics with a higher total news value obtain on average more space in the publication than topics with a lower news value. The term "news value" first occurred in 1922, when Walter Lippmann published his book *Public Opinion*, including some remarks on the news selection by journalists.

Østgaard (1965) was the first to classify news factors; he also postulated that a topic has to overcome a "news barrier" (a fictional value according to the used grade systems of news factors). Galtung and Ruge (1965), Norwegian peace researchers like Østgaard, codified twelve news factors in an analysis of the coverage of three foreign crises. The German communication researcher Schulz (1976) enlarged the catalogue of news factors up to 18 selection characteristics; he also operationalised the news factors and introduced a multivariate value scale, and he included formal aspects as position and extent of the publication. Staab (1990) extended the model by a back coupling effect: News factors are not just explaining news selection decisions. They are both results and a means to an end at the same time: When an editor has identified a certain news value, he may also have the tendency to underline this certain value in his article. As a result, the news value in the final article may be stronger than it was in the original event.

For more than three decades, the theory of news factors has been used for the analysis of news selection in the media's coverage especially of politics, foreign affairs or in non-specific news broadcasts. But this concept obviously does not describe the selection of *science* news sufficiently: For the last 20 years, journalism researchers have suggested several times that science journalists may have special selection criteria or at least that the criteria may be different in different sections (e.g., Hömberg 1987; Staab 1990; Ruhrmann 1990; Peters 1994a; Ruhrmann 1997; Milde and Ruhrmann 2006). However, empirical data are rare so far.

4.4.2 News Factors in the Context of Science Journalism

The probably most extensive considerations concerning news factors important for science journalism go back to Ruhrmann (1990, 1997). Ruhrmann examined especially the media coverage of genetic engineering where he identified three aspects: (1) the novelty of an event, respectively the probability of a risk, (2) the scientific-technical or social relevance of a technological development or the extent of its perils, (3) the common uncertainty of the gene technology's risk context.

[7] For the intensity, a grade system in four steps (between 0 and 3) is often used and was also applied in this work. Other authors use different scales, e.g., from 1 to 5 (Ruhrmann and Göbbel 2007).

Thus, the information value of the reported event and its associated risks should depend at least in part on the well known news factors *unexpectedness* ("exclusiveness of the event"), *potential harm* ("probability of damage"), and *establishment* ("duration since the incidence or observation of an event"), as well as *composition* ("variance of the event and topic"). At the same time, Ruhrmann indicated: The *recipients'* perception of risk communication uses other criteria, namely the criterion of the "social rationality". This includes "simplicity of events and risks" (the equivalent of Østgaard's news factor *simplification*), "personal relevance of risks" (equates the news factor *involvement/range*, measured by the number of people affected), and "credibility of persons and institutions" (which is the news factor *influence*).

Later, Ruhrmann (1997) hypothesised that a scientific event becomes news more easily the more the following conditions are given: the event is self-contained ("development of the event"), the event is sudden (i.e., news factor *unexpectedness*), the event is important and has consequences for the total population (i.e., news factors *damage*, *success*, and *involvement/range*), the event's consequences are negative (*damage*), the event is controversial and conflict-riddled (i.e., news factor *controversy*), and the scientific elite is involved (i.e., news factor *influence*, sometimes also *prominence*). According to this hypothesis, a distortion of the media reality is not astonishing: Science journalists report on "spectacular discoveries, laureates, and marketing opportunities but the daily routine of scientists, the merit of competitors, or the background of research promotion remain unknown" (Ruhrmann 1997). Relating to risk communication, this means: Rare and extraordinary risks are preferred, average risks of everyday life are neglected.

Although risk communication and genetic engineering are only a special segment in science journalism, many of his conclusions seem to be in line with the experiences in journalistic practice (see above). From this point of view, it can be predicted that many *conventional* news factors should indeed be transferable to the science coverage of the media (see Wormer 2010). Another indicator for an overall conformity of these factors could be the observation that science journalism is becoming more and more science *journalism* (instead of *science* journalism), i.e., practitioners see themselves as science *journalists* rather than *science* journalists (see Ruß-Mohl 1987). Especially in recent years, science journalism often has not been limited anymore to a kind of "nature protection area" (that is, for example, hidden in newspaper supplements) at the back end of the newspaper (see Elmer et al. 2008). Therefore, *non*-science editors have to deal with science news using their own (general) news criteria. The other way around is also true: All kinds of journalists, including science journalists, are socialised in editorial offices with a tendency to big newsroom concepts which limit the former frontiers between different departments. This again influences the selection process by science journalists (see Donsbach 2004).

Nevertheless, there is some evidence that some of the news factors in the existing catalogues may be rather unimportant to science journalists (see below). Vice versa it is natural to ask if science journalists use news factors which are more or less unique for the science section in the media. Of course, *conventional* news factors are obviously used by every journalist, independent from the field of coverage; these news factors are conventional exactly for this reason because they are cross-departmental. But what about news factors which do not make sense to journalists anymore behind the boundary of a certain editorial department?

This article aims at filling this research gap by combining the practitioners' perspective and the insights from the general news value theory with an empirical study of issue selection processes in science journalism. Our basic working hypothesis for the empirical study can be summarised as follows:

H1: In principle, the well established theory of news values should be applicable to science coverage, too (see Wormer 2010).

H2: The general theory of news values will *not* be applicable sufficiently to science coverage without a number of considerable adaptations of the existing framework. These could be in detail:

1. The exact definition of some news factors is incomplete because former study authors have only considered aspects which lie beyond the world of science but rather in the political, economic or cultural systems.[8]
2. Some news factors are hardly applicable because they anticipate an interpretation which is rather unusual for a science journalist.[9]
3. Some news factors such as "demonstration" are unnecessary because they refer to events which may belong to daily life in politics but not in science.
4. Some news factors specific to science media coverage may be missing in the general catalogue and cannot be included easily in existing news factors so that discrete new news factors need to be generated. Candidates are, for example, *scientific proximity* and *scientific relevance*.

4.5 Development of a Revised Catalogue of News Factors and a First Empirical Test

4.5.1 A Draft Catalogue of News Factors for Science Coverage

For the creation of a first draft catalogue of news factors specifically applicable to science journalism, we have chosen a systematic approach based on four maxims:

[8] For example, in the classical definition, the factor *influence* is defined according to "a person, group or organisation with political, economic or cultural power"; "scientific power" would not be taken into account.

[9] For example, applying the classical factor *reference to elite nation* science journalists would probably not have in mind the "military power" or the "foreign trade of the country where the reported event took place" but rather indicators such as the "scientific importance" of this country.

1. The draft catalogue should stay as close as possible to the common theory of news values and to the well established news factors.
2. The draft should consider Ruhrmann's (1990, 1997) thoughts about news factors especially important in science coverage.
3. The draft should be linked to the journalistic practice (see Section 4.3)
4. Each news factor should be operationable with appropriate standardized data, i.e., data for several countries that are collected by one institution with one definition following a common standard instead of information from different sources that are not comparable.[10]

Following these principles allows a traceable adaptation of existing catalogues as well as a highly intersubjective coding procedure for future studies applying this draft. Furthermore, the strict orientation on existing catalogues enables us to compare existing data in the literature with new data to be collected.

4.5.1.1 Analysis of Existing Catalogues of News Factors

First, an extensive literature analysis of the history and development theory of news values starting from Lippmann's publication in 1922 was performed. Then, a table with an overview of the most important and most popular versions of the theory of news values was generated (based on the model of Ruhrmann and Göbbel 2007). The table considered concepts of the theory of news values of the following scientists: Østgaard (1965), Galtung and Ruge (1965), Schulz (1976), Schulz (1977), Staab (1990), Eilders (1997), Ruhrmann et al. (2003), and Ruhrmann and Göbbel (2007). This chart showed when a news factor was renamed or separated into two distinct factors or newly introduced, etc. Every single news factor from the table as well as its extensive coding definition was surveyed in detail with regard to its applicability for science news. Afterwards, testing our working hypothesis we identified some factors which should be valid for science news with no or only a minor adaptation of the existing definitions in the classical catalogues. Examples include:

- *Geographical proximity* is one of the oldest news factors. As country borders are quite objective, no adaptation was necessary: It can be assumed that a political journalist as well as a science journalist in Berlin or London would prefer to select an event that took place in France instead of an event in Brazil.
- In its classical form, the definition of the news factor *influence* only includes "the political, economic, cultural or sportive power of a person, a group of persons or an institution" (see Ruhrmann et al. 2003; Ruhrmann and Göbbel 2007). By using

[10] Useful data collections are offered e.g., by the OECD or published in the CIA World Factbook which provides information on the people, government and further items for more than 250 world entities.

this definition, even influential scientists would not necessarily be considered. Hence, the definition had to be extended to *scientific power/influence*.[11]

- The classical news factor *range* is defined by the question whether "nobody", "a few people", "a professional group", or "all citizens of at least one nation" are directly affected by an event (see Ruhrmann et al. 2003). But for the coverage of medical issues this definition seemed to be too narrow. For example, which disease would affect every citizen personally at the same time? Thus, different diseases were classified according to their incidence or cases of death per year (with cardiovascular diseases and cancer on the top).[12]
- *Actuality* is also a classical news factor that has an additional dimension in science coverage: General news events which need a scientific explanation (e.g., natural disasters or political events) can trigger actuality. This seems to happen with increasing tendency (see Elmer et al. 2008). But genuine scientific events, such as journal publications or conferences, can also trigger (science) coverage.

Aside from such adaptations of existing news factors, some *news factors specific to science coverage* were proposed (mostly on the base of the practitioner's experience), e.g.:

- In addition to the news factors *geographical proximity*, *political proximity*, *economic proximity*, and *cultural proximity*, a new factor called *scientific proximity* was corollary. The new factor was defined – among other aspects – by considering scientific cultures and favourite research areas of a country.
- The news factor *composition* – so far only used by Galtung and Ruge (1965) – seemed to be important for science journalists, too: "We only have medical issues today, thus we also need some physics" is a well known argument in editorial conferences. In such cases, the news barrier would be reduced for certain topics.
- Another factor that seemed to be more important in science coverage than in the reporting on political topics is the *astonishing* aspect (see *surprise factor* in Section 4.3).

[11] As a consequence, the adapted definition of the news factor *influence* is given by: "political, economic, cultural, sportive or scientific power of a person, group, or institution." Our proposal to operationalise "scientific power" is as follows: 0 = no influence, e.g., a student; 1 = low influence, e.g., a PhD student; 2 = high influence, e.g., a professor or the scientific leader of a national research project; 3 = largest influence; e.g., leader of an international research group.

[12] By using this approach, the potential concern of a population regarding a certain disease is rated in a first approximation as equivalent to the mortality. This approach has the big advantage of being highly objective. However, in reality this selection process may be distorted at least in some cases by the subjective perception of a news editor. For example, editors are likely to over-estimate the impact of diseases such as HIV/Aids (about 650 deaths in Germany in 2008; Robert-Koch-Institut 2008) in comparison to diseases such as diabetes mellitus (20,000 deaths per annum; Statistisches Bundesamt 2009).

– Furthermore, the *intention* of the author, the aim that the author wants to achieve with the publication, is considered as a news factor for our empirical test.[13]
– Following Peters' (1994a) "four-stage hierarchy of access to the public", an essay written by a scientist or an interview could increase the total news value. Thus, the factor *expert's impact* on the publication is considered, too.[14]

In addition, the news factor *relevance* was re-introduced. This item occurred for the first time in the list of Galtung and Ruge (1965) and became interpreted as *range*, the number of affected people. We have defined *relevance* here as "intensity of damage or benefit" of an event (considering the outcome and not only the number of people involved). Initially, we had calculated this news factor with an index of the sub-criteria *political relevance*, *economic relevance*, *scientific relevance*, and *relevance to recipients and society*. However, it turned out in the pre-test that a high intensity in *scientific relevance* was often counterbalanced by non-existing *political* and *economic relevance* resulting in barely differing sums for different articles. Hence, it was decided that each sub-criterion should become an independent factor.

Finally, 29 news factors were derived inductively following the procedure described above, building a draft catalogue of factors for a subsequent empirical examination (Table 4.3).[15]

Table 4.3 News factors for the first draft of a prospective theory of news values specific to science journalism

Draft catalogue of news factors adapted on the basis of existing theories and journalistic practice	
Geographical proximity	Geographical distance from event country to publication country
Proximity of the politics of science[a]	Similarity between event country and publication country with regard to their political handling of science and research
Economic proximity	Similarity between event country and publication country with regard to their economic systems and their economic relations
Cultural proximity	Similarity between event country and publication country with regard to their language, religion, and culture
Scientific proximity[b]	Similarity between event country and publication country with regard to their scientific culture

[13] For example, the intention of an investigative report may be more attractive for a science journalist than a general news piece and thus increase the news value of this publication. The scale for the operationalisation follows Peters' (1994a) categories for the different kinds of science communication (popularization, clarification, and scientific controversy) and was complemented by the controlling function (Peters 1994b).

[14] With *expert's impact* we aim at the question whether a scientist appears in the media report just with a single quote or as an interview partner or even as a guest author.

[15] The entire codebook with all detailed definitions of the individual factors can be requested from the authors.

Table 4.3 (continued)

Draft catalogue of news factors adapted on the basis of existing theories and journalistic practice	
Reference to elite nation[a]	Status of the event country within the scientific community according to their science and engineering S&E article output
Reference to elite region[a]	Status of the event region within the publication country
Reference to elite person[a]	Political, economic, cultural or scientific power of a person, group, or institution ranked by its position in the hierarchy
Prominence[a]	Degree of notoriety of a person/institution independent from its power/position in hierarchy
Personalisation (reference to persons)	Inclusion of persons and importance for the reported circumstances
Controversy	Contrasting of differences in opinions
Aggression	Threat or use of violence with the aim to hurt or to damage
Demonstration (protest march)	Collective representation of goals
Unexpectedness	Extent to which an event was not expected
Range (number of affected people)[a]	Number of people participating in an event or affected by the event
Continuance	Establishment/period of time the media is already following a topic
Involvement of the publication country	Reporting about an event because it takes place with the participation of the publication country
Presentation of feelings	Display of human feelings via issuing gestures or facial expressions
Sexuality/eroticism	Verbal and pictorial presentation of sexuality/eroticism or allusion to it
Availability of graphical material	Extent to which an event becomes news just because pictures or figures are available
Scientific relevance[b]	Importance of an event for the scientific progress
Relevance to recipient/society[b]	Importance of an event for the recipient of the article or even the society in total
Economic relevance[b]	Importance of an event for the economy
Political relevance[b]	Importance of an event for politics or legislation
Composition[b]	Mix of topics within a distinct science page and the whole issue of a newspaper/broadcast etc.
Astonishment[b]	Extent to which an event causes amazed reactions ("Aah!")
Expert's impact[b]	Extent to which a scientist becomes involved in the publication and gets access to the public
Actuality[b]	Reason for the selection of an event at the present moment (coming from the general news situation, the research operation or both)
Intention[b]	Type of Science communication

[a] Definition of this classical factor slightly adapted to make it applicable for science coverage
[b] Newly introduced factor

4.5.2 Sample and Methods for the First Examination of the Draft Catalogue

To examine whether the developed model of news factors specific to science journalism is reasonable, a quantitative method and a qualitative method were used in a triangulation approach: a content analysis on the one hand, and guided interviews on the other.[16]

For the selection of the analysed media, the goal was to get a homogenous sample of comparable media that offer a notable amount of science coverage. Therefore, four German nationwide quality daily newspapers were chosen: *Frankfurter Allgemeine Zeitung* (*F.A.Z.*), *Süddeutsche Zeitung* (*SZ*), *Die Welt* and *Frankfurter Rundschau*. As the profile of the last one is slightly different, the *Frankfurter Rundschau* was just used for a pre-test of the applicability of our framework. Furthermore, in order to get a first hint for an international applicability in different media systems, the French daily *Libération* was selected as well.[17]

For the guided interviews, we included the responsible editor for the science sections in each of the selected newspapers. Thus, four plus one people were consulted for an approximately one-hour-interview each. During the conversation, the editors were first asked to which aspects they pay attention when selecting a topic. Afterwards, they were confronted with the draft of the adapted catalogue of news factors and its definitions in order to assign every item certain significance from their personal point of view. Finally, the science editors were asked whether they would see the need to go beyond the adapted catalogue by adding other factors which had not yet been included in the list. The recorded interviews were transcribed and then clustered with the text analysis software MAXQDA.

The final sample for the content analysis of the selected newspapers contained one stratified week randomly selected out of the first half of 2009, following the suggestions of Riffe et al. (1993). Every issue was scanned completely because a former study clearly indicated that a remarkable amount of science coverage can be found outside the science section. Every article containing scientific content in at least half of its length was encoded (following the "50+ percent scientific content" rule already applied in Elmer et al. 2008; see also Bucchi and Mazzolini 2003).[18] In total, 192 articles were classified as science journalistic coverage (*F.A.Z.*: 31, *SZ*: 59, *Welt*: 82, *Libération*: 20). These articles were encoded following the codebook.

---

[16] The empirical part is mainly based on a master thesis which was realised by F.B. and supervised by H.W.

[17] This approach has a kind of explorative character on the international level because, as a literature review showed, there is a lack of a scientific discussion about both news values and science journalism in France. Nevertheless, the French data will be considered only partly in the following analysis.

[18] Some older studies use definitions such as "An article is regarded as science coverage if a scientist or a scientific institution is mentioned in the first paragraph" (e.g., Böhme-Dürr and Grube 1989: 450). Although such definitions may be easier to apply, they are rather outdated because in the era of "narrative writing" many stories start with a colourful introduction, the story of a patient, etc. and switch to the scientific issue only in a later paragraph.

Afterwards, every news factor was analysed individually using the data analysis software SPSS. By combining the results of the content analysis with the information derived from the guided interviews, the extensive catalogue of 29 news factors was reduced to the most important factors.

4.5.3 Results

It could be confirmed *empirically* that some definitions of classical news factors were incomplete to describe the selection of science news. One example: Following the classical definition of the news factor *influence*, 41 of 192 (21.4%) articles noted people or institutions with (political, economic, or cultural) power. But within the 151 remaining articles in which none of the three categories of influence were noted, 116 dealt with people or institutions of notable *scientific* power. This is not surprising for science topics, but it is rather surprising that "scientific influence" has not been included in the theory before.

Another interesting question was whether the results of the guided interviews (self-perception of the editors) and the content analysis would be consistent. Here we found an ambivalent picture (see Table 4.4). In some cases, the most important news selection criteria were identical with the news factors that showed up the most or with the highest value in the analysed article. But in other cases they did not. On the one hand, the factors *unexpectedness* and *composition* were mentioned as two of the most important news selection criteria by the science editors as they were at the top of the ranking of the news values encoded in the content analysis. On the other hand, the editors stated the news factors *range* and *relevance to recipients/society* as one of the most important selection criteria while the average score of both news factors was low in the analysed articles. Furthermore, the news factor *astonishment* (derived from the practical approach) was rated high in both the top list of the science editors and the ranking of news values in the articles.[19]

All available information for each news factor was compared with the results of a survey of 43 executive news editors in the general news business (Ruhrmann and Göbbel 2007). Interestingly, the interviewed science editors often seem to focus on other news factors than the 43 news journalists in the former study. For example, *involvement of Germany* was the second most important criteria for the non-specialised journalists (Ruhrmann and Göbbel 2007); in contrast, the interviewed science editors in our study estimated the involvement of German scientists as "nice to have" at most.[20] *Prominence* and *geographical proximity* seem to be rather important in general news coverage, too, but not in the science coverage of nationwide quality newspapers. One exception from these differences was

---

[19] Although this result is striking it should be kept in mind that the sample was not representative for all kinds of media, which is especially true for the guided interviews.

[20] However, the French science editor interviewed in our study declared the involvement of France as a "must have" for an event to be selected.

Table 4.4 Comparison of the results derived from the content analysis and the guided interviews with a survey in the literature (including an accentuation of correspondence with the simple heuristic model and the principal component analysis)

	Opinion survey with 43 German executive news editors (Ruhrmann and Göbbel 2007)		Opinion survey with the head of science departments of F.A.Z., SZ, and Welt		Content analysis of science coverage of F.A.Z., SZ, and Welt	
Rank	News factor	Average score	News factor	Average score	News factor	Average score
1	Range (number of affected people)	4.26	Unexpectedness[a]	2.67	Unexpectedness[a]	2.41
2	Involvement of Germany	4.00	Range (number of affected people)[a,b]	2.67	Economic proximity	2.10
3	Negative consequences/damage/ failure	3.79	Scientific Relevance[a]	2.67	Composition	2.02
4	Unexpectedness	3.72	Relevance to the recipients/ society[a,b]	2.67	Reference to elite nation	2.01
5	Differences of opinion/controversy	3.67	Composition	2.67	Proximity of the politics of science	1.86
6	Positive consequences/benefit/ success	3.65	Trigger	2.67	Astonishment[a]	1.81
7	Prominence	3.33	Graphical material[a,b]	2.50	Scientific proximity	1.71
8	Geographical proximity	3.30	Intention[a,b]	2.50	Reference to elite region	1.70
9	Personalisation (reference to persons)	3.26	Controversy[b]	2.33	Trigger[a]	1.63
10	Continuance	3.26	Astonishment[a]	2.33	Reference to elite persons[b]	1.47
11	Visualise	3.15	Political relevance[a]	1.83	Prominence	1.46
12	Reference to elite region	3.14	Personalisation[a]	1.67	Involvement of Germany	1.45
13	Reference to elite nation	3.07	Continuance	1.67	Cultural proximity	1.44
14	Violence/ aggression	3.07	Economic relevance[a]	1.67	Geographical proximity	0.99
15	Demonstration (protest march)	3.07	Involvement of Germany	1.33	Scientific Relevance[a]	0.91
16	Availability of graphical material	2.95	Presentation of feelings	1.33	Range (number of affected people)[a,b]	0.84

Table 4.4 (continued)

Rank	Opinion survey with 43 German executive news editors (Ruhrmann and Göbbel 2007)		Opinion survey with the head of science departments of F.A.Z., SZ, and Welt		Content analysis of science coverage of F.A.Z., SZ, and Welt	
	News factor	Average score	News factor	Average score	News factor	Average score
17	Political proximity	2.91	Sexuality/eroticism[b]	1.33	Personalisation[a]	0.83
18	Presentation of feelings	2.86	Expert's impact[b]	1.33 (1.0)	Relevance to the recipients/society[a,b]	0.67
19	Economic proximity	2.81	Geographical proximity	1.00	Intention[a,b]	0.65
20	Reference to elite persons	2.70	Scientific proximity	1.00	Continuance	0.59
21	Cultural proximity	2.63	Reference to elite persons[b]	1.00	Graphical material[a,b]	0.57
22	Sexuality/eroticism	1.53	Prominence	1.00	Controversy[b]	0.52
23	–		Proximity of the politics of science	0.67	Expert's impact[b]	0.38
24	–		Reference to elite nation	0.67	Presentation of feelings	0.33
25	–		Reference to elite region	0.67	Economic relevance[a]	0.28
26	–		Demonstration (protest march)	0.67	Political relevance[a]	0.13
27	–		Economic proximity	0.33	Aggression[b]	0.09
28	–		Cultural proximity	0.33	Sexuality/eroticism[b]	0.05
29	–		Aggression[b]	0.33	Demonstration (protest march)	0.00

Ruhrmann and Göbbel (2007): $N = 43$ German journalists responsible for a general news media programme; Scale: 1 = news factor not important, 2 = slightly important, 3 = partly/partly, 4 = quite important, 5 = very important. Own Study: N = 3 interviewed science editors responsible for the science department and $N = 172$ articles; Scale: 0 = news factor with no or very low value, 1 = low value, 2 = high value, 3 = highest value

[a] News factors with a counterpart in the heuristic approach (importance, surprise, usability, actuality, possibility of visualisation, narrative factor; see Section 4.3)

[b] News factors important according to the principal component analysis (see Section 4.5.4)

found for the factors *unexpectedness* and *range (number of affected people)*: Those gained a high popularity both among the polled news editors (by Ruhrmann and Göbbel) and the science editors (our study). According to our interviews, the newly introduced news factors *scientific relevance*, *composition*, *actuality*, *intention*, and *astonishment* outperformed many of the well established news factors. Furthermore, following the explicit opinion of the news editors, no other specific news factors had to be added.[21] A comparative overview of the ranking by Ruhrmann and Göbbel (2007), the results of our guided interviews and the results of our content analysis are given in Table 4.4. In this overview, it is also highlighted which factors are strongly related to the simple heuristic model and which factors are exposed following a principal component analysis.

4.5.4 Reduction of the Draft Catalogue of News Factors

In order to simplify the application and the practicability of the concept, we examined whether the extensive catalogue of 29 news factors might be reduced to its most important criteria.[22] These criteria could be:

- news factors mentioned by the science editors as top news selection items in the list,
- news factors that showed up often or with high intensity in the content analysis,
- news factors that vary the most in their encoded intensity (identified with a principal component analysis[23] related to the results of the content analysis).

Building the intersection from these three perspectives, the following 14 news factors seem to be those with the highest impact on the news selection processes in science coverage (Table 4.5):

4.5.5 Summary and Limits of the Results of the Empirical Analysis

Although we have tried to synthesise the results of the guided interviews and of the content analysis into a reduced catalogue of news factors, there are considerable discrepancies between the results of both methods. In some cases,

---

[21] Asked whether they miss a criterion in the catalogue, the journalists mentioned rather an "anti-top list" of scientific fields: They only specified topics that *barely* have a chance to be selected for publication, e.g., "chemistry" or "research policy" instead of saying which topic will be selected *in any case*.

[22] According to Kepplinger (1998) renouncing a broad differentiation and specification of the catalogue of news factors would mean that the explanatory power of the theory of news factors would fall far short of its possibilities. However, it is a mistake to believe that a theory divided into small sections always delivers more knowledge; furthermore it will be less applicable with regard to the scientific and journalistic practice anymore.

[23] This method of multivariate statistics finds out which items (*news factors*) contribute the most to the total variance (*total news value*) (see Brosius 2006).

Table 4.5 The adapted and reduced catalogue

14 News factors with the highest impact on the selection process of science news (alphabetic order)	
Astonishment	Political relevance
Composition	Range (number of affected people)
Controversy	Reference to elite persons
Economic relevance	Relevance to recipients/society
Graphical material	Scientific relevance*
Intention	Actuality (Trigger)
Personalisation	Unexpectedness

*Scientific journals are the fundamental basis to encode the news factor *scientific* relevance in case the science coverage was triggered by the publication of a scientific paper. This was especially confirmed by the science editors in our survey. Thereby we may say that the reputation of a certain journal strongly determines the value of this news factor. Because of this close connection we renounced a factor of its own for "scientific journal."

these discrepancies may already be explained by a different self-perception of the science editors concerning their selection criteria and the real selection process itself (by themselves or another science editor). However, it also has to be kept in mind that a content analysis can only register "constructed news factors", i.e., the existence of a news factor or its intensity could be different in the newspaper (output) than it was in the original material (input) that the editors used in the selection process.[24] Furthermore, our sample was both rather small and highly specific to the genre of a "quality newspaper". The comparison of *F.A.Z.*, *SZ* and *Welt* on the one hand, and *Libération* on the other, already indicated that the data gained by encoding news factors are not sufficient for a complete interpretation of news selection: Contrary to the German science editors, the science editors of *Libération* do not have their own page and cannot decide on their own which scientific news will be published; it depends on the editor-in-chief of the newspaper who does not have the same estimation of certain news factors. Thus, one always has to keep in mind the context, the time dependent requests (see Section 4.3) and the country-specific or media-specific framework requirements.

Finally, it should be mentioned that the proposed catalogue of news factors is not a simple "check list": It will not automatically help science journalists to pick the "right" story; every medium has its own "pattern" of news factors. The theory of news values is a device for researchers to uncover this "pattern". However, many elements of such a catalogue are used like a list of ingredients in a cook book on journalism teaching or handbooks and therefore show up in the simple heuristic model as well.

[24] For example, the value of the news factor *influence* may be different in the published article and in the basic material if the science journalist has interviewed a further scientist during his investigation. When this scientist is higher in the hierarchy than the scientist mentioned in an original press release the news factor *influence* would become stronger in the reporting than it was before.

4.6 Conclusions and Forecast

In this work, we have analysed issue selection processes in science journalism in three steps: Starting with the practitioner's perspective and a simple heuristic model, in a second step, the classical theory of news values was adapted with regard to some specialities in science journalism. In a third step, this adapted theory was tested in two ways: by guided interviews with leading science editors as well as by a quantitative content analysis (including a principal component analysis). As a result, the proposed new catalogue could be refined and compared with the simple heuristic approach as well as with former classical catalogues. Our results clearly indicate that a certain adaptation of the classical theory of news theory for science journalism is reasonable. However, the question may be to what extent such an adaptation is useful.

The science section did not belong to the classical departments in the media. Therefore, it is not surprising that catalogues and codebooks of the classical news theory were developed only alongside departments such as politics, culture or economics. In our study, it could be confirmed that simply applying these catalogues and definitions may cause misleading results for science coverage. Furthermore, in our empirical tests, some newly introduced factors specific for science journalism outperformed many of the classical ones.[25] This brings us to the conclusion that the classical catalogues have to be regarded as incomplete and not sufficient to describe the selection processes in science journalism precisely. However, our first empirical testing also shows that *classical* news factors are relevant in science journalism, too. One explanation for these findings may be the fact that our content analysis has included articles that were triggered by the general news and not only by scientific events such as conferences or publications. A hypothesis for further research thus is that an only slightly adapted classical catalogue is suitable for the part of science journalism triggered by general daily news. In contrast, there is considerable benefit of an adapted and extended catalogue for describing the selection of classical science stories mostly triggered by scientific publications and conferences.

In this context it is an interesting question for further research how significant the factor *scientific relevance* is for the selection process in both types of science journalism. On the one hand, a recent publication or a scientific conference still strongly influences the selection process of science journalists. On the other hand, the perspective of the general audience seems to be of increasing importance for the selection by science journalists. This second tendency may be further amplified in the modern structures of the media (with all kinds of editors literally in the same

[25] It cannot be discussed here why the newly introduced news factors *scientific* relevance, *composition*, *actuality*, *intention*, and *astonishment* have overtaken lots of the well established factors (especially according to our interviews). However, two spontaneous explanations should be given: Science is often seen as a kind of entertainment (sometimes its editors are even part of the "miscellaneous" section in the media) which may explain the importance of the factor *astonishment*. One reason for the high attention to a good mix of topics (*composition*) might be that, in science journalism, usually fewer general topics are predefined than in the coverage of politics or sports.

"newsroom") which may level the difference between different sections. However, the stronger orientation to a broader audience is not only positive in terms of opening the world of science to science distant people: At the same time, it includes the danger that many really new and scientifically really important but not mass-popular science news will be ignored even by quality media and therefore will not get the public awareness they need. The question is then: Who will tell society what is really going on in science?

4.6.1 Further Research Needed for a Final Catalogue

Although we are convinced of the usefulness of an adapted and completed catalogue of news factors for science coverage, our data does not allow us to decide whether the proposed 14 factors should be the final choice. It is an open question whether the prognostic power of this catalogue would be significantly better than a prognosis based on the simple heuristic model. To answer such questions, we propose an input-output analysis with adjusted time dependent factors (see also Section 4.3.1). Therefore, in a first step, the theoretical news values of different topics in the press releases of leading journals such as *Science* and *Nature* could be analysed. In a second step, the number and amount of the related media publications as a result of the editorial selection processes could be measured.[26] As our current empirical data is still limited on quality daily newspapers, such examinations should also include other media. In this context, a further examination and differentiation of single factors would also be interesting. With regard to the practical perspective, it should be tested if a differentiation of *relevance for society* and *relevance for the individual ("usability")* in two discrete news factors would be useful.[27] Further attention should also be given to the factor *graphical material* which was regarded as very important especially in the guided interviews with the science editors. Interestingly, the editors mentioned a selection preference linked to the scientific field to which an event such as a publication belongs (which is in line with some of our ideas given in Section 4.2). It would be a challenge for further research to answer the question whether "a general news value estimation" could be constructed for different disciplines. Among others, that could be one tool to analyse in detail why there is a lack of some scientific topics in the media. In any case, the findings from the guided interviews are only one example for the benefit of the practical and the

[26] This is also reasonable because this work has not analysed the effect of the components "news factors" and "selection criteria" on the amount of coverage (the dependent variable in the model of two components).

[27] Both aspects were already separated in the simple heuristic model. One reason for that is our observation that science and medical reporting, on the one hand, has often a very personal component (i.e., an individual usability (*Nutzwert*), e.g.: "Where do I get this treatment? Is it harmful for me?"). On the other hand, the same treatment may be discussed in the context of exploding costs of the health system which is relevant for society as a whole but less important for the personal health question of an individual patient.

theoretical perspective going hand in hand in our work in which the practitioner's view was already the basis for the adaptation of existing classical catalogues of news factors.

4.6.2 Lessons to be Learned for Research Outside the Science Sections

Aside from the question of a specific selection process for science journalism, some of our results raise the question in how far the classical news theory is still up to date. Some aspects which we have identified as especially important for the selection process of science coverage may also be of growing importance in general. One example is the already mentioned factor *graphical material* which seems to be of increasing importance in a crossmedial world (from a picture gallery in the online edition to YouTube and other social media). Therefore, selection criteria for pictures and other graphical material and their influence on the selection of a story should be further analysed (maybe even thinking of a kind of "picture value theory" in the future). For the classical newspapers, there is evidence that it is of decreasing importance how "new" a selected topic really is. "Put more (elements) of a weekly newspaper into the daily newspaper!",[28] i.e., more background information instead of daily news, has been the motto of the editor-in-chief of the *Süddeutsche Zeitung* for more than 15 years (see Wormer 2006). Meanwhile, the front page of this paper's weekend issue has become remarkably similar to the German weekly *Die ZEIT*: Instead of political news, a rather timeless (often science) topic with a corresponding picture is featured prominently on the front page. That the front page of a classical *news*-paper renounces "news" (daily actuality) at the top may be taken as an indicator to question some aspects of the classical theory of news values: Is it still adequate to work only with a one-dimensional approach of "news"? Or should it be replaced by a multi-dimensional view with "first class fast news" reported in the online news section in the internet and "second class fast news" for the print edition? Different from the times of Lippmann (1922) and Galtung and Ruge (1965), editors nowadays do not only have to *select* a topic, but also need to decide on *which platform* the news should be placed (e.g., print or online?). Such "crossmedia strategies" considering print (TV, radio) and online together may also influence the news selection process for the classical media. These are some reasons why we think that a further engagement with issue selection in science journalism can inform and stimulate studies of other sections of the media.

Acknowledgments Both authors contributed equally to this publication. However, F.B. is especially responsible for the empirical part, H.W. primarily for the development from the practical perspective. We like to thank the alumni-society of our Institute (EX! e.V.) for financing the travel costs for the guided interviews.

[28] The original German quote is: "Mehr Wochenzeitung in die Tageszeitung!".

References

Bauer, M. (2000). "Science in the media" as cultural indicator: Contextualizing surveys with media analysis. In M. Dierkes et al. (eds.), *Between understanding and trust. The public, science and technology*. Amsterdam: Routledge, pp. 157–178.

Böhme-Dürr, K. and A. Grube (1989). Wissenschaftsberichterstattung in der Presse. *Publizistik*, 34, 448–466.

Brosius, F. (2006). *SPSS 14*. Heidelberg: Mitp.

Bucchi, M. and R. G. Mazzolini (2003). Big science, little news: Science coverage in the Italian Daily Press, 1946–1997. *Public Understanding of Science*, 12(1), 7–24.

Clark, F. and D. L. Illman (2006). A longitudinal study of the New York Times Science Times Section. *Science Communication*, 27, 496–513.

Donsbach, W. (2004). Psychology of news decisions. Factors behind journalists' professional behavior. *Journalism*, 5(2), 131–157.

Donsbach, W. (ed.) (2008). *The international Encyclopaedia of communication*. Oxford: Blackwell, Vol. II, 333–336; Vol. V, 1921–1925; Vol. V, pp. 1862–1868.

Eilders, C. (1997). *Nachrichtenfaktoren und Rezeption. Eine empirische Analyse zur Auswahl und Verarbeitung politischer Information. Studien zur Kommunikationswissenschaft*. Opladen: Westdeutscher Verlag.

Elmer, C., F. Badenschier, and H. Wormer (2008). Science for everybody? How the coverage of research issues in German newspapers has increased dramatically. *Journalism & Mass Communication Quarterly*, 85(4), 878–893.

Galtung, J. and M. H. Ruge (1965). The structure of foreign news: The presentation of the Congo, Cuba and Cyprus Crises in four Norwegian newspapers. *Journal of Peace Research*, 2, 64–90.

Hömberg, W. (1987). Auswahlkriterien für Wissenschaftsnachrichten. In S. Ruß-Mohl (ed.), *Wissenschafts-Journalismus. Ein Handbuch für Ausbildung und Praxis*. München: List, pp. 90–96.

Hömberg, W. (1989). *Das verspätete Ressort. Die Situation des Wissenschaftsjournalismus*. Konstanz: Universitätsverlag Konstanz.

Hömberg, W. (1996). Auswahlkriterien für Wissenschaftsnachrichten. In W. Göpfert (ed.), *Wissenschafts-Journalismus. Ein Handbuch für Ausbildung und Praxis*. München: List, pp. 88–93.

Illinger, P. (2005, October 6). Der Neandertaler schafft es auf die Titelseite. *Süddeutsche Zeitung. Beilage "60 Jahre Süddeutsche Zeitung"*, 230, 25.

Illinger, P. (2006). Fünf Fragen an Patrick Illinger. In H. Wormer (ed.), *Die Wissensmacher. Profile und Arbeitsfelder von Wissenschaftsredaktionen in Deutschland*. Wiesbaden: VS Verlag für Sozialwissenschaften, pp. 25–27.

Kepplinger, H. M. (1998). Der Nachrichtenwert der Nachrichtenfaktoren. In C. Holtz-Bacha (ed.), *Wie die Medien die Welt erschaffen und wie die Menschen darin leben. Für Winfried Schulz*. Opladen: Westdeutscher Verlag, pp. 18–39.

Kepplinger, H. M. and H. Weißbecker (1991). Negativität als Nachrichtenideologie. *Publizistik*, 36(3), 330–342.

La Roche, W. V. (1999). *Einführung in den praktischen Journalismus*. München: List.

Lippmann, W. (1922). *Public opinion*. New York: Macmillan.

Milde, J. and G. Ruhrmann (2006). Molekulare Medizin in deutschen TV-Wissenschaftsmagazinen. Ergebnisse von Journalisteninterviews und Inhaltsanalysen. *Medien und Kommunikationswissenschaft*, 54, 430–456.

Østgaard, E. (1965). Factors influencing the flow of news. *Journal of Peace Research*, 2, 39–63.

Peters, H. P. (1994a). Wissenschaftliche Experten in der öffentlichen Kommunikation über Technik, Umwelt und Risiken. In F. Neidhardt (ed.), *Öffentlichkeit, Öffentliche Meinung, Soziale Bewegungen (Sonderheft 34 der Kölner Zeitschrift für Soziologie)*. Opladen: Westdeutscher Verlag, pp. 162–190.

Peters, H. P. (1994b). Risikokommunikation in den Medien. In K. Merten et al. (ed.), *Die Wirklichkeit der Medien. Eine Einführung in die Kommunikationswissenschaft*. Opladen: Westdeutscher Verlag, pp. 329–351.

Rensberger, B. (2009). Science journalism: Too close for comfort. *Nature*, 459(7250), 1055–1056.

Riffe, D., C. F. Aust, and S. R. Lacy (1993). The effectiveness of random, Consecutive Day, and constructed week sampling in newspaper content analysis. *Journalism Quarterly*, 70, 133–139.

Robert-Koch-Institut. (2008). *HIV/AIDS in Deutschland – Eckdaten: Epidemiologische Kurzinformation des Robert Koch-Instituts*. Stand: Ende 2008. Hamburg.

Ruhrmann, G. (1990). Aidsmäuse und Schlimmeres. Risikokommunikation über Gentechnologie – ein systematischer Zugang. *Medium*, 20, 36–38.

Ruhrmann, G. (1997). Wissenschaft, Medien und öffentliche Meinung. In H. Hoebink (ed.), *Perspektiven für die Universität 2000*. Neuwied: Luchterhand, pp. 145–157.

Ruhrmann, G. and R. Göbbel (2007). *Veränderung der Nachrichtenfaktoren und Auswirkungen auf die journalistische Praxis in Deutschland*. Wiesbaden: netzwerk recherche.

Ruhrmann, G. et al. (2003). *Der Wert von Nachrichten im deutschen Fernsehen. Ein Modell zur Validierung von Nachrichtenfaktoren*. Opladen: VS Verlag für Sozialwissenschaften.

Ruß-Mohl, S. (1987). Was ist überhaupt Wissenschaftsjournalismus? In S. Ruß-Mohl (ed.), *Wissenschafts-Journalismus. Ein Handbuch für Ausbildung und Praxis*. München: List-Verlag, pp. 12–15.

Schäfer, M. S. (2007). *Wissenschaft in den Medien. Die Medialisierung naturwissenschaftlicher Themen*. Wiesbaden: VS Verlag für Sozialwissenschaften.

Schneider, W. (1986). *Deutsch für Profis. Wege zu gutem Stil*. München: Goldmann Verlag.

Schulz, W. (1976). *Die Konstruktion von Realität in den Nachrichtenmedien. Analyse der aktuellen Berichterstattung*. Freiburg, München: Alber.

Schulz, W. (1977). *Nachrichtenstruktur und politische Informiertheit. Die Entwicklung politischer Vorstellungen der Bevölkerung unter dem Einfluss des Nachrichtenangebots*. Unpublished Review by order of the Presse- und Informationsamt der Bundesregierung. Mainz.

Schütze, C. (1996). Der Wissenschaftsredakteur im Medienbetrieb. In W. Göpfert (ed.), *Wissenschaft-Journalismus. Ein Handbuch für Ausbildung und Praxis*. München: List Verlag, pp. 188–191.

Staab, J. F. (1990). *Nachrichtenwert-Theorie. Formale Struktur und empirischer Gehalt*. Freiburg, München: Alber.

Statistisches Bundesamt (2009). Gesundheit. Todesursachen in Deutschland [2007]. Fachserie 12, Reihe 4.

Ten Eyck, T. A., P. Thompson, and S. Hornig Priest (2001). Biotechnology in the United States of America: Mad or moral science? In G. Gaskell (ed.), *Biotechnology 1996–2000. The Years of Controversy*. London: Science Museum, pp. 307–318.

Ten Eyck, T. A. and M. Williment (2003). The national media and things genetic: Coverage in the New York Times (1971–2001) and the Washington Post (1977–2001). *Science Communication*, 25, 129–152.

van Rooyen, C. (2002). A Report on Science and Technology Coverage in the SA Print Media. http://www.saasta.ac.za/scicom/pdfs/setcoverage_printmedia.pdf (last accessed on September 07, 2010).

Wellershof, M. (2005). "Leben ist an sich ein Risiko." Interview with Frank Schätzing. Der Spiegel, 1, January, 3, 2005, pp. 114–115 (online at: www.spiegel.de/spiegel/print/d-38785544.html; last accessed on June 14, 2011).

Wormer, H. (2006). Wissenschaft bei einer Tageszeitung: Fragen zur Vergiftung von Ehegatten und andere Dienstleistungen. In H. Wormer (ed.), *Die Wissensmacher. Profile und Arbeitsfelder von Wissenschaftsredaktionen in Deutschland*. Wiesbaden: VS Verlag für Sozialwissenschaften, pp. 12–27.

Wormer, H. (2008). Science journalism. In W. Donsbach (ed.), *The international Encyclopaedia of communication*. Oxford: Blackwell, Vol. X, pp. 4512–4514.

Wormer, H. (2010). "Warum ist der Himmel blau?" Wie die Massenmedien Wissensthemen aufbereiten und verbreiten. In U. Dausendschön-Gay et al. (eds.), *Wissen in (Inter)Aktion: Verfahren der Wissensgenerierung in unterschiedlichen Praxisfeldern (Linguistik – Impulse & Tendenzen LIT)*. Berlin, New York: deGruyter, pp. 347–376.

Chapter 5
The Medialization of Regenerative Medicine: Frames and Metaphors in UK News Stories

Richard Elliott

5.1 The Nature of Medialization

The relationship between science and society is changing. Reports of the death of scientific "popularization" may have been grossly exaggerated, but we no longer accept without question the notion that expert scientists produce "true" knowledge before the mass media translate, transmit and corrupt it (by "hyping up" or "dumbing down") and "the public" passively and unthinkingly consumes it. We cannot ignore the scientific *raison d'être* – the search for "truth", or at least "truthful" verdicts on hypotheses derived from accumulated evidence – yet we acknowledge that science, the media and different publics construct their own realities according to idiosyncratic "validation criteria" (Weingart 1998) and "epistemic cultures" (Knorr-Cetina 1999). The medialization hypothesis suggests that, as the validation criteria and epistemic cultures of science are increasingly supplemented and challenged by those of other knowledge cultures and as the media becomes more influential in structuring public discourse, scientific agendas will adapt to become increasingly media-oriented (Weingart 1998; Schulz 2004; Peters et al. 2008a; Rödder 2009). The study of medialization, and of the structural changes and repercussions it may bring about, is therefore a study in contradictions. While journalists strive to objectively report scientific developments and offer balanced criticism, they must also sensationalise and scaremonger in order to attract an audience and "sell" newspapers, magazines, websites and television programmes. For non-specialist publics, the media are the dominant, and perhaps the only, source of information and debate about science, yet they are also a source of distortion, hyperbole and rhetoric to be viewed with suspicion (Nelkin 1995; Petersen 2001; Priest and Ten Eyck 2003; Peddie et al. 2009). For the scientific community, which has traditionally poured scorn on media "misrepresentations" of science (Salomone et al. 1990; Nelkin 1996;

R. Elliott (✉)
University of Nottingham
e-mail: LBXRE@nottingham.ac.uk

S. Rödder et al. (eds.), *The Sciences' Media Connection – Public Communication and its Repercussions*, Sociology of the Sciences Yearbook 28,
DOI 10.1007/978-94-007-2085-5_5, © Springer Science+Business Media B.V. 2012

Bucchi 1998), political pressure has made it increasingly important to engage with journalists in order to establish or reaffirm legitimacy and priority in the wake of widely-publicised moral panics and social controversies over issues like nuclear energy, BSE and genetic modification (House of Lords Select Committee on Science and Technology 2000). Furthermore, the practice of "science by press release" (e.g., in connection with the Human Genome Project or the *Darwinius* fossil named "Ida", see Chapter 18) has demonstrated that scientists are capable of exploiting the media as a political resource, especially in connection with scientific issues defined by competition or controversy (Persson and Welin 2008; Rödder 2009). Scientific reliance on non-scientific publics is nothing new, yet, driven by the need to secure funding in highly competitive fields like regenerative medicine, the practices and epistemic cultures of science communication appear to be changing in order to harness media narratives that will strengthen the public profile of research (Peters et al. 2008b, c). While the influence of public relations (PR) can vary wildly between different news outlets, over the last decade the conscious media management practices ordinarily associated with business and political journalism have become an increasingly important and unavoidable part of science journalism (Nelkin 1995; Bauer and Bucchi 2007; Williams and Clifford 2009). PR material, including speeches, websites, press releases, and press conferences is attractive to journalists, because it cuts the time needed to research a story. However, as these sources typically select specific pieces of information and frame them in particular ways, simply regurgitating them without additional verification may limit the independence of an article and help to establish an uncritical agenda for media coverage. Put simply, the privatisation and commercialisation of scientific research are contributing to conflicts between scientific claims to objective truth and the bottom-line logic of market returns (Bauer 2008).

> We went from famine to feast in the period between 1990 and say, well, now. You've gone from a period when you were scratching around for stories and you were glad if you found one, to one where you've always got a choice of four or five. [...] All the journals put their stuff on the web, and all the universities put their stuff on the web, and it's instantly accessible, which was never the case. So there's been a major change really. [...] And because it's all written in a form that you can just cut and paste, we're back to the old problem of churnalism. It's terribly easy to write perfectly coherent stories just by cutting and pasting press releases (Health journalist Nigel Hawkes quoted in Williams and Clifford 2009: 54).

Faced with the contradictions inherent in a media that serves both as entertainer and educator, critic and marketing tool, scientific ally and unreliable antagonist, it is important that we sharpen our definition of medialization. Rödder (2009) argues that we should assess medialization through the analytical differentiation of two key dimensions: the growing media appetite for scientific issues and an increasing scientific orientation towards the media. In this chapter I will argue that, as media appetites are somewhat ephemeral (often hinging on the presence of the proverbial "good story" behind the scientific jargon), the second dimension is the more interesting from a Science and Technology Studies perspective. For the sake of clarity, I would suggest that the scientific dimension of medialization be subdivided further to distinguish between (a) the modification of *science communication* practices and the *presentation* of scientific research to make science stories more interesting to audiences and (b) the modification of *scientific research agendas, practices and validation criteria* to make *science itself* more acceptable to audiences. To break down (a) further, we might also distinguish between (i) the *summary* and *simplification* of scientific findings and (ii) the *manipulation* or *misrepresentation* of scientific findings (whether by scientists, science communicators or the media). Whether it is possible to achieve (i) without also doing (ii) or to achieve (a) without also doing (b) is up for debate and, as already argued, the fulfilment criteria for each category are likely to be perceived differently by scientists, press officers, journalists and members of different publics. Nevertheless, these are important distinctions to make as they reflect discrete structural and ethical dimensions of the relationship between science and the media.

Schäfer (2009) associates medialization with a shift away from the popularisation and "deficit" models traditionally associated with the Public Understanding of Science (PUS) movement and toward improved dialogue and critical engagement with science under the banner of Public Engagement with Science and Technology (PEST). He suggests that medialization is defined by scientific coverage that is increasingly extensive, pluralised and controversial and argues that coverage of stem cells in the German media fulfils all three of these criteria. This chapter will use a similar case study to dispute these definitions and argue that medialization is more complex and ambiguous in nature, bringing with it greater challenges for science communication and public engagement in addition to greater rewards. As the other chapters in this volume suggest, scientific orientation towards the media (the second dimension of medialization, Weingart 1998; Rödder 2009) can be analysed profitably from a range of perspectives both within and external to the scientific community. Only one of these perspectives, specifically the content and construction of media articles, is explored here, necessarily limiting the scope of the conclusions drawn.

5.2 Media Communication of Science

Recent events in the UK, such as the controversial dismissal of Baroness Susan Greenfield as Director of the Royal Institution and of Professor David Nutt as chairman of the Advisory Council on the Misuse of Drugs have brought science communication and the medialization of science into the spotlight. Correspondents from two of Britain's leading daily newspapers recently used Greenfield's departure to criticise the traditional "patronising", "top-down style" of science communication practiced by an "obsolete" "old guard" of scientific "figureheads" – a select group of "grandstanding generalists" able to speak authoritatively on any topic (Blackmore 2010; Goldacre 2010; Henderson 2010a). In the current more "media savvy" climate, we are told, a greater number of "young, up and coming science popularisers" (Robbins 2010) and a diverse range of scientists, either with "nuanced specialist knowledge" or an "an obvious representative, financial or political role" are able to "see the value of explaining their research with all its subtleties and uncertainties" (Henderson 2010a, b).

These statements support the notion of a closer relationship between science and the media and suggest an improvement in the way that science is being communicated. However, they sit somewhat awkwardly with comments from other science journalists. Discussing the coverage of stem cell research, Tim Radford, former science editor of *The Guardian* newspaper admits that he, along with other "co-conspirators" throughout the British media, "employed a variety of outrageous images" and relied on "hype, spin, naiveté and ruthless headline grabbing" to "launch" a public debate and "push it through" to a vote in Parliament (Radford 2009: 148–149). His reasons for this are equally clear: firstly, the difficulty of communicating science ("a series of unemotional statements, hedged with caveat and festooned with proviso, couched in deliberately passive sentences, and phrased in wilfully opaque language...") and stem cell science in particular ("an expensive, laboratory-based technology of unproven merit guaranteed to lead to many years of frustration punctuated by small flashes of enlightenment") in a way that would excite the interest and hold the attention of readers (Radford 2009: 148–149). Secondly, he argues that journalists had the "willing and enthusiastic encouragement" of eminent scientists from academia, government and industry who "encouraged us to see their point of view and quite frankly enlisted our help: that was flattering, and flattery is a powerful weapon" (Radford 2009: 149). The role of scientists in sensationalising science is underlined by Mark Henderson, science editor of *The Times*: "[Journalists] do sometimes [sex-up science stories], but scientists sex-up science stories too. It's very difficult to sex something up that isn't without help. If we overplay something it is usually because somebody has overplayed something to us" (Sense about science 2006). While Radford acknowledges that the media "should have been more challenging" over stem cells, he insists that this would have been to "spoil a good story" (Radford 2009: 150) and, rather than single out this case as an aberration, he defends sensationalism as an essential journalistic tool, claiming that "the act of writing about something – to choose one topic from the hundred or so potential topics delivered every day in the scientific press – is to hype it" (Radford 2009: 147).

The idea that journalists sensationalise science is nothing new; however, the suggestion that they do so with the support and co-operation of scientists, while laying claim to a closer and more candid relationship with science, suggests a novel perspective on medialization and is worthy of investigation. The following discussion will develop this perspective and examine the relationship of scientists and journalists with an exploration of the frames and metaphors used to cover regenerative medicine and cell therapies in UK newspaper stories. This case is of particular interest since it not only touches on a broad range of developments in molecular and cellular biology (e.g., from improvements in cell and tissue culture to the creation of cloned and hybrid embryos and induced pluripotent stem cells) but also covers a range of ethically controversial issues related to the translation of basic research from "bench to bedside" (e.g., clinical trials, cord blood banking and embryo donation).

5.3 Regenerative Medicine in the Media

The mass media provides an inventory of "cultural elements", such as metaphors, frames and symbols, associated with particular issues (Peters et al. 2008c). In the context of media analysis, framing describes the rhetorical and metaphorical techniques used by journalists to create specific attitudes and expectations among their audience. Media frames organise stories around explicit and implicit themes and orientate audiences using a limited range of definitions, causal explanations and moral evaluations (Entman 1993). Although they may not be chosen consciously, frames represent the efforts of the journalist to convey their story in a direct and meaningful way (Iyengar 1987). Framing is also influenced by professional news value criteria (e.g., actuality, sensation, personalisation and locality) and by organisational factors such as the accessibility of information, editorial policies and strict print or broadcast deadlines (Weingart 1998; Schulz 2004). Consequently, frames not only draw on and reflect shared cultural narratives and social concepts, they also limit the range of public debate by encouraging audiences to interpret information according to the ideas and phrases of specific narratives (Petersen 2001; Tankard 2001; De Vreese 2004). Any frame can include pro-, anti- and neutral arguments and individual sources may disagree over an issue while sharing the same interpretive frame (Gamson and Modigliani 1989). However, frames may also promote one interpretation of events or issues over another by calling attention to "relevant" causes or consequences and singling out the actors responsible; simultaneously obscuring alternative points of view and tacitly identifying them as something to be ignored. In this way, the most effective frames set the agenda for (and persuasively package) complex issues, even when balanced with multiple perspectives (Gitlin 1980; Kitzinger 2007). Consequently, "breaking" media frames can be difficult because of the shared identities, traditions, history and culture that constitute their collective interpretive resources (Nisbet 2009).

Mass media coverage of cell and regenerative therapies began to climb significantly at the beginning of the new millennium (Petersen 2001; Kitzinger and Williams 2005; Haran et al. 2007), coincident with the "peak of the first wave" of regenerative medical research (Mason 2007: 11). Initial sampling of four UK daily newspapers from Nexis®, one of the world's largest archives of newspapers and printed documents, revealed many thousands of articles that mentioned stem cells in connection with some form of medical treatment or therapy. Tabloids were represented by *The Sun* and *The Daily Mail* and broadsheets by *The Times* and *The Guardian*. These particular papers were chosen for their average circulation statistics (among the highest in the UK) and for the variety of writing styles, political stances and publishing companies they represent. Existing studies of stem cells in the media have focused on particular methods, such as embryonic cloning, and limited their sampling to particularly intense periods of news coverage surroundings key events, such as government votes or the announcement of scientific "breakthroughs". In this case, a longitudinal sampling period, running from 1 January 2000 to 30 June 2008 and including periods of both high and low press coverage, was used to obtain a diverse range of articles and facilitate an analysis of the persistence and variation of frames and metaphors over time. The search terms *regenerative medicine*, *tissue engineering*, *cell therapy*, *cell treatment* and similar variants were used to narrow the sample size to a manageable 437 articles. Although not intended to be representative, this sample was chosen to afford a general picture of the many faces of regenerative medicine coverage. All of the articles in which regenerative medicine provided the focus or was mentioned in passing were included in the sample, even those in which references were brief or without a medico-scientific focus (e.g., business and sports articles).

Frames were identified by looking at the metaphors, phrases, exemplars and analogies found in the body, titles and sub-titles of each article. An understanding of the rhetorical construction and temporal patterns of each frame emerged gradually through two complete, chronological readings of the sample. On the first reading, *NVivo* software was used to flag the discourses and metaphors associated with particular themes and group them into relevant categories. Flagged extracts were then clustered into higher-order categories, moved between categories or removed from the hierarchy entirely in order to produce a tentative map of rhetorical frames. A second reading helped to refine the map by introducing additional extracts to reinforce or challenge the prominence and potency of particular frames. The sources cited by journalists and evidence of attempts to verify, disconfirm or present alternative information were also analysed, and key groups of sources were clustered together with their characteristic discourses to suggest coalitions of actors and ideologies.

Sampled articles occurred in the leader, news and features sections of newspapers, in addition to specialist science sections and supplements, and were written both by non-specialist news correspondents and specialist science and health reporters, with a handful of articles penned by scientists and members of the public. Most articles described regenerative medicine as a potential treatment for a huge variety of medical conditions: from the degenerative neurological diseases associated with aging to various forms of cancer and trauma. Many articles focused on specific developments connected with only one of these areas, often highlighting "breakthroughs" in basic or translational research or, to a lesser extent, the initiation and progress of clinical trials. Other articles reviewed a wide range of potential treatment areas to stress the scope and significance of regenerative medicine as a whole, thus broadening their audience appeal and enhancing their potential emotive impact. Scientists and clinical specialists were the sources referenced most frequently by journalists and had, in several cases, authored their own letters and articles. The recurrence of identical remarks from the same scientific sources in articles from different newspapers suggested that some statements had been recycled from pre-existing public relations material. Comparison of articles in which this was most evident with related press releases obtained from the *EurekAlert* science news service revealed several cases where the framing of press releases and media articles closely matched, often with the only key difference being that media versions of stories emphasised human interest elements. For example, in April 2006, articles in both *The Times* and *The Daily Mail* covered a story on the first human trial of tissue-engineered bladders and featured numerous sentences that closely matched a press release from Wake Forest University, including the following speculative extract on the future of regenerative medicine.

> The engineered bladders were grown from the patients' own cells, so there is no risk of rejection. Scientists hope that laboratory-grown organs can one day help solve the shortage of donated organs available for transplantation (Wake Forest University Baptist Medical Centre press release, April 3, 2006).

> Children and teenagers received bladders grown from their own cells, which means there is no risk of rejection. [...] U.S. researchers believe laboratory-grown organs could one day relieve the shortage of donated organs and save patients from a lifetime of anti-rejection drug therapy (*The Daily Mail*, April 4, 2006).

> The engineered bladders were grown from the patients' own cells, so there was no risk of rejection, [...] which scientists hope one day may help to solve the shortage of donors of hearts, livers, kidneys and other organs (*The Times*, April 4, 2006).

The somewhat technical nature of the search terms used in sampling may have weighted the sample towards more technical articles and articles from the broadsheet newspapers, which favour a more technical vocabulary than the tabloids. Nevertheless, elements of drama and "human interest" were amplified in regenerative medicine coverage with personal testimonies recounting the experiences and expectations of different patients and publics. Entire articles (occasionally authored by patients themselves) were given over to the medical narratives and disease histories of individuals, while letters pages covered the opinions of several different people on particular scientific or ethical issues. In addition, short quotations from patients, patient groups and celebrity patients (often advocating campaigns to find new therapies for particular diseases) were used to both "universalise" and "personalise" (Petersen 2001) articles focused on basic research.

A smaller proportion of articles covered debates surrounding legal campaigns and measures, such as the UK Human Fertilisation and Embryology Bill and attempts to secure legal permission to produce a "saviour sibling" for unwell children. Such articles produced the most obvious attempts to consolidate ethical concerns over human embryonic stem cells, embryonic cloning and other controversial issues in biomedicine with the "therapeutic promise" of regenerative medicine (Rubin 2008). Accordingly, they drew on a wider range of sources, including parliamentarians, civil servants, ethicists, pro-life groups and religious spokespersons. In addition, short updates on the fluctuating stock market prices of regenerative medicine companies and longer profile pieces, often in the business sections, helped to emphasise the commercial potential of regenerative medicine. Together with brief extracts within stories more directly focused on scientific developments, these articles stressed the economic significance of this emerging field, its potential impact on global markets and its relevance for international growth and competition.

5.4 Scientific Revolutions and Medical Miracles

Rather than portraying developments in regenerative medicine as merely the latest in a series of scientific steps, newspaper coverage relied on conceptual metaphors to position the field as a unique source of innovation. Articles focused on scientific research were routinely framed as *pioneering, cutting-edge* or *breakthrough* discoveries, each offering a *world-first* or *landmark* change, or representing a significant *milestone* or *step forward* for science. Accordingly, connections between regenerative medicine and existing biomedical areas, such as bone marrow transplantation or gene therapy, were routinely ignored by journalists.

The pervasiveness of the *breakthrough* metaphor, its use for generating support for scientific research by emphasising novelty and the extent to which it misrepresents the production and value of scientific knowledge are well recognised (Brown 2000; Petersen 2001; Wallis and Nerlich 2005; Wolvaardt 2007). However, coverage of regenerative medicine went further by framing this field as a *revolution* for medicine and healthcare. Direct quotes from scientific and clinical specialists emphasised the revolutionary nature of regenerative therapies for particular conditions:

> It will revolutionise our approach, which is largely palliative, to one that is truly regenerative (*The Daily Mail*, April 26, 2004 [quoting Professor Robert Kormos, University of Pittsburgh on a treatment for cardiac muscle]).

> To have an off-the-shelf skin replacement product that can be used in large numbers of patients will revolutionise the treatment of burned and skin-damaged patients *(The Daily Mail*, June 27, 2007 [quoting biologist Stephen Minger, King's College London]).

The rhetorical power of this frame was amplified by general statements in articles describing a more widespread medical revolution:

> Scientists claim that work on the embryonic 'stem cells' [...] will bring about a revolution in conventional medicine (*The Daily Mail*, August 28, 2002).

> The long-term implications of stem cell therapy could be a revolution in medicine (*The Times*, May 31, 2003).

In addition, the scientific progression from research with animal models to clinical products was often portrayed as something that would be rapid or easy. The time given as necessary for the translation of particular regenerative therapies ranged from several months to over 20 years, but estimates of around 5 to 10 years were common at both the beginning and the end of the eight-and-a-half year sample. In addition, numerous articles framed regenerative medicine as a more simplistic and more effective alternative to conventional therapies that would effectively eliminate harmful side effects or the need for intrusive surgery. Some articles, particularly those associated with medical tourism, positioned therapies as "life-saving", "overnight cures" or "just a jab" to emphasise the speed with which they would take effect, although this was challenged on several occasions in the statements of specialists:

> [...] she had pioneering treatment in the Netherlands where master cells from babies' umbilical cords were injected into her spine and transfused through a drip. Within hours, she was walking unaided (*The Sun*, July 20, 2005).

> The stories that one hears of patients in a wheelchair who are walking 10 minutes after receiving this treatment could not be explained by any stem cell therapy mechanism (*The Guardian*, March 20, 2006 [quoting Neil Scolding, professor of clinical neurosciences, Bristol University]).

Further articles, particularly those relying on patient narratives, framed the unique regenerative capability of cell therapies as *magical* and the therapeutic results of regenerative medicine as *miraculous*. Although such metaphors were more common in the tabloids than the broadsheets, both newspaper types channelled the rhetorical power of biblical miracles with descriptions of seemingly impossible feats (e.g., the restoration of sight to the blind and the ability to walk to paralysed patients).

> Within minutes of having the treatment, I felt better. It's a miracle I am walking again. No one can tell me it doesn't work (*The Sun*, March 31, 2006 [quoting multiple sclerosis patient Amanda Bryson]).

> Magic powder made my severed finger grow back (*The Daily Mail*, May 1, 2008 [Headline]).

The framing of regenerative medicine as *revolutionary* or *magical* appeared to mature over time and occurred less frequently after 2005, as scientific and legislative sources began to express concern that media hyperbole and "overblown promises" would generate impossibly high expectations among patients and seduce them into paying unscrupulous foreign vendors for expensive, unproven and potentially-dangerous therapies. Articles began to distinguish between research revolutions and their translation into improvements in clinical practice:

> The cloning revolution: A giant step forward for science, but quest for new medical treatments goes on (*The Guardian*, May 20, 2005 [Headline]).

> Stem cells promise a great deal, and may indeed prove to be the revolution in medicine that so many doctors and scientists believe is just round the corner. However, the harsh truth is that this revolution is not even on the horizon (*The Daily Mail*, August 31, 2006).

> Talking about a revolution, but which one? (*The Guardian*, June 18, 2008) [Headline]).

Accordingly, articles began to emphasise the *illusory* nature of regenerative medicine promises and the danger of being *tricked* into false hope.

> [...] stem cell therapy is now starting to be offered by the world's quacks and snakeoil salesmen as a magic cure or treatment for everything from multiple sclerosis to cosmetic surgery (*The Daily Mail*, August 31, 2006).

The twist in these "metaphor chains" (Koller 2003) reflects maturing scientific perspectives on regenerative medicine and the realisation by many researchers that regenerative therapies may be more difficult and more costly to develop, take longer to reach the market, have lower clinical utility and less impact on patient health than originally predicted (Wainwright et al. 2006; Rowley and Martin 2009). These realisations were stated infrequently in articles focused on scientific and legislative developments, which emphasised the potential obstacles to regenerative medicine in terms of ethics, safety, high cost (anywhere from £1,500 to over £100,000 per treatment), side effects, practicality and technical uncertainty. Nevertheless, similar frames, emphasising the clinical potential of regenerative therapies based on stem cells derived from hybrid embryos, resurfaced during UK parliamentary debate of the Human Fertilisation and Embryology Bill in 2008, leading to conflicting coverage:

> The Bill's clauses that allow experiments with human-animal embryos have been backed by research charities and patient groups, eight of which have written to MPs to urge them to vote for this research. It could 'fast-track' the development of new treatments, they said (*The Times*, May 19, 2008).

> Cells taken from 'cytoplasmic hybrids' or 'cybrids' – the main type of admixed embryos – are never likely to be transplanted into sick patients. Any insights that they might offer into diseases such as Parkinson's and Alzheimer's, too, are probably years away (*The Times*, May 20, 2008).

Such conflicts again appeared to reflect scientific perspectives, in this case the apparent lack of scientific consensus over the usefulness of hybrids, although this appeared very rarely as a frame within individual articles. One exception is the extract below, in which members of the scientific community describe their frustration with the language used by their own peers to frame the hybrid debate.

> [...] we also wish to caution against false optimism and unrealistic claims for as yet unproven avenues of research. It is irresponsible, unjustified and, especially, unfair to patients for researchers to claim without evidence that a refusal to fund, to license or to approve a particular research approach will 'delay treatments for incurable illnesses.' [...] we all believe that extravagant claims regarding the purported merits of human-non-human interspecies embryos are mistaken and misleading, and that such research would damage public confidence and support, to the detriment both of the cause of stem-cell science and, ultimately, of patients (*The Times*, May 16, 2008 [A letter jointly signed by 16 scientists and clinicians involved in stem cell research and regenerative medicine in the UK and abroad.]).

This extract suggests a division between scientists who are actively sensationalising hybrid research in order to advance regenerative medicine and scientists aiming to expose and challenge sensationalist frames. Both groups are engaging with and exploiting the media to achieve their goals but where the latter group summarises and simplifies (category i, above) the former manipulates and misrepresents (ii), or at least, that is the perspective of the 16 letter-writing scientists.[1]

5.5 Autonomy and Control

Newspaper coverage also framed regenerative medicine as a source of autonomy and empowerment and as a symbol of scientific control over human biology. The development of regenerative therapies was routinely positioned as a move towards *bespoke*, consumer-oriented products that could be *tailored* to meet the medical needs of individuals. Many articles suggested that autologous cell therapies would solve the problem of immunological rejection and facilitate the personalisation of healthcare and some stories even "telescoped" (Michelle 2006, 2007) these benefits to include allogeneic therapies (e.g., treatments using embryonic stem cells). Accordingly, articles used conceptual metaphors to frame the body as a *device* or *machine* that could be repaired and upgraded with an engineered *repair kit* or laboratory-grown stockpile of *spare parts*.

---

[1] Further discussion of medialization in relation to hybrid embryos can be found in Chapters 12 and 13.

> The promise is an inexhaustible supply of spare parts, tailor-made to match their intended recipient (*The Guardian*, May 21, 2005).

> [Stem cells] offer great hope as a 'repair kit' for the body, replacing dead and worn out cells in conditions from Alzheimer's disease to diabetes (*The Daily Mail*, June 7, 2007).

> Welcome to the new body shop (*The Times*, November 2, 2007 [Headline]).

This frame highlighted the ability of scientists to control and manipulate human cells, with the ultimate objective being a limitless supply of tissue and organs that one might choose to have fitted like a suit of clothes or a set of car tyres. Building on these ideas, the discourses of patients were used to emphasise personal independence and consumer choice over the medical suitability of a treatment option or the evaluation of related ethical issues, in a similar fashion to the direct-to-consumer marketing materials of private stem cell clinics (Lau et al. 2008). Stories of personal triumph over disease, community fundraising efforts for treatment abroad, self-improvement through cosmetic enhancement and the political advocacy of celebrities, such as Christopher Reeve and Michael J. Fox, positioned regenerative medicine as a patient-driven enterprise, at odds with discourses of caution and restraint from religious groups, pro-life organisations, regulators and scientists themselves.

> British docs told the grandmother of five there was no hope – and she would remain in a wheelchair. But then she heard of the success of a procedure pioneered at Rotterdam's Advanced Cells Therapeutics clinic in Holland. [...] The treatment is still illegal in the UK as researchers probe its safety. But Sue, from Wolverhampton, said: 'I researched the company and I was confident. I had to collect £ 13,000 through fund raising events so I didn't want to waste the money, but I really didn't have anything to lose' (*The Sun*, January 20, 2006).

While the "artefactualisation" (Parry 2008) of the human body and the reduction of the medical encounter to a personal campaign or commercial transaction highlighted the autonomy of individual "healthcare consumers" at a social level, other metaphors highlighted the power of the individual at a more basic biological level. Regenerative therapies were described as a *natural* part of the body's own healing and regeneration mechanisms or an *organic* aid to such mechanisms, while patients were positioned as active participants in their own treatment, by effectively *growing-their-own* tissues, organs or cosmetic enhancements. These metaphors added further weight to the idea of a medical *revolution* by performing implicit boundary work to demarcate regenerative medicine from the existing range of "unnatural" and "impersonal" medical techniques and biotechnologies.

> This is breathtaking stuff, signalling the body itself to grow its own remedies naturally (*The Daily Mail*, October 21, 2000).

> You grow your own in a dish, treat yourself and walk away, at ease with the new bit of you inside. There are no polymer implants, no alien grafts, no synthesised drugs on prescription. Just a fresh bit of you, fashioned by the same natural machinery that took you from a single fertilised cell to a whole human being in nine months (*The Guardian*, April 27, 2000).

> Patients help heal themselves (*The Times*, September 3, 2007) [Headline]).

Most articles associated increased autonomy and control over biology with the benefits of saving lives and reducing transplantation waiting-lists. Far fewer articles made a connection between scientific control and the misuse of scientific power and only rarely (e.g., in accounts of reproductive cloning and of the creation of "designer babies" as "saviour siblings") did journalists rely on characteristic metaphors, such as *slippery slope, playing god* and *Frankenstein's laboratory*, to suggest the transgression of moral boundaries related to embodiment (Turney 1998; Petersen 2005).

5.6 Regenerative Medicine and Medialization

The dominance of scientific and medical discourses within the hyperbolic framing of regenerative medicine in newspaper coverage suggests that an adaptation of *science communication practices to media agendas and value criteria facilitated* the construction of a sensationalist media narrative. Although articles were routinely located in the news and feature sections of newspapers (rather than being confined to designated science sections) they still tended to adopt positivist scientific perspectives and few offered genuine critical evaluation of the social context of regenerative medicine or connected this emerging field with existing biotechnologies. Instead, many articles offered an artificially balanced deliberation of the status of the embryo or other relatively simplistic moral choices where viewpoints could be polarised according to equal and opposite extremes. Over time I encountered a greater number of stories highlighting the risks, uncertainties and limitations of regenerative medicine, but, even towards the end of the sample, these were overwhelmed by scientifically-framed discussions of its benefits and the rhetorical contraction of lengthy and complex pathways from bench to bedside and from basic innovation to commercial success and international competition. Such framing closely reflects temporal and spatial contraction within UK scientific research agendas, which emphasise the direct translation of laboratory science into the clinic (Martin et al. 2008; Wainwright et al. 2008), and the "imagined geographies of regenerative medicine" reported in the discourses of scientists (Williams and Wainwright 2008). Accordingly, articles tended to focus on and emphasise the novelty and revolutionary therapeutic potential of research and development "breakthroughs", concealing the processes of research, development and translation; exaggerating the extent of scientific control over human biology and "telescoping" characteristics associated with specific developments to emphasise the magical and miraculous ease-of-use, simplicity and "naturalness" of a vast range of potential end products. Consequently, coverage helped create an impression that biomedical research is principally oriented towards the resolution of social problems and the needs of publics and patients rather than the realisation of basic scientific objectives (Peters et al. 2008c).

The dominance of scientific and clinical "experts" within articles, reflects the complexity and specialisation of research on regenerative medicine and consequent journalistic reliance on "reputable" sources of information and clarification (Nelkin 1995; Conrad 2001). However, it also suggests that the discursive frames of these specialists were being regurgitated with a bare minimum of critical assessment or contextual review, effectively allowing them to set the agenda for science stories (Anderson 2002; Nisbet and Lewenstein 2002; Michelle 2006). Journalistic emphasis of the independence and autonomy of individuals (through anecdotal personalisation, dramatic narratives, emotive celebrity endorsements and their association with campaigns against disease, luddism and narrow-mindedness) helped simplify and universalise scientific discourses. As a result, articles were more exciting, engaging and potentially of greater relevance to readers, but may, from a scientific perspective, have risked overgeneralisation (Conrad 1997; Petersen 2001; Nisbet et al. 2003). These findings are consistent with reports of growing economic and institutional pressures on journalists, which leave them with less time to research stories and check facts and lead to greater reliance on public relations materials and a limited pool of sources (Williams and Clifford 2009).

The framing and organisation of coverage identified here, which suggest a close coupling and interaction of science communication and media value criteria, appear to conflict with those described by Schäfer (2009) as the symptoms of medialization. Schäfer argues that medialization is described by scientific coverage that is more extensive, features a greater diversity of actors and content and provides considerable journalistic scrutiny and challenges to scientific authority. He also associates medialization with movements toward increased dialogue and PEST and away from "deficit" models of science "popularisation" associated with the PUS movement. While Schäfer's criteria – extensiveness, plurality and controversy – are significant for any study of the relationship between science and the media, they are, by themselves, insufficient to account for the changes to this relationship associated with medialization. As these findings suggest, an increase in one or even all three of these criteria does not necessarily correlate with the medialization of a particular scientific issue.

I would argue that medialization is more complex and subtle, and that an overlap of science communication and journalistic value criteria does not necessarily equate with greater engagement of science and society. While coverage of regenerative medicine was extensive and appeared to increase throughout the limited sample, this cannot be attributed to medialization. The willingness of scientists to give up their time to engage with the press undoubtedly helped to promote regenerative medicine stories, but, as the comments of journalists themselves suggest, the extensiveness of coverage can also depend on the existence of a "good story" (Radford 2009: 150), which scientific encouragement only helps to package and sell to an audience. On the contrary, a closer relationship between scientists and journalists can even help to reduce the coverage of particular stories, by persuading journalists that they are irrelevant or inaccurate:

> Rather than hectoring journalists for covering stories that their editors tell them to cover, we instead find them leading experts willing to offer quotable comments rubbishing the story that can be included in their copy. Reassuringly, we have countless examples of how these comments often do help specialist journalists to persuade their news editors not to run the story (Fiona Fox, Director of the Science Media Centre, a unit which connects scientists and journalists and helps promote the views of the scientific community (2009: 118)).

Similarly, the medialization of regenerative medicine appears to be associated with the prioritisation and emphasis of scientific discourses within coverage, rather than their pluralisation amid statements from a more diverse pool of sources and spokespeople. While the appearance of science articles in the news and features sections of newspapers hint at greater plurality, contextualisation and critical reflection on coverage, the opposite is often true, since it is recognised that non-specialist news reporters are far more likely to over-simplify, sensationalise and unfairly "balance" scientific content than specialist science journalists (Fox 2009; Williams and Clifford 2009). Moreover, these findings suggest that the media tends to rely on scientific validation criteria, such as reputation, consensus and peer-reviewed publication, when these help to frame a story but may de-emphasise or disregard them when they do not (e.g., stories covering medical tourism against the advice of UK medical professionals or non-peer-reviewed "breakthroughs" such as finger regrowth using "pixie dust").

While medialization challenges the notion of science communication as a linear popularisation of science, this does not necessarily go hand in hand with greater evaluation and dispute of scientific perspectives, improvements in the diversity of scientific discussion or even increases in coverage. Rather, it describes a complex web of converging and conflicting agendas, interests and organisational structures that gradually and sporadically unite to bring science and the media closer together. Therefore, while medialization may offer certain improvements over past forms of science communication and bring with it exciting media narratives and renewed political enthusiasm for scientific endeavours, the closer coupling of scientists and journalists does not guarantee eager public engagement with science or a stronger relationship between science and society.

5.7 Conclusions

The medialization hypothesis describes the growing importance of the media for communication within society and highlights the causes and repercussions of the structural changes that contribute to the increased orientation and coupling of social systems to the media (Weingart 1998; Schulz 2004; Rödder 2009). The growing popularity of books, internet blogs and television programmes reflecting on the power of media narratives to amplify scientific predictions and selectively frame information (e.g., layscience.net; Goldacre 2008; *Newswipe* 2010) suggest that medialization is of increasing interest not only to scholars but also to journalists and the different "publics" that constitute their audiences. This chapter has used a longitudinal qualitative analysis of the frames and metaphors used in media coverage of regenerative medicine, one of the most celebrated and controversial scientific developments of recent years, to explore and help define the concept of medialization and to discuss its potential implications for science communication.

Individual cell therapies and regenerative medicine as a field were routinely sensationalised by newspaper articles that overemphasised the significance and simplicity of basic scientific developments and extrapolated from the narratives of individuals for greater emotional resonance. The prevalence of scientific sources and stories that drew heavily on scientific public relations material and the consequent reiteration of scientific frames provides further evidence to support the medialization hypothesis and the notion of closer coupling between the validation criteria and knowledge cultures of scientists and journalists. These findings suggest that science communicators organised their inputs according to media parameters of drama and relevance and that, because the "regenerative medicine revolution" represented a good story, journalists encouraged scientists to set the agenda for coverage, amplified the rhetorical power of scientific frames through anecdotal personalisation, hyperbole and blanket coverage and regularly omitted the sorts of contextual information that might enable a more critical evaluation of scientific perspectives. These findings, together with the comments of both scientists and journalists reflecting on regenerative medicine stories, suggest that an increased orientation of science towards the media and growing media interest in science do not necessarily correlate with greater diversity or critical engagement within science coverage and instead may contribute to greater exaggeration and sensationalism.

This chapter offers another indication, or "fingerprint of the phenomenon" as Weingart (1998) puts it, to suggest that media validation criteria (such as sensation, circulation and personalisation) are influencing the presentation and communication of science. The framing of regenerative medicine appears to reflect the increased commercialisation and politicisation of science and increasing scientific awareness that only scientific stories with a powerful social and political context (those with implications for healthcare in this case) are likely to receive significant coverage (Nelkin 1995; Weingart 2002). To return to the dimensions of medialization defined in the introduction, the case of regenerative medicine appears to fulfil category (a), suggesting that *science communication* practices and the *presentation* of scientific research have changed to make science stories more interesting to the media. Furthermore, it is arguable that the frames used to describe regenerative medicine go beyond category (i) the mere *summary* and *simplification* of scientific findings and in certain instances verge on (ii) their *manipulation* or *misrepresentation*, although whether this is attributable to scientists, science communicators or the media is difficult to ascertain and may vary from one news story to another. Without also studying in detail the research and communication practices of scientists, science communicators, public relations representatives and journalists, it is not possible to say whether the framing of scientific stories such as regenerative medicine can also be associated with (b) the modification of *scientific research agendas, practices and validation criteria* to make science itself more acceptable to audiences. Further studies of the practical interface between science and the media are warranted to determine whether the changes associated with medialization are limited to the communication of science or involve also the modification of core scientific values.

One can only speculate as to the influence medialization is likely to have on the most vulnerable audiences (patients in this case) and other publics. We must credit audiences with enough insight to know not to believe everything that is printed in the newspapers or broadcast on television (Peddie et al. 2009). Yet, the amplification and relentless bombardment of particular frames and metaphors may facilitate their tacit acknowledgement and incorporation into the background to future debates and certainly help to concentrate public attention and interest (Nerlich and Halliday 2007). Misleading or contradictory coverage creates the potential for confusion, false hope or frustration, which could help to nurture public apathy and suspicion of science in the media and of scientists themselves as one more "interested" voice among many, forcing scientists to work even harder in order to achieve priority and legitimation (Weingart 1998, 2002).

It is not the job of the media to educate the public about science, unless as Radford suggests, education occurs "by the way, and as a kind of happy accident" (2009: 152). However, neither is it the job of journalists to perform uncritical public relations work for the scientific community. The process of scientific research, and the numerous scientific, ethical and political perspectives from which it is considered, are often uncertain or conflicted, but objective reporting benefits from acknowledging and exploring this subtext rather than presenting successive scientific developments as unique or isolated incidents. It would be naive to expect popular science stories to account for the full complexity of the scientific process and, in any case, greater distinction between developments in basic laboratory science and applied technologies is but one of the missing components. Nevertheless, as Murcott (2009) and Nisbet (2009) suggest, improving the communication of science is all about nuance and approach – providing context, values-based reasons, background and relevance rather than simply reporting the results.

Despite changes to formats and means of distribution, the objectives of the mass media remain much the same – to capture the attention of audiences and maintain it for as long as possible by whatever means necessary – and therefore it can be tempting to argue that there can be no "adequate" presentation of science in the media sufficient to satisfy scientists (Weingart 2002). Consequently, changes to the practices and validation criteria of science represent the most significant aspects of medialization and it is scientists and science communicators who must assume a greater responsibility for the content and character of media engagement and science coverage. In the past, scientists have preferred to think of themselves as "speaking truth to power" (Price 1965) but not entering into policy debates. Today scientists must not only engage with the media but must also reflect on the values and interests underlying their science communication practices if they wish to maintain roles as neutral "arbiters" or transparent "brokers" of knowledge instead of becoming one more group of "stealth issue advocates" among many (Pielke 2007). However, the predicted decline of generalist science "figureheads" in favour of specialist scientists who will engage with the media from the "bottom-up", suggests that there could soon be an even greater number of competing voices clamouring for support and funding and thus greater potential for exaggeration and sensationalism.

References

Anderson, A. (2002). In search of the Holy Grail: Media discourse and the new human genetics. *New Genetics and Society*, 21(3), 327–337.

Bauer, M. W. (2008). Paradigm change for science communication: Commercial science needs a critical public. In D. Cheng, M. Claessens, T. Gascoigne, J. Metcalfe, B. Schiele, and S. Shi (eds.), *Communicating science in social contexts*. London: Springer, pp. 7–25.

Bauer, M. W. and M. Bucchi (2007). *Journalism, science and society: Science communication between news and public relations*. Oxon: Routledge.

Blackmore, S. (2010). Goodbye to a not-so-good scientist. *The guardian – Comment is free*. Available at: http://www.guardian.co.uk/commentisfree/2010/jan/12/susan-blackmore-royal-institution-science (last accessed 13 January 2010).

Brown, N. (2000). Organising/disorganising the breakthrough motif: Dolly the Cloned Ewe meets Astrid the Hybrid Pig. In N. Brown, B. Rappert, and A. Webster (eds.), *Contested futures: A sociology of prospective techno-science*. Aldershot: Ashgate Press, pp. 87–110.

Bucchi, M. (1998). *Science and the media: Alternative routes in scientific communication*. New York: Routledge.

Conrad, P. (1997). Public eyes and private genes: Historical frames, news constructions, and social problems. *Social Problems*, 44(2), 139–154.

Conrad, P. (2001). Genetic optimism: Framing genes and mental illness in the news. *Culture, Medicine and Psychiatry*, 25(2), 225–247.

De Vreese, C. H. (2004). The effects of frames in political television news on audience perceptions of routine political news. *Journalism and Mass Communication Quarterly*, 81(1), 36–52.

Entman, R. M. (1993). Framing: Toward clarification of a fractured paradigm. *The Journal of Communication*, 43(4), 51–58.

Fox, F. (2009). Science communication and ethics – Trying to get it right: The science media centre – A case study. In B. Nerlich, R. Elliott, and B. Larson (eds.), *Communicating biological sciences: Ethical and metaphorical dimensions*. Farnham: Ashgate, pp. 109–128.

Gamson, W. A. and A. Modigliani (1989). Media discourse and public opinion on nuclear power: A constructionist approach. *American Journal of Sociology*, 95(1), 1–37.

Gitlin, T. (1980). *The whole world is watching: Mass media in the making and unmaking of the New Left*. Berkeley, CA: University of California Press.

Goldacre, B. (2008). *Bad science*. London: HarperCollins.

Goldacre, B. (2010). A clarification: Why people have been concerned by Baroness Greenfield. *Guardian journalist Ben Goldacre's Posterous blog*. Available at: http://bengoldacre.posterous.com/a-clarification-why-people-have-been-concerne (last accessed 13 January 2010).

Haran, J., et al. (2007). *Human cloning in the media: From science fiction to science practice*. London: Routledge.

Henderson, M. (2010a). Baroness Greenfield and the Royal Institution are both out-of-date. *The Times*. Available at: http://www.timesonline.co.uk/tol/comment/columnists/guest_contributors/article6984214.ece (last accessed 13 January 2010).

Henderson, M. (2010b). Baroness Greenfield, the Royal Institution and science communication. *Times Online – Eureka Zone blog*. Available at: http://timesonline.typepad.com/science/2010/01/baroness-greenfield-the-royal-institution-and-science-communication.html (accessed 13 January 2010).

House of Lords Select Committee on Science and Technology (2000). Third Report, Science and Society. London: Parliament Stationery Office. Available at: http://www.parliament.the-stationery-office.co.uk/pa/ld199900/ldselect/ldsctech/38/3801.htm (last accessed 15 January 2010).

Iyengar, S. (1987). Television news and citizens' explanations of national affairs. *American Political Science Review*, 81(3), 815–831.

Kitzinger, J. (2007). Framing and frame analysis. In E. Devereux (ed.), *Media studies: Key issues and debates*. London: Sage, pp. 134–161.

Kitzinger, J. and C. Williams (2005). Forecasting science futures: Legitimising hope and calming fears in the embryo stem cell debate. *Social Science & Medicine*, 61(3), 731–740.

Knorr-Cetina, K. (1999). *Epistemic cultures: How the sciences make knowledge*. Cambridge, MA: Harvard University Press.

Koller, V. (2003). Metaphor clusters, metaphor chains: Analyzing the multifunctionality of metaphor in text. *Metaphorik.de*, 5, 115–134.

Lau, D., et al. (2008). Stem cell clinics online: The direct-to-consumer portrayal of stem cell medicine. *Cell Stem Cell*, 3(6), 591–594.

Martin, P., et al. (2008). From bedside to bench? Communities of promise, translational research and the making of blood stem cells. *Science as Culture*, 17(1), 29–41.

Mason, C. (2007). Regenerative medicine 2.0. *Regenerative Medicine*, 2(1), 11–18.

Michelle, C. (2006). Media (ted) fabrications: How the science-media symbiosis helped 'sell' cord banking. *Communication & Medicine*, 3(1), 55–68.

Michelle, C. (2007). Human clones talk about their lives: Media representations of assisted reproductive and biogenetic technologies. *Media, Culture & Society*, 29(4), 639.

Murcott, T. (2009). Journalism and society. In B. Nerlich, R. Elliott, and B. Larson (eds.), *Communicating biological sciences: Ethical and metaphorical dimensions*. Farnham: Ashgate, pp. 93–107.

Nelkin, D. (1995). *Selling science (revised edition)*. New York: W.H. Freeman and Company.

Nelkin, D. (1996). An uneasy relationship: The tensions between medicine and the media. *The Lancet*, 347(9015), 1600–1603.

Nerlich, B. and C. Halliday (2007). Avian flu: The creation of expectations in the interplay between science and the media. *Sociology of Health & Illness*, 29(1), 46–65.

Newswipe (2010). Series 2, Episode 1. TV programme. BBC 4. 24 January, 30 minutes.

Nisbet, M. C. (2009). The ethics of framing science. In B. Nerlich, R. Elliott, and B. Larson (eds.), *Communicating biological sciences: Ethical and metaphorical dimensions*. Farnham: Ashgate, pp. 51–73.

Nisbet, M. C. and B. V. Lewenstein (2002). Biotechnology and the American Media: The policy process and the elite press, 1970 to 1999. *Science Communication*, 23(4), 359–391.

Nisbet, M. C., et al. (2003). Framing science: The stem cell controversy in an age of press/politics. *The Harvard International Journal of Press/Politics*, 8(2), 36–70.

Parry, B. (2008). Entangled exchange: Reconceptualising the characterisation and practice of bodily commodification. *Geoforum*, 39(3), 1133–1144.

Peddie, V., et al. (2009). 'Not taken in by media hype': How potential donors, recipients and members of the general public perceive stem cell research. *Human Reproduction*, 24(5), 1106–1113.

Persson, A. and S. Welin (2008). *Contested technologies: Xenotransplantation and human embryonic stem cells*. Lund: Nordic Academic Press.

Peters, H. P. et al. (2008a). Science-media interface: It's time to reconsider. *Science Communication*, 30(2), 266–276.

Peters, H. P. et al. (2008b). Interactions with the mass media. *Science*, 321(11), 204–205.

Peters, H. P. et al. (2008c). Medialization of science as a prerequisite of its legitimization and political relevance. In D. Cheng, M. Claessens, T. Gascoigne, J. Metcalfe, B. Schiele, and S. Shi (eds.), *Communicating science in social contexts: New models, new practices*. New York: Springer, pp. 71–92.

Petersen, A. (2001). Biofantasies: Genetics and medicine in the print news media. *Social Science & Medicine*, 52(8), 1255–1268.

Petersen, A. (2005). The metaphors of risk: Biotechnology in the news. *Health, Risk & Society*, 7(3), 203–208.

Pielke, R. A. (2007). *The honest broker: Making sense of science in policy and politics*. Cambridge, UK: Cambridge University Press.

Price, D. K. (1965). *The scientific estate*. Cambridge, MA: Harvard University Press.

Priest, S. and T. Ten Eyck (2003). News coverage of biotechnology debates. *Society*, 40(6), 29–34.

Radford, T. (2009). A workbench view of science communication and metaphor. In B. Nerlich, R. Elliott, and B. Larson (eds.), *Communicating biological sciences: Ethical and metaphorical dimensions*. Farnham: Ashgate, pp. 145–152.

Robbins, M. (2010). Susan Greenfield sacking: Now the Royal Institution can focus on science. *Guardian Science Blog*. Available at: http://www.guardian.co.uk/science/blog/2010/jan/11/susan-greenfield-sacking-royal-institution (last accessed 13 January 2010).

Rödder, S. (2009). Reassessing the concept of a medialization of science: A story from the "book of life". *Public Understanding of Science*, 18(4), 452–463.

Rowley, E. and P. Martin (2009). *Barriers to the commercialisation and utilisation of regenerative medicine in the UK*. Nottingham: Institute for Science and Society.

Rubin, B. P. (2008). Therapeutic promise in the discourse of human embryonic stem cell research. *Science as Culture*, 17(1), 13–27.

Salomone, K., et al. (1990). Science and the news: A question of quality: How journalists and news sources evaluate coverage of environmental risk. *Journal of Communication*, 40(4), 117–131.

Schäfer, M. S. (2009). From public understanding to public engagement: An empirical assessment of changes in science coverage. *Science Communication*, 30(4), 475–505.

Schulz, W. (2004). Reconstructing mediatization as an analytical concept. *European Journal of Communication*, 19(1), 87–101.

Sense about science (2006). *Interview with Mark Henderson*. Available at: http://www.senseaboutscience.org.uk/index.php/site/about/100 (last accessed 13 January 2010).

Tankard, J. W. (2001). The empirical approach to the study of media framing. In S. D. Reese, O. H. Gandy, Jr., and A. E. Grant (eds.), *Framing public life: Perspectives on media and our understanding of the social world*. Mahwah, NJ: Lawrence Erlbaum, pp. 95–106.

Turney, J. (1998). *Frankenstein's footsteps: Science, genetics and popular culture*. New Haven, CT: Yale University Press.

Wainwright, S. et al. (2006). From bench to bedside? Biomedical scientists' expectations of stem cell science as a future therapy for diabetes. *Social Science & Medicine*, 63(8), 2052–2064.

Wainwright, S. et al. (2008). Shifting paradigms? Reflections on regenerative medicine, embryonic stem cells and pharmaceuticals. *Sociology of Health & Illness*, 30(6), 959–974.

Wallis, P. and B. Nerlich (2005). Disease metaphors in new epidemics: The UK media framing of the 2003 SARS epidemic. *Social Science & Medicine*, 60(11), 2629–2639.

Weingart, P. (1998). Science and the media. *Research policy*, 27(8), 869–879.

Weingart, P. (2002). The moment of truth for science: The consequences of the 'knowledge society' for society and science. *EMBO Reports*, 3(8), 703–706.

Williams, A. and S. Clifford (2009). *Mapping the field: Specialist science news journalism in the UK national media*. The Risk, Science and the Media Research Group, Cardiff University School of Journalism, Media and Cultural Studies. Available at: http://slb.cf.ac.uk/jomec/resources/Mapping_Science_Journalism_Final_Report_2003-11-09.pdf (last accessed 15 January 2009).

Williams, C. and S. Wainwright (2008). From basic science to clinical medicine: Reflections on stem cell translational research. *Society for the Social Studies of Science (4S) and European Association for the Study of Science and Technology (EASST) Joint Annual Meeting*. Rotterdam, the Netherlands.

Wolvaardt, E. (2007). Extra! Extra! New science breakthrough! *Cosmos Magazine*. Available at: http://www.cosmosmagazine.com/features/online/1003/extra-extra-new-science-breakthrough (last accessed December 2008).

Chapter 6
Medialization and Credibility: Paradoxical Effect or (Re)-Stabilization of Boundaries? Epidemiology and Stem Cell Research in the Press

Arlena Jung

6.1 Introduction

It is a commonly held position in science communication studies that the relationship between science and its social environment is undergoing a fundamental change. As expressed in the characterization as a "knowledge society," it is assumed that scientific knowledge increasingly permeates all areas of social activity (Gibbons et al. 1994; Stehr 1994; Willke 1998; Nowotny et al. 2001; Weingart 2001, 2005). One feature of this "tighter coupling" between science and its social environment is the medialization of science (Weingart 2001, 2005). In this paper I focus on one aspect of medialization: How does the increasing presence of science in the news affect the credibility of science? With the increasing importance of science in society, reporting on science has intensified both quantitatively and qualitatively. Not only is science given more space, coverage also increasingly goes beyond merely reporting on the results of scientific research (Bauer et al. 1995; Nelkin 1995; Elmer et al. 2008). The scientific process itself as well as the relationship between science and its social environment is subjected to observation by the mass media. As a result, much of what previously took place in the seclusion of the ivory tower comes into the field of public vision. Thus, more science in the news also means more scientific dissent, uncertainty, fraud, misconduct and more instrumentalization of science in the news (Nelkin 1995).

This quantitative and qualitative rise in reporting on science is in part due to a greater interest in science. The increasing importance of science raises both public interest and the democratically legitimated expectation of accountability. Mediated by mass media, science becomes subjected to the continual observation by the public and this on all levels of the scientific process. The attention given to the precarious

A. Jung (✉)
Social Science Research Center Berlin (WZB)
e-mail: arlena.jung@wzb.eu

S. Rödder et al. (eds.), *The Sciences' Media Connection – Public Communication and its Repercussions*, Sociology of the Sciences Yearbook 28,
DOI 10.1007/978-94-007-2085-5_6, © Springer Science+Business Media B.V. 2012

nature of scientific findings, shady characters and ethically disputable uses of science has to do with the nature of mass media and their weakness for the dramatic and the negative (Galtung and Ruge 1965; Schulz 1976; Nelkin 1995). Why restrict coverage to scientific findings when dissent, fraud and the instrumentalization of science for political and economic goals provide ample opportunities for reporting on conflicts and scandals? The mass media and the public are, however, not the only parties involved in changing how science is covered. The mechanisms endorsing the attention media give to uncertainty and dissent also have to do with the changing relationship between science, politics and economics.

Mass media is not only an impartial observer or platform for spreading information of public interest. It is also the stage on which power struggles are enacted, won and lost. And as the importance of science in society increases, science becomes intimately involved in these power struggles. It is not only increasingly relevant as a basis of political decisions and economic production processes, but also as a means of legitimating and selling decisions and products. Thus, whether in competition for the better argument or in an attempt to solve policy problems or make money, politicians as well as actors from the economic field increasingly turn to the forefront of scientific developments and thus to areas where scientific knowledge is unstable and contested. As different parties resort to scientific knowledge as a source of legitimacy, expertise is countered with counter-expertise, exposing the uncertainty of scientific knowledge to the public eye (Jasanoff 1987, 1990: 197ff; Peters 1994; Nelkin 1995; Weingart 2001; Kinchy and Kleinman 2003). Furthermore, policy-makers often deliberately set the stage so that science appears as unable to provide clear answers, thereby justifying the right of politicians to interpret science and base their decisions on their "personal" interpretations of scientific findings (Jasanoff 1987: 199). Politicians are, however, not the only ones to have discovered the media as an instrument in battles for monetary, political and legal resources and for legitimacy (Gieryn 1983, 1999; Nelkin 1995; Bucchi 1996; Simon 2001; Weingart 2001; Peters et al. 2008; Chapter 11; Haran and Kitzinger 2009). Scientists, too, "enlist journalists as vital allies" (Gieryn 1999: 200).

As shown, we can piece together a relatively coherent narrative about the mechanisms changing the quality and quantity of science coverage in the media. When it comes to the *effects* of these changes, we are, however, confronted with what at first seem to be rather strong contradictions. Here, too, we encounter a coherent narrative: The theoretically well founded assumption that more dissent, misconduct and instrumentalization of science in the news leads to a loss in credibility. At the same time, we are, however, faced with numerous empirical studies showing the stability of the credibility of science. Thus, in an attempt to explain and resolve these contradictions, we are well advised to begin by distinguishing between areas of consent and dissent. Uncontested is that the credibility of science rests on the special epistemological status of science as a source of knowledge impartial to particular interests, be they political, economic or of any other nature. Uncontested is also that autonomy functions as a prerequisite for impartiality and objectivity. Thus, both the mobilization of legal and economic support for scientific undertakings and the acceptance of the autonomy of science depend on the credibility of science as a source of objective and certain truth (Gieryn 1983; Jasanoff 1987, 1990; Luhmann 1990; Weingart 2001, 2005; Kinchy and Kleinman 2003). In this regard, the concept of credibility is surprisingly indifferent to differences in theoretical perspective.

How then do compatible definitions of medialization and of credibility allow for such different understandings of the effects of medialization? Let us begin with what Peter Weingart calls the paradoxical effect of proximity (Weingart 2001, 2005). As a result of the increasing importance of science in its social environment, the presence of science in the news increases. And, as shown, more science in the news means more dissent, fraud and instrumentalization of science in the news. Thus, on the one hand, science loses its credibility as being politically neutral and impartial to particular interests (Daniels 1967; Jasanoff 1987, 1990; Habermas 1992: 426; Weingart 2001, 2005: 52ff). Or as Kinchy and Kleinman (2003) put it: The ideals of "value neutrality" and "utility" come into conflict. On the other hand, increasing dissent and uncertainty in the news erodes the image of science as a source of certain and objective truth. Fuelling this process, scientists, in an attempt to fulfil the expectations society places on them, end up making knowledge claims that cannot be upheld and thereby further inflate the value of scientific truths (Luhmann 1990: 623ff). Following this line of thought, the effect of medialization on credibility seems obvious: The increasing presence of science in the news leads to a loss in credibility.

From a difference theoretical perspective, (1) distance between science and its social environment appears as a condition for legitimacy and hence for autonomy and (2) medialization is equated with a loss in distance. Tearing down the ivory towers, the media ends up making the incompatibility of social expectations towards science (such as utility and certainty) with the logic of the scientific process apparent. In this perspective, transparency means a loss in autonomy. Although the loss in distance caused by medialization is seen as posing a threat to the autonomy of science, distance is, however, not equated with difference. The special epistemological status of science has its roots in the differentiation of science from other forms of communication and knowledge attainment. The development of a logic *specific* to the scientific subsystem is what allows science to produce *comparatively* objective and certain knowledge. With medialization, science is, however, increasingly confronted with concrete expectations of objectivity and certainty it is not able to meet. As a result, the incompatibility of the logic of the scientific knowledge production process with the logics and expectations of other social systems becomes apparent. Or in other words: Science inevitably appears as continually disappointing the expectations of society. Thus, the increasing importance of science in its social environment paradoxically erodes the very foundations on which the special authority and status of science is based.

It is when we confront this conceptually sound reasoning with the results of empirical research that we begin to lose our footing. No one disputes the fact that the increasing presence of science in the news goes hand in hand with more "negative" science in the news. What is called into question is the assumption that this leads to a loss in credibility. Numerous studies show how science immunizes itself against the negative connotations of fraud and misconduct by publicly expelling the "black sheep" from the scientific community (Gieryn 1983, 1999; Nelkin 1995; Haran 2007; Franzen et al. 2007; Haran and Kitzinger 2009). In addition, Collins shows how uncertainty is framed not as failure but as part of normal science, justifying more science (Collins 1987). "We are shown only a small 'window of uncertainty' set within walls of certainty that extend into the past and the future" (Collins 1987: 692).

As the term "boundary work" implies, this can be attributed to the effectiveness of scientists as gatekeepers (Gieryn 1983, 1999; Schäfer 2008; Gerhards and Schäfer 2009; Haran and Kitzinger 2009). Scientists are not passive observers but rather actively influence how science is perceived. Another explanation is "that the repeated drawing of boundaries along similar lines [...] reflects the historically resonant, and consequently, taken-for-granted character of the discourses on which actors draw" (Kinchy and Kleinman 2003). In a similar vein, Gieryn observes: "So secure is the epistemic authority of science these days, that even those who would dispute another's scientific understanding of nature must ordinarily rely on science to muster a persuasive challenge" (Peters 1994: 181; Gieryn 1999: 3).

While neither the "paradoxical effect" explanation, nor the "boundary work" and "cultural patterns" explanations are wrong, each is in its own way insufficient for understanding the effect of medialization on credibility. The assumption that more "negative" science in the news *necessarily* leads to a loss in credibility underestimates the influence of framing. Only then can dissent, misconduct and instrumentalization be equated with negative science. And only then can distance be seen as a condition for legitimacy and thus for autonomy. The "boundary work" and "cultural pattern" explanations, on the other hand, do not provide a plausible explanation for *why* scientists have such easy play defending their boundaries. Why is it that the epistemic authority of science is so secure that – in spite of so much dissent, misconduct and instrumentalization – there seems to be no valid alternative to countering scientific expertise with counter-expertise (Peters 2008: 141)?

In order to better understand the effects of medialization on the credibility of science, it is, I argue, necessary to differentiate between normative and cognitive expectations. While cognitive expectations are based on *observations* about how things function, normative expectations are based on *judgements* about how things *should* function (Luhmann 1987: 436). Thus, while cognitive expectations change with contradictory empirical evidence, normative expectations are rooted in cultural patterns and moral values. As a result, the *experience* that my good friend Charlie always comes too late will change my cognitive expectation, while it is unlikely that my normative expectation that he *should* be punctual will be affected. Thus, normative expectations tend to be more stable than cognitive expectations.

Distinguishing between normative and cognitive expectations allows us to identify three possible effects of medialization: (1) a *loss* in credibility, (2) a *tension* between normative and cognitive explanations, and (3) a *re-stabilization* of cognitive and normative expectations. A *loss* in credibility means that science is no longer given credit for what it once was. Here, the public may, for example, "learn" that science is not to be understood as a source of objective and certain truth, but rather as one of many different ways of constructing reality, and that scientists are always "blinded" by their particular perspective on the world. In this case, we would be facing both a cognitive and a normative change in expectations. Science would neither cognitively nor normatively be expected to function as a source of certain and objective truth. When the credibility of science is discussed in science communication studies, what scientists are usually talking about is whether a loss in credibility can be observed. Thinking in categories that permit only two alternatives – a loss or the reproduction of credibility – has made science communication studies blind to two further possibilities: that the "negative" coverage of science does not lead to a *tension* between normative and cognitive expectations or to a re-stabilization of the relationship between the two. In the first case, the *normative expectations* placed on science remain stable. One may, for example, agree that scientists *should* act as impartial observers, while at the same time being aware of the fact that career interests often play a decisive role in influencing the judgement of scientists. In this case, one would *cognitively expect* scientists to continually disappoint the *normative expectations* placed on them. In the second case, medialization leads neither to a tension nor to a loss in credibility, but rather to a *re-stabilization* of the expectations placed on science. In this case, science would maintain its special epistemic status, while the expectations of *how* science attains knowledge may change. Uncertainty may, for example, come to be viewed not as failure but as a normal part of science. What we would then be observing is an adjustment of cognitive expectations while normative expectations remain untouched.

Although a tension between normative and cognitive expectations refers to a credibility problem in the present, both a loss of credibility and a re-stabilization of credibility refer to a change over time. A re-stabilization of credibility can, however, be operationalized as the production of an image of science in which the "negative" aspects of science covered in the media are accounted for, without creating a tension between cognitive and normative expectations. A loss in credibility can, in turn, be operationalized as an image of science in which the validity and objectivity of scientific knowledge is not seen as being superior to other knowledge forms.

Using this analytical distinction, I interpret the implications of the image of science produced in the coverage of epidemiology and stem cell research in the German Press on the credibility of science. Here, I show that what at first sight may be interpreted as successful boundary work is rather a re-stabilization of credibility or even the stabilization of a tension between normative and cognitive expectations. In the conclusion, I then argue that these effects must be seen in context of the nature of mass media as a communication form. I argue that although mass media can and do allow for learning in the sense of a re-stabilization of expectations, they would not be able to produce such a fundamental cultural change as the loss of the special epistemic authority of science. However displeasing it may be to scientists, the continual reproduction of a tension between normative expectations and cognitive observations is, on the other hand, precisely the democratic function the press should be fulfilling.

6.2 Epidemiology and Stem Cell Research – Between Routine Coverage of Science for Policy and Science Policy for Breakthrough Science

In the following, I present the results of an analysis of the coverage of epidemiology and stem cell research in the German national press (*Das Handelsblatt*, *Der Spiegel*, *Die Tageszeitung*, *Die Welt*, *Die Zeit*, *Frankfurter Allgemeine Zeitung* and *Süddeutsche Zeitung*) between 2000 and 2006. The goal of this analysis was to reconstruct the images of science produced in the media.[1] Epidemiology and stem cell research were chosen as two areas that represent "visible science," both enjoying a high degree of media attention while, at the same time, differing in ways that are likely to influence the effect of proximity on credibility. Health is an "area in which science, politics and the media interface [...] and in which regulatory measures are connected to the actual state of knowledge and are changed in accordance with its progress" (Weingart 1997: 605). Thus, in the case of epidemiology, "the process of decision-making places unusual strains on science" (Jasanoff 1987: 195). Epidemiology is, however, at the same time, an area in which reporting is to a large extent a routine matter because it has to do with an issue of universal concern (health!). In contrast, stem cell research is breakthrough science both in the sense of being seen as being at the outer most frontiers of scientific development (Nelkin 1995) and in the sense of "violating" cultural categories of the natural, thus disturbing the existing social/moral order (Bloomfield 1995). In addition, while epidemiology is almost only covered as *science for policy,* stem cell research is largely reported on in terms of *science policy*. Thus, on the one hand, reporting on *science for policy* is compared with reporting on *science policy* while, on the other hand, *routine reporting* on science is compared with *reporting on breakthrough science*.

6.2.1 Epidemiology

6.2.1.1 Method and Material

The articles were analyzed with the qualitative interpretation method developed by Ulrich Oevermann: objective hermeneutics (Oevermann et al. 1979; Oevermann 1991, 2000; Reichertz 2004). This is a method that takes the contextuality of meaning seriously. Two principles serve to ensure that this constructivist understanding of meaning is respected in the interpretation processes. Instead of focusing on parts of a text that might seem particularly interesting, the researcher goes through the text step by step, analyzing each sequence in its particular context. Each sequence is analyzed in terms of all possible meaning contexts in which the selection of precisely

[1] This study was part of the project "Integration of scientific expertise into media-based public discourses (INWEDIS)" (see Peters et al. 2008) which was supported by a grant from the German Federal Ministry of Education and Research (BMBF) in the research programme "Knowledge for Decision-making Processes – Research on the Relationship between Science, Politics and Society."

this form of constructing reality would make sense. As one progresses in the analysis, more and more of the many possible contexts can be eliminated so that in the end one has only one coherent "reality" – in our case one coherent image of science. This meticulous approach usually greatly restricts the amount of material that can be analyzed. Since the image of science produced in the coverage of epidemiology proved to be very stable, it was possible to analyze a comparatively large number of articles (120).

Based on a keyword search in the Database Lexisnexis, a stratified random sample of 40 articles was taken from each of the following three areas of epidemiology: BSE, cancer and epidemiology excluding BSE and cancer. Restricting the analysis to cancer and BSE served to neutralize possible context variables. In the case of cancer, the purpose was to assure that reporting was on an area of epidemiology that enjoys continual and high media presence. BSE is an area of epidemiology in which media coverage was not a matter of routine reporting. Rather, it took place in a situation that can safely be described as a crisis both in terms of policy and in terms of the relationship between science and politics. No variation of the image of science was found in the coverage of cancer, BSE and other areas of epidemiology.

6.2.1.2 Empirical Results[2]

A very consistent image of science is created in the media coverage of epidemiology. Politicians are expected to be informed on the latest research results and to base their decisions on this knowledge. Policy that does not take the latest scientific discoveries into account is framed as not well founded and irrational, thereby reaffirming the epistemic status of scientific knowledge.

> Those who demand cannabis for everyone are not familiar with current research results and with the dramatic epidemiological data of the last two to three years (*Spiegel*, "Kick from the bong," 33/2002).

> With the slogan 'the fattening television' the eager Minister Künast wants to drive kids away from the TV screen. At the same time, a recent study [. . .] couldn't affirm the often claimed relationship between television consumption and overweight (*Zeit*, "Round and healthy," September 30, 2004).

The objectivity of scientific knowledge as the legitimate basis of policy decisions is contrasted with the influence of interest groups.

> The scientific findings on the dangers of passive smoking should have long ago been translated [. . .] by a responsible politics into effective laws and regulations. [. . .] The dependence on and the fear of lobby groups has [. . .] played an overwhelming role in tobacco prevention or rather non-prevention [. . .] (*Süddeutsche Zeitung*, "More protection for non-smokers," March 11, 2003).

---

[2] Since the interpretation method objective hermeneutics produces very long texts, the quotes used serve to illustrate the results of the analysis.

The image of science as the legitimate basis of policy decisions and as a source of objective truth is reinforced by the expert role of scientists as educators and policy advisors.

> 'In this regard we are politically not on the right path,' says epidemiologist Niklaus Becker from the German Cancer Research Centre Heidelberg. Although both the educational campaigns of the government and its Mammography project [. . .] are positive, 'the behaviour of Germany in the fight against tobacco is a disgrace' (*Süddeutsche Zeitung*, "The lifestyle that makes you sick," April 4, 2003).

> Andreas Krause from the rheumatism clinic in Berlin-Wannsee summarizes the prejudices of the nation as follows: 'It is seen as an old people's alignment against which one cannot do anything.' But lay people are mistaken (*Zeit*, "Smouldering fire in the body," July 29, 2004).

As expected, dissent, uncertainty and the instrumentalization of science were recurring themes in the media coverage of epidemiology. How did this affect the credibility of science? In the empirical material analyzed, two contexts were found in which the credibility of science becomes an issue: (1) the inadequacy of science for solving concrete problems (utility) and (2) the instrumentalization of science. In the following, I will analyze each of these contexts separately.

Science and Utility

The coverage of science in contexts in which the lack of scientific knowledge that is needed for solving concrete problems is an issue, the credibility of science is at least implicitly called into question. One common means of solving this credibility problem is by re-defining the problem. Knowledge gaps are attributed to the complexity of the subject matter rather than to deficiencies in science. As can be seen in the article cited below, more science thus appears as the only means of solving the problem.

> Nothing can conceal the fact that the knowledge gaps concerning cancer promoting or cancer inhibiting effects of the natural and synthetic ingredients of our nutrition are immense. Well founded knowledge on these complex relationships could help countless people to attain a healthier life. Detailed research into carcinogenic and anti-carcinogenic effects of food should therefore be more intensely driven forward than they have been till now (*Frankfurter Allgemeine Zeitung*, "Cancer and nutrition – A never ending story," February 28, 2001).

This article begins by calling the utility of science into question. The title "Cancer and nutrition – a never ending story" evokes an impression of futility. As we quickly learn, the "story" referred to is scientific research on the relationships between cancer and nutrition. We are shown that science is unable to find clear answers to questions concerning cancer inhibiting and cancer promoting effects of nutrition. Describing these scientific uncertainties as "a never ending story," of course, implies that science, at least in this field, cannot be given credit for attaining certain and objective truth. The utility for preventing cancer is called into question. Ongoing research thereby appears as futile and hence as a waste of time and resources. This image is, however, quickly contradicted in the running text. The text begins by distinguishing between areas of scientific consent and dissent: The fact that nutrition has both positive and negative effects on cancer development is undisputed. What is difficult to understand, we are told, is what effects which substances have. These difficulties are not attributed to scientific deficiencies but rather to the complexity of the subject matter. The only means of attaining "well founded knowledge on these complex relationships" is more "detailed research." The implication is that scientific research is the only means of attaining "well founded knowledge" and that the uncertainties of current scientific findings can be resolved by more "detailed research." At the same time, research is defined solely in terms of utility ("help countless people attain a healthier life").

In short: While the article begins by addressing a credibility problem in its title, the text serves to (1) re-define the problem as a problem that is caused by the complexity of the subject matter and not by deficiencies in science and (2) thereby stabilize the credibility of science as the only means for solving the problem: "Detailed research into carcinogenic and anti-carcinogenic effects of food should therefore be more intensely driven forward than they have been till now." Thus, as in Collins' "window of uncertainty," scientific uncertainty is framed as being a temporary state set within "walls of certainty." The credibility problem addressed in the title can be defined as a tension between normative and cognitive expectations. The purpose of science appears as attaining knowledge concerning the relationship between nutrition and cancer. The impression is, however, evoked that this is a futile mission. In the text of the article, this impression is implicitly repudiated and thus framed as being a false impression. In this sense, what we observe in this article can be defined as a re-stabilization of credibility. It addresses and corrects an "existing" credibility problem. The credibility of science as a source of objective truth and the legitimacy of science in terms of utility is upheld (1) in that uncertainty and dissent are framed as a normal part of science and (2) in that science alone appears as able to resolve the conflict, close the gaps and thereby replace uncertainty with certainty.

Another means by which the tension between the normative expectation placed on science to solve concrete problems and its inability to meet these utility demands is resolved is by: (1) addressing the *incompatibility* of science process and knowledge acquirement and with the demands of utility, in terms of solving concrete problems as they arise –while at the same time – (2) annihilating the difference between the scientists as researchers and policy advisors and policymakers as the ones making collectively binding decisions.

> It is as if we were sailing a boat which we are still building. We are still missing unbelievable much knowledge. But we still have to continuously make decisions that can be a question of life and death – and at the same time about a lot of money (*Spiegel*, "International alliance of virus hunters," 19/2003).

Instead of scientists as policy advisors, we encounter a collective engaged in a common endeavour of both building (attaining knowledge) and steering (making decisions). In this context, an understanding of science in terms of its utility is pushed to its extreme. Although the process of knowledge acquisition (building) is distinguished from its application (steering), no distinction is made between those responsible for building and those responsible for steering. In this context, scientists are characterized as people mobilizing all possible resources and energy to find the knowledge needed. The inadequacy of knowledge is explained by the complexity of the issue at hand and by the nature of the scientific process, or more precisely, the incompatibility of the nature of science and the nature of policy. While decisions must be made as problems arise, science must adhere to scientific methods of knowledge acquisition. Thus, policymakers are faced with the dilemma of having to steer a boat that is still being built. In terms of credibility, here, too, we see a re-stabilization of the image of science in the sense that science is neither normatively nor cognitively expected to be able to provide the knowledge needed for decisions *in the time frame* set by the necessity of action. The consequence that science is inadequate in terms of what is needed to solve concrete problems is avoided by annealing the difference between science and policy in terms of relevance structures. The distinction between scientists and policymakers disappears and scientists are portrayed as part of an international collective working for the common good.

In contexts of concrete utility expectations, scientific uncertainties are not only addressed in terms of knowledge gaps, but also in terms of contradicting information. Here, the credibility of science as a source of certain and objective truth is called into question. The tension between the expectation of objectivity and the observed uncertainty is resolved by framing existing contradictions as a result of knowledge gaps. Scientific dissent, we are told, can, must and will be resolved by further research. Thus, the credibility problem that scientific dissent implies is re-defined as a problem of missing knowledge. The lack of scientific certainty is then, as shown in the above examples, not attributed to deficiencies in science, but rather to the complexity of the subject matter. Thus, here, too, one can see a very consistent pattern of re-stabilizing the credibility of science. The following example shows how the coverage of dissent and uncertainty can create an image of science that at first may appear as a form of successful boundary work. The credibility of science as a source of objective and certain truth is re-stabilized. Upon closer examination one can see, however, that, at the same time, a tension between the normative and cognitive expectations placed on science is stabilized on another level.

> *Spiegel*: Professor Michaelis, the findings of your discipline are causing an epidemic of fear. The basis are epidemiological studies which make the consumption of coffee responsible for pancreas cancer, fatty foods for heart disease and anti-baby pills for thrombosis. Often enough a little later studies claim the exact opposite.
>
> Michaelis: Contradictions are often caused by epidemiological studies collecting extensive data. [...] Then, in the analysis of all kinds of things, one has the statistical problem that relationships can always be seen which are merely coincidental results of the numbers. [...] The perhaps coincidentally observed results have to be – and this is the point – researched methodologically in a further study. [...] That is where the epidemiologists often make the mistake that they want to increase the value of their study and sell still uncertain coincidental findings as definite results (*Spiegel*, "Fishing for data is widely spread," 15/2001).

In his "question," the journalist of the magazine *Spiegel* prompts his interview partner, an epidemiologist, to justify the behaviour of "his field." Both the expectation of utility and of science as a source of certain and objective truth is called into question with the metaphor "an epidemic of fear." Instead of helping to prevent and heal diseases, science is portrayed as being a source of illness and suffering. Framing science as the cause of an "epidemic of fear" not only reverses the role of science from helper to ill-doer. The term "epidemic" also refers to a highly contagious, rapidly spreading disease and therefore to a threat or danger, implying a loss of control. Thus, instead of functioning as a source of certainty allowing for rational decisions, science is portrayed not only as causing "fear" as an irrational emotion, but as causing this emotion to be passed from one member of the community to the next like an uncontrollable disease. The credibility problem addressed, thus, concerns both the epistemological status of science as a source of certain and objective truth and the utility of science. By beginning this description of epidemiology with the phrase "the findings of your discipline" a collective is defined. The journalist does not speak of the finding of individual epidemiologists but rather of "your discipline." Hence, the implication is that the discipline as a whole, as a collective, is responsible for "causing an epidemic of fear." By calling it "your discipline" the interview partner is identified as a member of this collective and thereby given responsibility for "the epidemic of fear." Without formulating a question the journalist clearly calls on his interview partner to justify the behaviour of *his* discipline.

The interview partner begins by locating the cause of contradiction in what seems like sound scientific practice: "epidemiological studies collecting extensive data." This sound practice is then shown to bear a problem: "Relationships can always be seen which are merely coincidental results of number." Then, a norm of good scientific practice is formulated: "The perhaps coincidentally observed results have to be researched in a further study." It is only in a next step that we are told what the true source of contradictions is. It is not the extensive data that cause the contradictions or the problem of distinguishing between coincidental and real relationships. With the formulation of the norm of good scientific practice we have already been told what the scientific solution to this problem is: more research. The cause is the "selling of still uncertain coincidental findings as definite results." The implication of "*still* uncertain findings" is that these momentary contradictions will eventually be resolved – the means of resolving the contradictions being further research. The deviation from good science is attributed to personal career interests: "Epidemiologists want to increase the value of [the] study and sell uncertain finding as definite results." So here, too, we see familiar mechanisms of re-stabilizing the credibility of science at work: (1) the distinction between good science and bad science and (2) the setting of uncertainties within walls of certainties.

While the "setting of uncertainties within walls of certainties" re-stabilized the epistemic status of science, the distinction between good and bad scientific practice functions somewhat differently here than in the contexts considered so far. It both re-stabilizes the credibility of science and stabilizes a tension between normative and cognitive expectations. In order to understand this double effect, let us compare Michaelis' epidemiologists with the "black sheep" we meet in Gieryn's studies (Gieryn 1983, 1999). The "culprits" we encounter in this example differ from Gieryn's "black sheep" in three regards. (1) The deviation from the norm of good science is not described as fraud or misconduct but rather as a "mistake." There is a great difference in the moral implications of the term "mistake" and the terms "fraud" or "misconduct." Accusing somebody of "fraud" or "misconduct" bears a clear moral judgement expressing both a breach of deeply-rooted values and contempt for the person or at least his behaviour. Classifying behaviour as a mistake, in contrast, places it safely beyond the reach of moral judgement. Not only does everybody make mistakes. Mistakes are not a result of ill intentions or moral indifference but rather of bad judgement or lack of knowledge. Therefore, even if the "mistakes" are interpreted as mistakes in moral judgement, it would be inappropriate to sanction them with disdain let alone such drastic measures as the expulsion from the scientific community. (2) The deviation from the norm is not framed as an exception. Both the title ("Fishing for data is *widely spread*") and the text ("epidemiologists *often* make the mistake") emphasize that what is being described is common practice. The implication is that this form of behaviour is not only common, but also that it does not breach deeply-rooted norms in the scientific community but is rather tolerated as an understandable mistake – and this in a context, where the detrimental effects of this behaviour on society are described in dramatic terms: "causing an epidemic of fear." (3) The mistake in judgement is made possible by a characteristic of the scientific research process itself: "One has the statistical problem that relationships can always be seen which are merely coincidental results of the numbers." By framing this as an objective "problem" that exists for all researchers ("*one* has the problem." "*always* can be seen"), we are given the impression that the sale of "*perhaps* coincidental relationships" as clear research results is a "mistake" caused not only by the career interests of individual scientists, but also by the nature of this kind of research. Instead of despicable personality traits being responsible for the crossing of the boundary from good science to bad science, we are shown a kind of slippery slope. Combined with the nature of epidemiological research and the normal, morally acceptable motivation to further one's career interests, the mistake of "fishing for data" becomes common practice.

The image of "black sheep" both serves to immunize science by making misconduct and fraud appear as an exception to the norm and by re-affirming the compatibility of social and scientific norms. Here, far from being described as "black sheep," the culprits are described as scientists making a by no means uncommon "mistake." The possibility for wrong interpretations, though of temporary nature, appears as something inherent to the scientific process and selling research results over value as an impulse that is both common and tolerated within the scientific community. At the same time, the detrimental effects of the resulting bad scientific practice for society – the "epidemic of fear" – are not called into question. As a result, scientific norms appear as being in conflict with basic social interest on a very fundamental level. Thus, although the repertory used to address the credibility problem is not extended, the implications for the credibility of science are very different. The immunizing effect is restricted to the epistemological status of science and as a means of attaining certain and objective truth in the *long run*. Science, we are told, is in principle able to function as a source of reliable information with practical implications. One can, however, expect that "in reality" any particular statement might very well be based on "bad" science.

Instrumentalization of Science

In articles in which the instrumentalization of science is an issue, the credibility of science is stabilized in two regards: (1) The instrumentalization of science is framed as a breach of the norm of scientific autonomy. (2) Scientists appear as key actors fighting for the autonomy of science. In both cases, the credibility of the autonomy of science is reinforced. The credibility problem we encounter here does not pertain to the validity of science as a knowledge form or to its autonomy, but rather to the relationship between science and its social environment. Decisive for whether credibility is re-stabilized or a tension between cognitive and normative expectations is stabilized is whether the breach in norms appears as common practice or as a scandal in the true sense of the word. The following example shows how the coverage of the instrumentalization of science can re-stabilize the credibility both of science and of the relationship between science and its social environment.

> One is otherwise only used to this kind of influence in totalitarian states: Researchers weren't allowed to freely participate in conferences, their findings were censored. British clerics and politicians picked out those statements from their findings that served to calm down the population. Warnings were kept concealed. Instead, British agricultural minister John Gummer went in front of the camera with his daughter and asserted, eating a hamburger: Beef is safe (*Süddeutsche Zeitung*, "Living with the risk," November 28, 2000).

By framing the instrumentalization of science as something "one is otherwise *only* used to in totalitarian states," the incident is marked as a scandal in the true sense of the word. It is not only something which should never have happened; it is also something lying beyond normal experience. The comparison with totalitarian states, at the same time, legitimates the autonomy of science in two different ways: The autonomy of science appears as a condition for utility. Scientific knowledge is framed as something the public both has a right to and needs for making rational judgements. Autonomy from politics appears as the condition for open communication about scientific research. The autonomy of science is, however, not only justified in terms of utility, but also – as the comparison with totalitarian states implies – in terms of freedom and thereby as something of inherent value. Thus, the instrumentalization of science is constructed both as a breach of democratic norms (freedom of science, the right of citizens to information on matters of public concern) and as a threat to the social function of science. Eating beef on camera is framed as being a form of propaganda based on emotional manipulation and is implicitly contrasted with the legitimate means of evaluating risk: objective, scientific knowledge. The instrumentalization of science described here is framed not only as a breach of the norms of a legitimate relationship between science and politics, but also as something breaking with previous experience – at least in democratic states: "One is otherwise *only used to* this kind of influence by totalitarian states." Thus, what we see here is a re-stabilization of the credibility of the autonomy of science both as the legitimate and the normal form of the relationship between science and its social environment.

> Because the study didn't find anything new it wasn't published, the company defends itself. In addition, it was faulty anyways. Frits Rosendaal of the University of Leiden does not accept this excuse: 'Whether the study is worth it, is decided before one starts,' he says.

> 'Even if nothing new comes of it, the result is of interest. [. . .] This is selective publication.' The case thus shows what scientists often criticize. Studies that fit the concept are published, others kept back (*Süddeutsche Zeitung*, "New debate on the third generation pill," March 20, 2001).

As in the previous examples, here, too, we can observe how a breach in norms is defined, by distinguishing between good and bad practice ("selective publication"). What is wrong with bad practice becomes clear through the more or less implicit comparison with good practice. And once again, we see that how responsibility for crossing the boundary between good and bad practice is decisive for the implications of this distinction for the credibility of science. The "defence" of the company is framed as being an "excuse." Unlike an "explanation," an "excuse" always bears the negative connotation of trying to explain unacceptable behaviour on false grounds. With the phrase "does not accept this excuse" the scientist cited is framed as being an expert for judging the validity of the "defence" and thus for distinguishing between what is good and what is bad science communication. At the same time, his judgement is implicitly marked as being correct. It would obviously be wrong to accept an excuse. Why the "excuse" is unacceptable is shown by comparing good science-communication with the communication practice of the company. Good science-communication is described as an objective, existing norm with the phrase "*is* decided before *one* starts." The existing norm of communicating research results appears as the legitimate basis for assuring objectivity and impartiality. The bad communication practice described here is attributed to the economic interests of the company and thus to motives ulterior to the scientific process itself. Scientists and science, thus, appear as impartial and objective and scientists both as defenders of impartiality and victims to the power and influence of political and economic parties.

In contrast to the previous example, the bad practice described here is, however, framed as being common. We are told about an ongoing "censorship" of which the current incident is but one example ("*often* criticize," "Studies that don't fit the concept *are* published, others kept back"). Thus, while stabilizing the epistemological stability of science and the credibility of scientists as impartial and objective observers, a tension between the normative and cognitive expectations placed on the relationship between science and economics is stabilized.

6.2.2 Stem Cell Research

6.2.2.1 Method and Material

The articles on stem cell research were selected based on two criteria: (1) the phase within the stem cell debate and (2) the method of maximal case contrast. The stem cell debate was divided into three phases: the coverage prior to the decision of the lower house of the German parliament on the import and use of human embryonic stem cells (January 2001–August 2001), the debate in the immediate context of this decision (August 2001–March 2002), and a third phase that I understand as a re-warming of the debate (May 2005–May 2006). Forty articles were selected from each of these three phases based on a keyword search in the Database Lexisnexis. Within these three phases, the articles were chosen using the method of maximal case contrast. The purpose of this method is to allow for a maximal representation of a field in studies where relatively few cases can be analyzed. Here, too, the image of science was reconstructed based on the interpretation method of objective hermeneutics. After analyzing the image of science in an article, I formulated hypotheses as to what context factors might be responsible for the production of precisely this image – such as the political orientation of the paper, the column in which the article was published or the particular issue at hand. The articles were selected with contrasting context factors – for example, articles dealing with a different topic or with a different political orientation. Once again, relatively little variation was found in the images of science, allowing for the analysis of a comparatively large number of articles (120).

6.2.2.2 Empirical Results

With the analysis of the coverage of stem cell research in the news, we come to an area of science in which one would expect the credibility of science to become an issue in a very different sense. Aside from being a prominent example of visible science, stem cell research shows interesting parallels to epidemiology. Here, too, science is reported on in a context where utility appears as the criterion legitimating scientific endeavours. And as in the coverage of epidemiology, sound policy is seen as being based on sound science. Thus, here too, we see politicians as being faced with an insoluble dilemma. They must make decisions based on judgements about the potential uses of science in areas where the knowledge needed for these judgements is not available. In the case of stem cell research, there are, however, three additional layers to the dilemma. On the one hand, politicians must make science policy decisions that will affect the ability of scientists to gain the knowledge needed for judging the utility of stem cell research. The research needed for closing the knowledge gaps is in this case precisely the ethically contested stem cell research about which policy decisions must be made. On the other hand, politicians must rely in their science policy on the knowledge of scientists who have a vested interest in the topic. Thus, scientists appear as political actors. In addition, stem cell research is an area of breakthrough science. Not only is science seen as posing an at least potential threat to the social/normative order of society (Bloomfield 1995). It also produces "monsters" and is thus compatible with the discourse of fear used both in science fiction and politics to discredit science (Haynes 1994, 2003). Thus, stem cell research is an area of visible science in which one would expect the credibility of science to be at risk.

As in the reporting on epidemiology, science consistently appears as a source of certain and objective truth and scientific research as a process by which knowledge gaps are gradually closed and uncertainties reduced. Thus, the epistemological authority of science remains intact. In the reporting on stem cell research, there is, however, an additional layer of "reality." Here we find three different images of science: (1) science as a "sport", (2) science as a "guild" and (3) science as "hubris." In the "hubris" image, and in part in the "guild" image, the credibility of the autonomy of science becomes problematic.

In the following section, I describe the key features of these images in terms of credibility. While the quotations used here serve to illustrate these features, the full objective hermeneutical analysis of all three images is available in "Mediale Konstrukte von Wissenschaft in den Bereichen Stammzellforschung und Epidemiologie" (Jung 2009).

6.2.2.3 Science as "Sport"

A frequent pattern of portraying science is as a "sport." In this context, the goal of science is framed as winning in a competition between nations. Thus, the success of German scientists is a German success, and the success of scientists from other countries is a national set back.

> The success of Asiatic clone laboratories shows scientists that they not only have lost the chance of competing in the champions league of biotechnology but are in danger of being handed down to the regional league. [...] South Korea is on the move [...] As a result, Germany is in danger of irrevocably falling behind (*Zeit*, "Stem cell researchers are pessimistic," May 25, 2005).

Consequently, the importance of a victory or defeat is neither measured in terms of the epistemological value of scientific findings, nor their medical relevance. Rather, it is qualified in terms of the region in which the finding was made for the first time and the medial attention the finding aroused.

> Before it became publicly known yesterday that South Korean researchers have for the first time succeeded in the therapeutic cloning of cells of patients, scientists of the University of Newcastle reported on Thursday evening that they have cloned human embryos for the first time in Europe (*Welt*, "Clone successes cause new ethics debate," May 21, 2005)

Doing something "for the first time in Europe" appears as a success, doing something for the first time in the world as an even greater success. Mirroring the shifting possibilities of regional identification, the English success is framed as a European victory and the Asian success as a setback for Germany.

The flip side of the construction of science in the context of a competition between nations and world regions implies the normative expectation of internal co-operation.

> In a (for Germany unimaginable) closing of ranks politicians of the two biggest [South Korean] parties have founded a committee. Their goal: the Noble Prize for Hwang Woo Suk (*Zeit*, "Stem cell researchers are pessimistic," May 25, 2005).

Society, and in particular national politics, is framed as being responsible for assuring that science is given the support it needs to attain a high ranking and play in the champions league.

What are the implications of this form of constructing science in terms of credibility? The expectation of internal cooperation is stabilized, while the implied criticism of the "reality" of the relationship between science and politics can be seen as a form of successful boundary talk. Here, too, however, the legitimacy of science is upheld at the cost of the creation of a tension between normative and cognitive expectations. The credibility of German scientists is upheld in the face of what is constructed as a national defeat by criticizing the relationship between science and politics: The failure of German scientists is not a result of any inherent inferiority. Rather, it is a result of unfair competition conditions. While South Korean scientists can rely on the support of politicians and the nation, German scientists are faced with a political landscape deeply split in their standing on stem cell research and the even drier prospect of ending up with a political party in power, which is categorically against stem cell research. As the phrase "in a for Germany unimaginable closing of ranks" shows, the expectation that the normative expectation concerning the relationship between science and politics/society be fulfilled is framed as being "unrealistic."

6.2.2.4 Science as a "Guild"

Another frequent image is that of science as a "guild." In contrast to the image of science as a "sport," the success or failure of German scientists is not equated with the success or failure of Germany. Rather, individual scientific successes or failures are seen as the successes or failures of an internationally, or rather globally, organized community. Science appears as a self-governing community, sharing common interests and norms and capable of strategic action. The implications for the credibility of science are somewhat more complex than the implications of the "sport" image. Two different forms of this image can be distinguished. In the first image, the goals of science appear as being at least potentially in conflict with the interests and values of society, and the values and interests of society are only insofar of relevance to scientists as they hinder or promote the realization of their goals.

> While society, at least in its political sphere, is dedicatedly debating the acceptability of research on stem cells [...] science is– though for now elsewhere – unflinchingly going its way (*Frankfurter Allgemeine Zeitung*, "Stem cells – the next step," March 3, 2001).

> Stem cell researchers are thus eagerly looking for new ways, because they see their work hindered in many countries by ethical hurdles [...] (*Süddeutsche Zeitung*, "Politically correct stem cells," October 18, 2005).

Scientists appear as being unscrupulous, bound neither by the moral values nor by the legal norms of nation-states and oriented solely to the interests of their own scientific community. Although the credibility of science as a knowledge form remains intact, the credibility of science as an institution is called into question, the implication being that science should be subjected to the control of society or the political system.

In the second form of this image, science is portrayed as a self-governing international community adhering to a moral code which is compatible with fundamental social norms and values.

> Those who let themselves be caught in massive deceit don't get a second chance in the scientific community. No one cooperates with them, no one supports their research proposals,

no one appraises their work. The rigorous exclusion for caught deceivers is the only protection against charlatans in one's own herd. [...] But because researchers are humans, cases of fraud occur time and time again. It is the exaggerated ambition, the pressure to publish and the concern for one's own position and occasionally hubris and thoughtlessness that drive some to dishonesty (*Zeit*, "Hero or charlatan," December 21, 2005).

The stringency with which the scientific community enforces its moral code appears as evidence of the morality of science and the functionality of scientific self-control. That fraud is possible is attributed to factors ulterior to science per se.

The wonderful thing about science is that the moment, in which trust is lost, its true strength comes to light. Religious leaders can make unfounded statements and cause millions of trusting people to bash up each other's heads without any possibility of proving that they are lying. For the man, however, who used the hopes of the seriously sick part of humanity for his own fame colleagues and instruments are now available to verify or disprove his work. The less wonderful aspect of current science is that Hwang could rise to such heights with his statements. The South Korean state has made a star of Hwang in order to bring itself onto the world stage. [...] The success addict Hwang didn't leave any room for the possibility of failure. No scientists should work in that kind of pressure chamber. Some undesirable developments should now be ended thanks to the self-control of science. [...] If Hwang turns out to be a forger science will spit him out like a cold fish (*Frankfurter Allgemeine Zeitung*, "Trust gambled away," December 17, 2005).

Here, too, neither the legitimacy of scientific norms nor the autonomy of science as an institution is called into question. The comparison with religion implies that although science, like religion, is an area with a great deal of power, science is capable of self-control. The sentence "If Hwang turns out to be a forger, science will spit him out like a cold fish" not only illustrates how science governs itself, but also, as in the example above, how important this moral code is in the scientific community. Thus, the reaction implies that forgery is not a common and by no means an accepted practice. And here, too, the occurrence of forgery is explained by factors external to science per se such as the role of the state or Hwangs personal character traits. Once again we find the credibility of science is maintained at the cost of the stabilization of a tension between normative and cognitive expectations. An image of science is produced in which society creates an atmosphere in which scientists are subjected to undue pressure. So here the tension is between the normative expectations placed on the relationship between science and society and the reality of this relationship. In the case of the Hwang scandal, this tension is largely restricted to the relationship between the South Korean state and its science (see also Haran and Kitzinger 2009).

6.2.2.5 Science as "Hubris"

Another recurring image found in the coverage of stem cell research is that of science as "hubris." Here, the goal of scientific endeavours appears as the realization of megalomaniac fantasies.

Worldwide, so at least the prophets of the new era announce, new human ES cells will be ripening to liver, heart and nerve tissues. Alzheimer, Parkinson, heart attacks and cancer – Brüstle and his co-fighters hardly leave out any of the great captivators of humanity when they list the sicknesses that one day will supposedly be healed with the help of ES cells (*Spiegel*, "We are better than God," May 14, 2001).

Although science is portrayed as *claiming* to have a noble purpose, what is promised appears as being so ambitious – freeing humanity of almost all its great captivators – that it seems utterly unrealistic. In addition, scientists are depicted as being prophets and visionaries, not only comparing themselves with God but even appearing in their self-evaluation as the winners in that comparison. The implication of this characterization is that their predictions and promises are not based on professional expertise and a rational evaluation of the potential utility of research.

Scientists are not only portrayed as being irrational, but amoral and down right manipulative.

It is an amazing fact of contemporary history that scientists always remember the sick and the weak of the world precisely at those moments when acceptance is needed for future technologies (*Spiegel*, "Cell workshop," January 1, 2001).

The delusions of grandeur possessing scientists appear as posing an at least potential threat to the moral and social order of society.

If human nature is technologically changed, this affects the very idea of man on which constitutional law is founded. It is part of human dignity that the imprinting given to man by nature remains unchanged. [...] If and for how long the scientific community allows itself to be tamed by these consideration remains unclear (*Handelsblatt*, "Science won't let itself be permanently tamed," August 6, 2001).

In this context, we encounter the allusions to "monsters" ("human nature technologically changed") threatening the existing social and moral order typical of the coverage of breakthrough science (Bloomfield 1995). And scientists seem indifferent to these consequences. Thus, the "hubris" image of science picks up on the "Frankenstein discourse" of science fiction (Haynes 1994, 2003).

Scientific cooperation with politicians and entrepreneurs is consequently framed as a "devilish alliance" in which the allies are at best victims, if not succumbed to megalomaniac fantasies of their own, themselves knowing and willing partners in the manipulation of the public.

With his unfailing sense for symbols Schröder understood: A new type of researcher had entered the scene. Their message was epic, their promise a revolution. [...] Schröder wanted to partake in the glamour (*Spiegel*, "We are better than God," May 14, 2001).

What we find here is a tension both between the normative and cognitive expectations placed on science and on the relationship between science and the rest of society. Central to the credibility problem science faces here is that the values and goals of science are framed as being in conflict with fundamental values and norms around which society is organized. The implication is that both scientists and science as an institution must be subject to political and social control. Thus, here, the legitimacy of the autonomy of science is called into question. It is, however, important to note that even in the construction of science as "hubris", both the credibility of the *normative* expectation that science function as a source of objective and certain truth and the epistemological status of science are not called into question.

6.3 Conclusion

Medialization means not only more science in the news, but also more "negative" science in the news. What, however, are the effects on the credibility of science? Both common sense and theoretical consideration give us a rather clear cut answer: More "negative" science in the news leads to a loss in credibility. With the increasing importance of science in society, science becomes subject to continual observation by the media – and this on all levels of the scientific process. The result is what Peter Weingart calls the paradoxical effect of medialization: What previously took place in the seclusion of ivory towers or backstage is now a matter for public viewing (Weingart 2001). Dissent, uncertainty, fraud, misconduct and the instrumentalization of science for political and economic interests are not only topics the media size upon due to their affinity for the dramatic. They are also matters the public has a right to be informed about. Thus, with the increasing importance of science in society, its special authority as a source of certain and objective knowledge – the bases of the increasing importance of science in society – is, paradoxically, undermined. We are, however, also confronted with ample empirical evidence, showing how the epistemic status of science is stabilized. Culprits are identified and expelled from the scientific community as black sheep; scientific uncertainties are set within walls of certainty, etc. Here, the media appear as a stage on which the credibility of science is successfully defended. How can these seemingly incompatible findings be explained?

The distinction between normative and cognitive expectations allows us to identify three possible effects of medialization: (1) a *loss* in credibility, (2) a *tension* between normative and cognitive expectations and (3) a *re-stabilization* of the relationship between normative and cognitive expectations. Exemplified with the results of an analysis of the coverage of epidemiology and stem cell research in the German press, it was shown that in the coverage of fraud, uncertainties or the instrumentalization of science one of two things happen: Either a tension between normative and cognitive expectations is constructed or the relationship between the two expectations is re-stabilized. A loss in credibility could not be observed. Thus, some of the tension between the conceptually convincing expectation of a loss in credibility and the empirical observation of the stability of the epistemic status of science was resolved. It was shown that, although the epistemic status of science is not called into question, what at first may be mistaken for "successful" boundary work in fact often (re)produces a tension between normative and cognitive expectations.

In order to fully understand the effects of medialization on the credibility of science, we must, however, also take the logic of mass media into account. The function of mass media can be described as an integration function. In highly differentiated societies, mass media fulfils an integration function by co-ordinating the mutual expectations and expectations-expectations of different social spheres. Mass media function as a mirror, both reflecting the worlds of different social spheres to the public and to each other and reflecting these reflections back to the respective social spheres. Thus, they make a continual adjustment of mutual expectations and expectations-expectations possible (Marcinkowski 1993, 2002; Blöbaum 1994;

Esposito 1995; Kohring 1997, 2005; Sutter 2002, 2005). And this is exactly what happens when the coverage of science leads to a re-stabilization of the expectations placed on science. Social learning takes place.

As Shirley Ramsey has shown, reporting on science intensifies when the credibility of science is on the line (Ramsey 1994). Only then can science slide from the science pages to political and cultural columns. Only then is a more differentiated reporting on science possible, allowing for a "more realistic" understanding of science. Only when reporting on science takes place in a more extensive debate, in which science itself becomes an issue, can the real and necessary tensions between the logics of application – be they economic, political or of an everyday nature – and the logic of research be addressed. Thus, more science in the news does not only mean more dissent, uncertainty and misconduct in the news. It also means that the conditions for a "more realistic" re-alignment of cognitive and normative expectations are created. By intensifying the relationship between science and the media, the respective expectations can be better co-ordinated. Or put more succinctly: Distance is not necessary for maintaining credibility. Once the ivory towers have been torn down, the medialization of science is the functional and necessary counterpart to the politicization and economization of science as well as to the increasing permeation of scientific knowledge into all areas of human activity.

The resulting re-stabilization of the relationship between cognitive and normative expectations is but one dimension of the integration function of mass media. The mass media also fulfil a co-ordination function by continually confronting the "reality" of different social spheres with the normative expectations placed on these spheres. The mass media thus forces politics, the economy and, as in our case, science to adjust to the normative expectations of their social environment. Although the adjustment often seems to be restricted to rhetorical strategies and symbolic acts, such as the expelling of black sheep, the orientation to the normative expectations of a generalized public has a long-term integrating effect. In order to maintain their credibility, the respective social spheres are continually forced to make their actions at least *seem* compatible with the normative expectations of their social environment. Thus, both the re-stabilization of expectations and the continual (re)production of a tension between normative and cognitive expectations serve to fulfil the integration function of mass media – a function that can also be described in normative terms as a democratic function of transparency and accountability.

A loss in credibility is in contrast an effect that is not compatible with the function of mass media. Addressing a large and heterogeneous public, journalism is not a communication form, in which fundamental cultural patterns, such as the epistemological status of science, can be radically undermined. In order to serve as a mirror for and to different publics and diverse social spheres, mass media has to appeal to the lowest common denominator, or more precisely: to shared meaning structures. Thus, it is not a communication form, in which a fundamental transformation of deeply-rooted cultural patterns is to be expected. Rather, long held assumptions and norms tend to be reproduced. Insofar as medialization does indeed have the paradoxical effect of causing a credibility problem, it does so by (re)producing a tension between normative and cognitive expectations. Criticizing the way our institutions function in light of existing norms is not only an important part of the democratic function of journalism. It is also a means of breaking with routine reporting while at the same time reinforcing existing social norms. Just as political and economic scandals do not undermine the normative legitimacy of democracy and capitalism, uncertainty, misconduct and the instrumentalization of science do not undermine the normative expectation that science function as a source of certain and objective truth.

References

Bauer, M., J. Durant, A. Ragnarsdottir, and A. Rudolfsdottir (1995). Science and technology in the British press, 1946–1992. *The Media Monitor Project, Vol 1–4, Technical Report*. London: Science Museum and Wellcome Trust for the History of Medicine.

Blöbaum, B. (1994). *Journalismus als soziales System. Geschichte, Ausdifferenzierung und Verselbständigung*. Opladen.

Bloomfield, B. P. (1995). Disrupted boundaries: New reproductive technologies and the language of anxiety and expectation. *Social Studies of Science*, 25(3), 533–551.

Bucchi, M. (1996). When scientists turn to the public: Alternative routes in science communication. *Public Understanding of Science*, 5(4), 375–394.

Collins, H. M. (1987). Certainty and the public understanding of science: Science on television. *Social Studies of Science*, 17(4), 689–713.

Daniels, G. H. (1967). The pure-science ideal and democratic culture. *Science, New Series*, 156(3782), 1699–1705.

Elmer, C., F. Badenschier, and H. Wormer (2008). Science for everybody? How the coverage of research issues in German newspapers has increased dramatically. *Journalism and Mass Communication Quarterly*, 85, 878–893.

Esposito, E. (1995). Interaktion, Interaktivität und die Personalisierung der Massenmedien. *Soziale Systeme*, 1, 225–260.

Franzen, M., S. Rödder, and P. Weingart (2007). Fraud: Causes and culprits as perceived by science and the media. *EMBO Reports*, 8(1), 3–7.

Galtung, J. and M. H. Ruge (1965). The structure of foreign news: The presentation of the Congo, Cuba and Cyprus crises in four Norwegian newspapers. *Journal of Peace Research*, 2(1), 64–91.

Gerhards, J. and M. S. Schäfer (2009). Two normative models of science in the public sphere: Human genome sequencing in German und US mass media. *Public Understanding of Science*, 18(4), 437–451.

Gibbons, M., H. Nowotny, and C. Limoges (1994). *The new production of knowledge. The dynamic of science and research in contemporary societies*. London et al.: Sage.

Gieryn, T. (1983). Boundary-work and the demarcation of science from non-science: Strains and interests in professional ideologies of scientists. *American Sociological Review*, 48(6), 781–796.

Gieryn, T. (1999). *Cultural boundaries of science: Credibility on the line*. Chicago: University of Chicago Press.

Habermas, J. (1992). *Faktizität und Geltung. Beiträge zur Diskurstheorie des Rechts und des demokratischen Rechtsstaates*. Frankfurt/M.: Suhrkamp.

Haran, J. (2007). Managing the boundaries between maverick cloners and mainstream scientists: The life cycle of a news event in a contested field. *New Genetics and Society*, 26(2), 203–219.

Haran, J. and J. Kitzinger (2009). Modest witnessing and managing the boundaries between science and the media: A case study breakthrough and scandal. *Public Understanding of Science*, 18(6), 634–652.

Haynes, R. (1994). *From Faust to Strangelove*. Baltimore and London: The Johns Hopkins University Press.

Haynes, R. (2003). From alchemy to artificial intelligence: Stereotypes of the scientist in Western literature. *Public Understanding of Science*, 12(3), 243–253.

Jasanoff, S. (1987). Contested boundaries in policy-relevant science. *Social Studies of Science*, 17(2), 195–230.

Jasanoff, S. (1990). *The fifth branch: Science advisers as policymakers*. Cambridge, MA: Harvard University Press.

Jung, A. (2009). Mediale Konstrukte von Wissenschaft in den Bereichen Stammzellforschung und Epidemiologie. In H. P. Peters (ed.), *Medienorientierung biomedizinischer Forscher im internationalen Vergleich. Die Schnittstelle von Wissenschaft und Journalismus und ihre politische Relevanz*. Jülich: Forschungszentrum Jülich, pp. 177–226.

Kinchy, A. J. and D. L. Kleinman (2003). Organizing credibility: Discursive and organizational orthodoxy on the borders of ecology and politics. *Social Studies of Science*, 33(6), 869–896.

Kohring, M. (1997). *Die Funktion des Wissenschaftsjournalismus: Ein systemtheoretischer Entwurf*. Opladen: Westdeutscher Verlag.

Kohring, M. (2005). *Wissenschaftsjournalismus. Forschungsüberblick und Theorieentwurf*. Konstanz: UVK.

Luhmann, N. (1987). *Soziale Systeme*. Frankfurt/M.: Suhrkamp.

Luhmann, N. (1990). *Die Wissenschaft der Gesellschaft*. Frankfurt/M.: Suhrkamp.

Marcinkowski, F. (1993). *Publizistik als autopoietisches System. Politik und Massenmedien. Eine systemtheoretische Analyse*. Opladen.

Marcinkowski, F. (2002). Massenmedien und die Integration der Gesellschaft aus Sicht der autopoietischen Systemtheorie: Steigern die Medien das Reflexionspotential sozialer Systeme? In K. Imhof, R. Blum, and O. Jarren (eds.), *Integration und Medien*. Mediensymposium Luzern, Bd. 7, Opladen: Westdeutscher Verlag, pp. 110–121.

Nelkin, D. (1995). *Selling science. How the press covers science and technology*. New York: Freeman.

Nowotny, H., P. Scott, and M. Gibbons (2001). *Re-thinking science: Knowledge and the public in an age of uncertainty*. Cambridge: Polity.

Oevermann, U. (1991). Genetischer Strukturalismus und das sozialwissenschaftliche Problem der Erklärung der Entstehung des Neuen. In S. Müller-Doohm (ed.), *Jenseits der Utopie: Theoriekritik der Gegenwart*. Frankfurt/M.: Suhrkamp, pp. 267–336.

Oevermann, U. (2000). Die Methode der Fallrekonstruktion in der Grundlagenforschung sowie der klinischen und pädagogischen Praxis. In K. Kraimer (ed.), *Die Fallrekonstruktion*. Frankfurt/M. Suhrkamp, pp. 58–156.

Oevermann, U., T. Allert, E. Konau, et al. (1979). Die Methodologie einer 'objektiven Hermeneutik' und ihre allgemein forschungslogische Bedeutung. In H. G. Soeffner (ed.), *Interpretative Verfahren in den Sozial- und Textwissenschaften*. Stuttgart: Metzler, pp. 352–434.

Peters, H. P. (1994). Wissenschaftliche Experten in der öffentlichen Kommunikation über Technik, Umwelt und Risiken. In F. Neidhardt (ed.), *Öffentlichkeit, öffentliche Meinung, soziale Bewegungen*. Opladen: Westdeutscher Verlag, pp. 162–190.

Peters, H. P. (2008). Scientists as public experts. In M. Bucchi and B. Trench (eds.), *Handbook of public communication of science and technology*. New York, NY: Routledge, pp. 131–146.

Peters, H. P., et al. (2008). Medialization of science as a prerequisite of its legitimization and political relevance. In D. Cheng et al. (eds.), *Communicating science in social contexts*. Dordrecht: Springer, pp. 71–92.

Ramsey, S. (1994). Science and technology: When do they become front page news? *Public Understanding of Science*, 3(1), 71–82.

Reichertz, J. (2004). Abduction, deduction and induction in qualitative research. In U. Flick et al. (eds.), *Companion to qualitative research*. London: Sage, pp. 159–165.

Schäfer, M. S. (2008). Diskurskoalitionen in den Massenmedien. Ein Beitrag zur theoretischen und methodischen Verbindung von Diskursanalyse und Öffentlichkeitssoziologie. *Kölner Zeitschrift für Soziologie und Sozialpsychologie*, 60(2), 367–397.

Schulz, W. (1976). *Die Konstruktion von Realität in den Nachrichtenmedien. Analysen der aktuellen Berichterstattung*. Freiburg: Alber.

Simon, B. (2001). Public science: Media configuration and closure in the cold fusion controversy. *Public Understanding of Science*, 10(4), 383–402.

Stehr, N. (1994). *Arbeit, Eigentum und Wissen. Zur Theorie von Wissensgesellschaften*. Frankfurt/M.: Suhrkamp.

Sutter, T. (2002). Integration durch Medien als Beziehung struktureller Kopplung. In K. Imhof, O. Jarren, and R. Blum (eds.), *Integration und Medien*. Reihe: Mediensymposium Luzern, Band 7, Wiesbaden, pp. 122–136.

Sutter, T. (2005). Vergesellschaftung durch Medienkommunikation als Inklusionsprozeß. In M. Jäckel and M. Mai (eds.), *Online-Vergesellschaftung? Mediensoziologische Perspektiven auf neue Kommunikationstechnologien*. Wiesbaden: VS Verlag für Sozialwissenschaften, pp. 13–32.

Weingart, P. (1997). From 'finalization' to 'Mode 2': Old wine in new bottles? *Social Science Information*, 36(4), 599–613.

Weingart, P. (2001). *Die Stunde der Wahrheit? Zum Verhältnis der Wissenschaft zur Politik, Wirtschaft und Medien in der Wissensgesellschaft*. Weilerswist: Velbrück.

Weingart, P. (2005). Die Wissenschaft der Öffentlichkeit. Essays zum Verhältnis von Wissenschaft, Medien und Öffentlichkeit. Weilerswist: Velbrück.

Willke, H. (1998). Organisierte Wissensarbeit. *Zeitschrift für Soziologie*, 27(3), 161–177.

Part IV
Scientists' Attitudes to Media Visibility

Chapter 7
Re-ordering Epistemic Living Spaces: On the Tacit Governance Effects of the Public Communication of Science

Ulrike Felt and Maximilian Fochler

In 2008 *Science* featured an article entitled "Interactions with mass media" with the following teaser: "A survey reveals that media contacts of scientists in top R&D countries are more frequent and smooth than was previously thought." Stressing their astonishment and satisfaction, the authors – a consortium of researchers from Germany, France, the UK, the US and Japan – further report that "the scientists most involved in these interactions tend to be scientifically productive, have leadership roles, and [. . .] perceive the interactions to have more positive than negative outcomes" (Peters et al. 2008: 205). Apparently the often repeated saying that scientists and journalists are "like oil and water" no longer holds. Neither does the long-held opinion that too much media presence harms the reputation of scientists. This, the study suggests, is valid "across the five countries, the basic patterns [being] surprisingly similar. The functional necessity of public science communication may be a global phenomenon in democratic knowledge societies" (ibid.: 205).

Indeed, in recent years academic scientists have been increasingly trained to talk to journalists and write press releases, they are asked to speak in schools, and science weeks, festivals or nights are organized to allow "ordinary citizens" to get in touch with "their" scientists. FameLabs are staged for young researchers to showcase their talent as infotainers and prizes have been established to reward excellent communicators. Examples of an increased public presence of scientists are numerous. Beyond these communication activities aimed at broader publics we find a flurry of other efforts such as glossy brochures, annual reports, web pages, video clips, and many other professionally produced media, which are directed at selected communities of potential funders, supporters or collaborators, but also used for recruiting purposes.

Thus, science is no longer only performed at the lab bench, but needs to be staged in an ever-increasing variety of places and contexts. A possible reason for this could be the growing competition for public attention at a time when pictures and stories about science and technology proliferate. Another reason could be the need to rehearse and thus stabilize certain foundational narratives to assure

U. Felt (✉)
University of Vienna
e-mail: ulrike.felt@univie.ac.at

S. Rödder et al. (eds.), *The Sciences' Media Connection – Public Communication and its Repercussions*, Sociology of the Sciences Yearbook 28,
DOI 10.1007/978-94-007-2085-5_7, © Springer Science+Business Media B.V. 2012

an unquestioned position of authority for science and technology in a world where publics are no longer seen as unconditional admirers of science, but much more as potential dissenters (Felt et al. 2007; Felt and Fochler 2010).

A broad body of literature has dealt with these phenomena, in particular analyzing the presence of science and scientists in classical media and the representations of research and of being a researcher that they create (e.g., LaFollette 1990; Nelkin 1994). Weingart's (1998) concept of a "medialization of science" characterizes science as increasingly media-oriented, and addresses the consequences this has for the public perception of science. Developing this notion further, Rödder (2009) points at the need to analytically differentiate between two dimensions of medialization: One "characterizes a phenomenon on the level of media content [that] can be regarded as part of the changing environmental framework of science" (ibid.: 453); the other hints at structural changes within science, meaning "that scientific institutions as well as scientists increasingly orientate themselves towards public and media attention rather than the truth" (ibid.: 454).

This paper focuses on the second dimension, because little reflection has been offered so far on how the proliferation of media formats and contexts in which communication takes place affects research and scientific practice. Other than the initial proponents of medialization (Weingart 1998; Rödder 2009), we do not situate our analysis on the level of systems and their changing dynamics. For example, we are not asking whether media orientation is replacing truth as the functional code of the science system. Rather, our approach starts from a person-centred perspective, and asks what consequences the proliferation of communication activities and of media representations of research has for academic scientists and their ways of living and working in research. How does scientists' increasing engagement in communication activities feed back into research and influence their identities and practices as scientists? How do these communication activities affect the social and symbolic orders that define what it means to do research today? And how do they both tacitly govern research environments as well as researchers' self-understanding? By starting from these questions about how researchers' ways of living and working in science are affected by medialization on a micro-level we aim to work towards conclusions about the more systemic effects of medialization, both in science as well as in its relation to other societal actors and systems.

Our aim in analyzing these questions is to grasp the plurality of contexts and formats in which academic scientists orient themselves towards public attention. Our starting hypothesis is that medialization does not only take place in relation to the classical mass media, but that the ongoing process of classical medialization as described by Weingart (1998) has spawned a number of heterogeneous contexts which are governed by similar logics, but do not immediately relate to the mass media. Although they may not be directly reported by the media, the glossy brochures and accounts produced by scientists and scientific institutions, along with instances of more direct public engagement, such as when a life science department invites pupils to learn about "doing real science" in a visitors' lab, are thought and designed in such a way that they could (and should) attract public attention. They are at least partially governed by the logics of the media, and hence need to be understood as sites where medialization takes place. Our working definition of medialization is thus broader than that of many other contributions in this volume.

We will start by addressing some of the more recent debates on the changing relations between science and society and introduce the key-concept of epistemic living spaces (Felt 2009). This will help us to reflect on the articulations between wider societal changes and the way they impact the social and epistemic lives of researchers. After presenting the empirical research that serves as the basis for this article, we will elaborate our findings. In the first part of our empirical section, we will analyze life scientists' accounts of the role medialized interactions play in their professional lives. In the second, we will ask which tacit governance effects medialization has on the development of research fields and on the career imaginations and decisions of young scientists. In conclusion we will reflect on the ways in which medialization processes affect the symbolic and social orders of contemporary research and the role they have come to play in tacitly governing research, while critically pointing to the fact that this role is disconnected from any responsibility.

7.1 Eroding Demarcations Between Science and Society and the Consequence for Research(ers)

From a macro perspective, several key contributions have argued that the cultures, practices and contexts in which research is done have significantly changed over the last decades – be it due to a progressive intertwining of academia, industry and the state (Etzkowitz and Leydesdorff 2000), or to a new relation between science and society in which knowledge is more often produced and validated in contexts influenced by extra-scientific rationales (Nowotny et al. 2001). Though these approaches vary in their emphases, they all point at a convergence of different societal spheres and rationales, and at the dynamic erosion and re-definition of boundaries previously thought to be stable.

In the framework of co-evolution, science is characterized as being increasingly contextualized by societal rationales. This is considered the main process changing research cultures and practices. While in the literature context is mostly thought of in terms of different social groups or market structures, in our argument the proliferating public images and narratives of science, as well as the equally proliferating spaces in which scientists and publics interact, also need to be considered as contextualizing scientific knowledge production. In their communication activities beyond the scientific community, researchers can be shown to constantly engage in boundary work, both tearing down and simultaneously re-erecting boundaries between science and society (Felt and Stöckelová 2009).

While macro-diagnoses such as those by Nowotny and co-authors or Etzkowitz and Leydesdorff are much rehearsed, surprisingly little work has been done to empirically address how these changes affect the actual cultures and practices of research, and how researchers perceive their work and their own identity. In our perspective, this is a major lacuna, because, paraphrasing Jasanoff (2004), the ways in which researchers live and situate themselves in society at large, in particular epistemic cultures and in concrete institutional and social settings, is inseparable from the kinds of knowledge they produce. Epistemic and social orders are always co-produced, and the multiplication of interactions with the media and the growing attentiveness of research(ers) towards media representations are major factors influencing the processes in which this co-production takes place in contemporary research.

To address this, we will use the notion of "epistemic living spaces" (Felt 2009). By epistemic living space, we mean researchers' individual or collective perceptions and narrative re-constructions of the structures, contexts, rationales, actors and values which mould, guide and delimit their potential actions, both in what they aim to know as well as in how they act in social contexts in science and beyond. This concept directs our attention to the efforts of researchers to stabilize, extend or protect the space they occupy socially and epistemologically, as well as institutionally. It resonates with Knorr-Cetina's notion of epistemic cultures, which aims to analyze the "knowledge machineries of contemporary sciences" (Knorr-Cetina 1999: 3), and their technical, symbolic and social dimensions. However, we do not intend to reconstruct and map stable cultural assemblages and their differences, but instead we focus on individual and collectivized perceptions in order to analyze the changes, heterogeneities and fluidities in today's research landscape, and to link individual and collective experiences to more global systemic changes. Hence, elements which are less important for studying tightly-bound epistemic cultures are key for understanding more fluid epistemic living spaces and how they are transformed by the framing of research in policy discourses, the societal imaginaries which influence research, and the manifold normative symbolic regimes which govern contemporary research practices. The concept thus allows us to address the inextricable interdependence of epistemic practices, institutional rationales, individual biographical decisions, and political frameworks, which characterizes the lived experiential realities of researchers today. One particularly salient example for this interdependence is the rise of New Public Management in scientific institutions. This ideology sets out to introduce corporate management techniques into academic institutions, to reform the efficiency of decision-making in public institutions by basing decisions on "objective", quantitative criteria, and to make these institutions more favourable to policy-makers and the general public. Medialization is strongly linked to these institutional changes, as, for example, academic PR departments increasingly shift to selling science in a corporate model of public relations (see Chapter 11), and as scientists' media contact is monitored and quantified as an indicator of societal impact.

Epistemic living spaces can be characterized along at least five dimensions: an epistemic, a spatial/material, a temporal, a symbolic, and a social dimension. While making a distinction between these five dimensions makes sense for analytical purposes, at the same time they are inextricably intertwined.

The first and most central dimension of epistemic living spaces is their epistemic structure. They are framed by specific assumptions about which kinds of research questions are central, how knowledge should be produced, and which properties and procedures constitute good knowledge. As opposed to the many other human practices besides science that involve the production and use of knowledge, producing knowledge is seen as the prime aim by the actors populating the epistemic living spaces called academic research, and hence reflecting on the guiding values, means, and methods of their knowledge production is perceived to be of key importance in these spaces. One crucial axis of recent change in research cultures that is related to this, for example, is whether researchers see the purpose of their work as following mainly academic curiosity, or whether contexts of application matter more in how they frame and answer their research questions (Hakala 2009; Nowotny et al. 2001).

The second dimension is the spatial/material dimension. On the one hand it points to how research work is enabled or constrained by material artefacts such as technologies or the more or less convenient spatial and architectural arrangements in which everyday work is done (Felt et al. 2010b). On the other hand, it also addresses the simultaneously geographic and symbolic maps researchers use to orient themselves. This implies that we need to analyze researchers' ideas of proximity and distance to what they perceive as central nodes in the system, as well as their "tacit geography", which informs how they understand their own place within this system (see Felt and Stöckelová 2009). For example, while the development of scientific careers and interaction with peers is often seen as international[1] in scale, making mobility a pre-requisite for scientific careers, in individual career decisions this has to be reconciled with more localized rationales, be they those of specific institutions or of the individuals' private relations.

Third, the temporal dimension addresses (1) the perceived tempo of academic work (2) the relation of research work (most often organized into projects) to other time regimes, such as the rhythm of institutional evaluations or of private lives and (3) the different forms of time which play out in academic work (Garforth and Cervinkova 2009). For the latter, recent studies (Ylijoki and Mäntylä 2003) highlight that the perceived time pressure in research work is strongly linked to a shift in the relation between different categories of time. As project research increasingly constitutes the *modus operandi* of academic work, "timeless time", i.e., time which is not a priori dedicated to a specific task becomes a scarce and precious resource in relation to time units already dedicated to the production of a specific output.

The fourth dimension, the symbolic dimension, is of specific relevance to this paper, as it touches on the values and modes of ordering seen as crucial in governing research work, and the related virtues and qualities expected of the individual researcher (Shapin 2008). In recent years a range of powerful normative distinctions and values around notions such as excellence, mobility and accountability have emerged, which play a central role in guiding both researchers' epistemic practices and their career decisions (Felt and Stöckelová 2009). For example, the rituals of academic audit (Power 1997; Strathern 2000) lead researchers to conceptualize their output in terms of units, such as publications, which can be counted and should

[1] Even though this supposed globality may often turn out to be a circulation in a sphere comprising Europe and North America.

be maximized. In relation to this, other aspects of academic work such as support or articulation work (e.g., work needed to set up interdisciplinary collaboration) that are not as easily translated into quantifiable outputs, are increasingly devalued and hence less attractive for researchers to pursue.

Fifth, the social dimension scrutinises which forms of togetherness researchers imagine as specific to their epistemic living spaces. On the one hand this relates to which forms of collaborative knowledge production exist. On the other hand, sociality is also about expectations for other individuals and groups who share the epistemic living space, for example, whether and in which moments these others are perceived as peers or competitors. And it is always linked to the question whether or not they are sharing the same perception of the values and symbolic modes of ordering characteristic of a particular epistemic living space – for instance whether they buy into the audit and quantification logic, or whether they see themselves committed to pursuing other goals.

This last argument points to a crucial specificity in individual perceptions of epistemic living spaces: the different degrees to which the perception of the actual epistemic living space corresponds with one's own normative expectations about life and work in this space. Our hypothesis is that the rapid changes the research system is currently undergoing create frictions and dissonances between normative ideals and perceived realities, which can be voiced in different rhetorical forms, such as "not feeling at home" or an academic nostalgia for the "good old times" (Ylijoki 2005). This again points to two specific kinds of work academic researchers have to engage in. The first is orientation work, which refers to the activity of making sense of the shape, the boundaries and the central logics governing their own epistemic living space. Second, there is the boundary work researchers engage in to re-shape their epistemic living space to fit better with their own expectations, or to resist shifts and changes that are perceived as negative. As we intend to show and discuss, medial representations of science play a crucial role in actually shaping epistemic living spaces, as well as orientation and boundary work.

7.2 Material and Methods

Our empirical argument draws on data collected in a number of projects conducted in Austria between 2001 and 2010. Broadly speaking, these projects fall in two categories. On the one hand, we are building on several projects concerned with the relations of sciences and their publics, and in particular, with how scientists experience science/public interactions. On the other hand, a major part of our arguments derives from intense fieldwork in two projects on the changing research cultures and practices in the life sciences. We will briefly describe each project category.

Concerning science/public relations, our argument first draws on three evaluations of science communication events we carried out in the early 2000s.

The evaluation of both the 2001[2] and 2002[3] Austrian science weeks and of the "Discourse day on genetic diagnosis" in 2002[4] were all projects which involved both participant observation as well as an elaborate collection of qualitative data in the form of interviews and focus groups with participating scientists and members of the publics. Secondly, we will draw on data from the project "Let's talk about GOLD. Analysing the interactions between genome research(ers) and the public as a learning process".[5] The key methodological feature of this project was a series of six whole-day Round Table discussions between scientists working on a genome research project and 14 members of the public, on the topic of the ethical and societal dimensions of this project and genome research in general. The media was a recurrent theme in these discussions, and one Round Table was explicitly devoted to discussing "science and the media". Besides the transcripts of the Round Table discussions, we will refer to interviews conducted with all participants before and after the series of Round Tables.

The second category of projects we are drawing on was concerned with changing research cultures and practices in the life sciences, and in particular with how societal rationales such as ethical considerations, public debates or media logics influence these changes. The European project "Knowing – Knowledge, Institutions and Gender. An East-West Comparative Study"[6] compared the research cultures in molecular biology and sociology in five European countries. For this paper, we are mainly referring to interviews and focus groups conducted with molecular biologists in Austria. The project "Living Changes in the Life Sciences. Tracing the Ethical and Social within Scientific Practice and Work Culture"[7] developed an innovative interview method called the "reflexive peer-to-peer interview."[8] In more than

---

[2] Funded by the Austrian ministry of research. http://sciencestudies.univie.ac.at/research/completed-projects/scienceweek-2001/?L=2.

[3] Funded by the Austrian ministry of research, the ministry of traffic, innovation and technology, and the Austrian council for research and technological development. http://sciencestudies.univie.ac.at/research/completed-projects/scienceweek-2002/?L=2.

[4] Funded by the Austrian genome research programme GEN-AU. http://sciencestudies.univie.ac.at/index.php?id=57585&L=2.

[5] Funded by the Austrian genome research programme GEN-AU as an ELSA project. http://sciencestudies.univie.ac.at/index.php?id=57575&L=2.

[6] Funded by the European Commission, FP6. http://www.knowing.soc.cas.cz/.

[7] Funded by the Austrian genome research programme GEN-AU as an ELSA project. http://sciencestudies.univie.ac.at/research/living-changes-in-the-life-sciences/?L=2.

[8] These two to three hour qualitative interviews were structured by different question blocks in which the life scientists we interviewed talked about their personal professional development, about the epistemic directions of their work and how they have changed over time, and the institutional contexts they have worked in. Toward the end of the interview they were asked to give their impression of a series of key terms and catchwords currently used in academia, such as mobility or excellence. During the interview, the interviewer invited the interlocutor to add a reflexive dimension to his or her narration, either by asking him/her to relate the different blocks of the interview – such as epistemic orientations to institutional framings, and/or by asking him/her to compare his/her story to the stories of others, e.g., to prior generations. The interviewer/interviewee "peer to

fifty qualitative interviews of 2 to 3 h, we explored how life scientists narrate the relationships between biographical, epistemic and institutional rationales in their personal biographies. Science-society relations and the relation of scientists to the media were a part of these interviews. The scientists interviewed covered all career stages from PhD students to full professors, as well as the major research areas in the life science landscape in Austria.

We thus draw on material resulting from nearly 10 years of research on the life sciences and on science/society relations in Austria. Besides more formally recorded data, our dense personal experiences and observations in working with life scientists in these and in other project contexts are another important resource for our arguments. We have been involved in one of the largest genome research consortia in Austria as ELSA/social science partners for more than 5 years, and our co-operative work with some of the scientists included in the focus groups, Round Tables and interviews dates back considerably before this. Hence, virtually all of the arguments that we illustrate with single quotes can and have been cross-validated in other interviews or more informal contexts.

7.3 Tracing Medialization

7.3.1 Keeping Society Close but Outside – Scientists' Narratives on Sciences' Strategic Use of the Media

For most scientists we talked with in our field work, the idea that scientists should communicate with the public had become almost self-evident. Rather, a crucial question for them concerned what is to be gained by communicating to whom. Here, society and policy were often described as separate communication partners, with policy unsurprisingly attributed with considerably more emphasis than society. Particularly for resource-intensive fields such as the life sciences, attracting media attention was perceived as central, because policy attention was thought of as following at the media's heels. Senior scientists talked about medialized stories about research as particularly useful for securing the resources distributed through policy, as policy-makers would be able to use the promises made in these narratives to justify their actions more persuasively than other arguments offered by scientists.

> If I say, I will create a centre for molecular medicine and will use 80 million Euro – then you have to argue that it will have an immediate impact on society, because otherwise no politician will stand in front [. . .] and cut the green ribbon. If I say that I want an institute

peer"-relationship differed from most other types of qualitative research, as both conversation partners were conceptualized as different types of experts on the issue at hand, as well as colleagues affected by different issues touched upon in the interview (e.g., the pervasiveness of audit logics), albeit in very different disciplinary contexts. This peer-to-peer relation allowed for building trust and to explore the discussed issues in considerable depth, but it also needed to be reflected in analyzing the interviews, in particular with regard to meanings taken for granted by both interviewer and interviewee.

> for basic research in order to have my tranquillity and publish two Cell papers every year – that will not get anything off the ground (lab leader, male).

From another perspective, making arguments about the immediate impact of research in the media was also seen as fulfilling science's need to be accountable for how it uses societal resources. In this context, the media was not only conceptualized as an arena to convince policy to provide resources; it was also seen as a possible ally in protecting science from the downside of policy logics, such as the assumption that immediate impact is the only criteria for deciding how research benefits society. If politicians are as oriented towards direct societal benefits as imagined in the quote, then they may just as quickly decide that societal resources should not be spent on basic research at all – a debate, which has been constantly going on in Austria in recent years. Hence, a number of scientists we talked to argued that science needs to use the media to foster a different image, one which is not as dependent on promises of short-term applicability. Consider the following quote, in which a scientist sums up a small group discussion between several researchers on science and the media:

> I find it essential that research is recognized as a cultural good. [This] means that it is not merely utility which is at the centre of considerations, that the question of what can be practically achieved does not come up immediately (project manager, female).

Hence, slightly paradoxically, the media was seen both as a means to instrumentalize policy and as a potential ally to protect science from policy, as keeping policy "close but outside". Similar patterns can be identified in how scientists talk about the role of the media in negotiating science's wider relation to society. On the one hand, the media is seen as a space to win societal support like in the "science as a cultural good" argument. However, it was also conceptualized as an arena where precautionary measures are necessary to avoid societal conflict on potentially controversial issues.

> Especially in our case, when it is about genetic research, it is important to also deliver arguments to a wider public to shape their opinion. There was an initiative in Switzerland. They wanted to ban [animal] experiments. This was debated broadly, and many scientists went out into the public and initiated debates on TV, radio, everywhere, Nobel Prize winners, what have you; and they tried to convey the importance of such a decision. And that's why the decision has turned out positively for science. Something similar should also happen in Austria (lab leader, male).

To achieve this science-friendly coverage, two factors were seen as crucial: On the one hand, the involvement of respected senior scientists in this kind of media work is necessary. But on the other hand, "scientising" the media is of high strategic importance. This means influencing science journalists, science journalism training, or at best encouraging people with training in the life sciences to become science journalists themselves.[9] Particularly this last option was seen as a very promising way to assure an adequate "understanding" of the life sciences and their needs in the media.

---

[9] In a related rationale, study programmes in the life sciences at Austrian universities have offered courses and trainings in science communication.

Within academia, we encountered a quite coherent vision of who should interact with the media, and who does not need to do so. Doing this kind of boundary work was conceptualized as the senior scientists' task, in order to assure that their junior colleagues can "just do science". Playing the media game in the public eye and hence, at least rhetorically, buying into it was seen as necessary pre-condition for achieving the resources to do "science as usual" on the back stage.

7.3.2 *Press-Packaging Science? Why and How Life Scientists Relate Their Work to the Media*

When life scientists talked about their work, their projects and their careers in the different fieldwork settings that we analyzed, the media was a recurring theme. For making a successful academic career, the media, and in particular scientists' ability to skilfully deal with it, was seen as crucial. Consider the following quote from a post-doc researcher in response to the question of what aspect of scientific careers has changed most compared to previous generations:

> I suppose [. . .] that it is increasingly important to also sell yourself to the outside. On the web or anything else, newspapers, [. . .] to really show off. Probably this also existed in the past, but I suspect to a lesser extent (post-doc, male).

Successfully dealing with the media requires skills, which have also become of key importance in other aspects of scientific careers, as the interviewee goes on to explain. For him, this is mainly about being able to construct and tell a convincing story in order to capture the interest of the audience, and to be able to do so succinctly, in both writing and in speech. Whether one is chatting with a potential collaborator or employer, giving a scientific presentation, or talking to a local newspaper, in the end it is all about being able to deliver short and exciting tales about one's work and its outcomes.

Most scientists we talked to would not see as clear connections between the different contexts where scientists have to communicate effectively as this particular post-doc. However, his story points to an increasingly important form of talking and thinking about one's work we found in many narrations, and which we label "press-packaging research": to communicate one's research in a brief form adapted to and attractive for a specific public. The example given also suggests that the contexts in which these skills are required have multiplied and go beyond the classical instances of science/media interactions. Offering a press-packaged version of one's research is seen as crucial in competing for attention in very different situations.

Being a good "press-packager" was also seen as likely to foster an individual's chances of receiving funding, in particular in the context of excellence-programmes, which distribute large amounts of money and hence are also reported on in the media. A researcher who had won one of the most prestigious prizes in Austria, which is awarded for basic research on the senior post-doc level, commented that in his view the projects of all individuals who received this prize shared a focus on potential applications, and hence could be easily "sold" to the public by the funding agency, even though the scientists situated themselves in the basic science realm. He argues that the fields that are more likely to be funded are those that one

> can sell to the wider public as exciting, I believe. Well, I can imagine, that this plays a role, when [funding] choices are made (lab leader, male).

This argument points to the ambivalent perception of the funding process that many of our more senior interviewees in particular shared. On the one hand, they saw the funding process as an arena for performing internal academic competences, as the decision would ultimately be based on the opinion of peers. On the other hand, as the quote implies, research funders were also seen as needing to legitimize their funding to the public, particularly with regard to their most prestigious prizes and programs. Hence, presenting one's research not only to one's peers, but also at least indirectly to wider audiences was seen as crucial for succeeding in the funding game.[10] In quotes such as these, two different kinds of value seem to be relevant to the decision of which projects to fund: the value of the specific project or publication within science and the value of its presentation to a wider public in terms of societal relevance (see Chapter 2). An interesting line of further inquiry would be to empirically study how these two kinds of value are related and negotiated in the actual decision making processes and, following from this, how far funding institutions inhabit a hybrid position between science and other societal rationales.

For the purposes of this paper, scrutinising the rhetorical form of press-packaging has high analytic salience. In a wider context, scientists' attempts to align their work with particular societal concerns have been analyzed as part of an "economy of promises" (Felt et al. 2007), in which promises of applicability themselves become the central currency. This is nicely illustrated by a senior researcher as he talks about how he presents his work in grants and to wider publics. Here he points at the form these promises need to take within a medialized logic:

> I have come to understand from my experience with journalism that you only are allowed to say two sentences: 'Here we do xx-research.' is the first. But this does not thrill anybody. So you need to have a second one: that in doing this, you found a solution for this problem; then it is cool (lab leader, male).

Scientists assumed that to be reported on by the media, they would have to position their work as crucially contributing to a certain societal issue, if not directly solving that very problem. The brevity required by the medium and its temporal logics amplifies the problem. In the further context of the quote, the scientist goes on to argue that his relation to the media would be much more satisfactory if he was given more time and space to actually explain in how far and in which precise ways his discovery contributes to addressing an issue, and in how far it does not. These other modalities, however, are sacrificed for the sake of the brevity that is required by media communication, leaving just the decontextualized and unqualified promise –

[10] Depending on the topical structure of public discussions, making these kinds of arguments can be harder for some fields than for others. See also Section 7.4.1 of this paper. Further our argument resonates with the findings reported by Peters (Chapter 11).

and a scientist who is ready to make this promise to foster his research and career. This points to a general ambivalence scientists are facing: on the one hand, being successfully represented in the media is tantamount to having a large number of media citations. On the other hand, this quantity comes at the price of what is seen as quality, as most of this coverage tends to be (too) short.

As the general thesis of an "economy of promises" (Felt et al. 2007) suggests, promissory logics are by no means exclusive to science/media interactions. As press-packaging is also seen as crucial for achieving funding, similar rationales govern the process of grant writing, as the same scholar argues:

> Everybody writes, and so do I, as first sentence in the [grant application]: 'metabolic diseases is a major burden on humanity.' As if, with my grant, if I manage to get it through, I would solve that problem. [laughs] But of course it's actually not like that (lab leader, male).

This quote points to a particular image of scientific practice and its relation to societal problems that is characteristic of press-packaging-practices more broadly: in factual statements such as this, a direct linear relation between societal problem and scientific practice is established, leaving no room for any complexities, uncertainties, or alternative constructions of problems and solution. This also renders the modalities of this form of statements particularly hard to "unpack" for their respective recipients.

While most senior scientists we talked to, in particular, would agree to the basic tenet that relating one's work to particular societal expectations is part and parcel of playing the "funding game" today, quite a number of them expressed concern that these promises, by their very definition, create unrealistic public expectations about the potential and the speed with which science can address societal needs. While there is worry that this "promise bubble" might burst and negatively affect the public image of science, scientists in our interviews hardly ever reflected on a second type of expectation that might be generated by these promises. It seems more than likely that the rehearsal of these promises in doing boundary work also affects scientists' orientation work, and hence also changes the perceptions and expectations of the scientists themselves, or others working in the institutional spaces created by successful promises.

7.3.3 Their Stories About Us and Our Stories About Ourselves. Researchers' Perception of Their Work Between Media Narratives and Everyday Practice

In this section, we will explore scientists' perception of the relation between their everyday work practice happening backstage and medialized representations of this very work performed on the front stage. In our fieldwork, a number of life scientists complained that the media, as well as press-packaged stories produced e.g., by institutional public relations offices, would very often misrepresent the temporalities of scientific work. In particular, interviewees believed that the media focused its reporting on what scientists themselves perceived as long term goals that are not within their direct influence (such as curing a specific disease). Researchers argued that the concrete discoveries that they themselves saw as important breakthroughs were considered as much less important in constituting "news value". Of course, top publications in journals such as *Science, Nature* or *Cell* would be reported on, but in these contexts the reporting also stressed the long term goal rather than actually explaining what was discovered; and it suggested that this particular discovery would be a much larger step towards these long term aims than the scientists themselves would argue. Hence, the temporal pathways researchers conceptualized for themselves were seen as being shrunk in media reporting, with long term goals turning into short term aims for which a breakthrough is imminent, and in which their "real" short term aims were of little interest or relevance. Consider the following quote from a scientist referring to her negotiations with a TV channel about whether they would cover a recent *Science* publication:

> If we would have claimed to have found 'the gene for x', and that hence x wouldn't be something to worry about anymore, then they would have done the feature right away (project manager, female).

Later, she goes on to explain that it's not that she would object to talking about any application of her research in principle, but instead she would like to stress that

> possibly there isn't an immediate application in months, years or even decades; but rather that there's a long time horizon, and that basic science is necessary to achieve important changes also later on (project manager, female).

Researchers often expressed that they felt uneasy about this type of reporting, because it completely bypasses the complex translation logics of a basic science discovery into societal applications, and hence raises false societal expectations about how and most of all how quickly science can contribute to solving societal problems. In relation to this argument, we witness an interesting contradiction: The mode of reporting criticized here of course follows precisely the same rationale scientists themselves employed in the press-packaging practices analyzed in the preceding section. At another point in the very same discussion that the quotes are taken from, the very same scientist employed this type of argument to justify her own work to the citizens present. From an analytic point of view, this shows how deeply medialized rationales have entered scientific practices, and that they, as in this case, create tensions between scientists' self-perception of their work and its medialized representation, which also includes self-representation.

However, it was not only the temporal dimension of scientific work in relation to societal aims that was seen as misrepresented. Scientific practice itself and the nature of everyday work in science was seen as distorted in such medialized accounts. From scientists' perspective, several crucial aspects of scientific practice were made invisible. First, they criticized the media's nearly exclusive focus on success in science. Comparing science to sports, they argued that the public was only interested in "highlights and gold medals" (lab leader, male), but not in the hard path which leads to these successes and particularly not in the many experiments and attempts which turn out to be unsuccessful. Second, the media featured labs "where it goes boom, where it is loud and smoky" (project manager, female) rather than the often boring everyday routines of life science work, misrepresenting the majority of scientific practice. In particular, researchers were critical of the media's, and also their own press-office's, propensity to consciously omit any aspect of research which might be controversial in the public sphere, such as whether experiments used transgenic animals. In a discussion with citizens on science and the media, a senior researcher half-jokingly suggested having a TV-reality show on science – "then we would have time for all those stories which aren't told" (lab leader, male). In fact, the group of citizens he was talking with had just rejected a tour of the public labs that are usually shown to visiting school classes, and asked to see the places where "real work" is done.

This again leads us to an interesting contradiction. Researchers criticized the media for not representing scientific practice as they see it, and instead omitting the tiring and routine aspects of scientific work and presenting research work as if it will always quite quickly lead to successful results, without any uncertainties involved. Possibly the very same researchers, however, as directors of their respective departments or even institutions, would play important roles in designing "open labs" to fascinate pupils and invite them to choose a career in science. Currently, in the Austrian as well as in other European contexts, these open labs are only a small part of a burgeoning culture of public initiatives created to interest young people in choosing science as a career by producing glossy brochures about science or by offering summer internships in laboratories. However, the image of science presented in these brochures, or in the "open lab" strongly resembles the one criticized by the researchers above. Here, science is also portrayed as fascinating and colourful; and of course the "open lab"-experiments may be completed within a few hours, and are successful by default – in order not to "bore" pupils with the exigencies of actual lab work. When these pupils decide to go into science they will probably realize quickly how different the reality of research actually is.

In this section, we have tried to sketch the deeply ambivalent relationship between media representations of scientific practice, scientists' self-perception of this very practice, and the medialized representations of their own practice that scientists produce in different contexts. In a nutshell, our argument is that consciously or unconsciously, scientists re-produce the very same things they criticize as misleading representations in the media when they produce their own medialized accounts of their practice. This may be at least part of the reason why most of the life scientists that we talked to saw dealing with the media and engaging in medializing their own work as deeply troubling, despite the fact that they also saw it as "part of the game" of being a successful scientist today. However, also in line with this ambivalence, playing this game too successfully was often also seen as some form of treason, such as in the frequently expressed opinion that highly visible scientists in the media were actually rather mediocre, and engage in media activities to compensate for their lack of scientific skills.

> And often there's the argument, I always see that guy on TV, he must be a pretty good scientist. Even though he's doing some kind of crap, and just able to sell himself well. Of course, one doesn't foster one's own reputation by doing that (project manager, female).

7.4 Tacit Governance Effects of the Medialization of Science

Most life scientists we talked to strongly objected to the idea that the media in any way impinges on how they do their research. In their argument, media work is boundary work done on the front stage, while the backstage activities of actual research remain untouched by media logics and rationales. Already in the preceding sections, we have argued that this distinction only holds rhetorically, and that medialized ways of thinking and presenting their work have become deeply entrenched in research cultures and practices at least in the life sciences, but most likely in research more generally. In this section, we will discuss two issues for which the tacit governance effects of the medialization of science are particularly strong: (1) the development of research fields, in our case in the life sciences in Austria and (2) the career imaginations and decisions of young scientists.

7.4.1 The Impact of Media and Societal Framings on the Development of Research Fields in the Austrian Life Sciences

The Austrian media coverage of the life sciences is highly polarized. After some initially more heated discussions about red biotechnology in the 1990s, today the coverage of medical biotechnology and related life science research is uncontroversial, in comparison to other countries. Much of the reporting buys into the "economy of promises" (Felt et al. 2007) outlined earlier and reports breakthroughs together with their projected social benefits. For green biotechnology, nearly the reverse is true. From the 1990s onwards, any agricultural or food-related use of biotechnology has been deeply controversial, with media coverage and especially public opinion being predominantly negative. There were intense protests against field experiments involving genetically modified plants in the 1990s, as well as a very successful petition to parliament in 1997 – whose central slogans were "No food from the gene laboratory" and "No field trials involving genetically modified organisms". Today, the slogan "gene-free Austria" has permeated deep into everyday culture, and the label "gene-technology free" may be found on many agricultural products in every supermarket.

Most life science researchers both from the red and green field are very critical of this public debate, in particular because it renders "genome researcher" one of the least popular lines of work to mention in a pub or any other informal conversation. Beyond this annoyance, most scientists working in red areas of the life sciences felt mostly untouched by this debate. For people doing research on plants however, it tacitly governs how they think about, how they describe and what they (can) do in their research.

When asked about the direct impact of the 1997 petition on research in this field, senior researchers in plant genomics stressed that this public debate has "at the very least prevented [this kind of] biotechnology to develop more strongly" (lab leader, male). In the wake of the protests against field trials, institutions stopped entire lines of research, and even partially dissolved the research groups working on these topics. Today, green biotechnology research in Austria is mostly basic research, and hardly ever dares to rhetorically or actually venture beyond the confined and protected spaces of its laboratories.

But even these basic researchers would argue that Austrian public opinion and discussion still influences and impedes their work. First, getting legal clearance to do even extremely limited field experiments is extremely difficult, especially in comparison to the regulations for the animal experiments that affect researchers working in medical fields. Consider the following quote by a female group leader: "What an effort, if you want to plant 20 transgenic maize plants, unbelievable, only because we investigate soil bacteria." Second, in relation to the "economy of promises" that seems so crucial for acquiring funding, plant researchers clearly feel at disadvantage in relation to their colleagues working in the medical domain. Making promises about future applications is seen as pointless, if not dangerous, in a societal context which clearly rejects these applications, however well argued they may be. This deprives an entire research field of one of the most crucial rhetorical resources in the tough competition for institutional funds and grant money.

> I simply can't argue on the basis of the potential applications of these things, because there is no political support for these applications here. That means we can't do what the basic researchers in the medical field do, which is claim that they have a therapy for XY in five years time (lab leader, female).

As this senior researcher goes on to argue, the negative societal image of the field is not only an issue in acquiring financial resources, but it also impedes the social re-production of the field, because it renders it a less attractive career choice for young scientists. Also for the purpose of recruiting good students, the ability to offer possible societal applications is seen as central, on the one hand, and, on the other hand students are seen as more likely to choose a field they can relate to based on their prior experiences. As public discussion about green biotechnology is predominantly negative, most life science students lack these experiences.

> Of course, if I can promise my students that they will work on the problem of lung cancer in my lab, [. . .], then that's of course a challenge. If I say it's about raising the salt resistance of plants, then that's far more removed from their personal experiences in the first place (lab leader, female).

Finally, press-packaging is a more difficult activity for "green" life scientists in the Austrian context, not only because they cannot mobilize certain rhetorical resources, but also because they constantly need to avoid sensitive issues and buzzwords when writing or talking about their research. However, not talking about their research to the media is also not an option, because institutions require their researchers to issue press releases and to engage in media activities. Hence, things sometimes go wrong, such as when a project that was using genomic methods to study root development in a common tree in Austria issued a press release and was suddenly attacked for genetically altering the very same tree. "And then of course I'm at a consortium meeting, and am being attacked for the way we communicate.

But the initial press release didn't say that at all" (lab leader, female). Hence, whoever has a choice in the Austrian life sciences would rather not touch any topics related to plant genomics. In this respect, "white" biotechnology, life science research on micro-organisms, is an interesting case, because there is no public discussion on it, and because it can potentially relate to the rhetorical domains of both medical and agricultural uses. Not unexpectedly, the researchers we talked to in this field painstakingly sought to avoid any visual or rhetoric reference to agricultural genomics in their arguments.

As we have shown, media framings govern, at least indirectly, the development of research fields in the Austrian life sciences. Due to the controversial public framing, life scientists working on plants describe themselves as being at a disadvantage in relation to their colleagues working in the medical field, which receives very positive societal and media attention. This disadvantage does not only apply to direct contact with the media or societal actors, but can also be seen in the limited repertoire of the arguments they can deploy when applying for funding and also in the difficulty of attracting prospective students.

7.4.2 On Becoming a Scientist – Media Images and the Career Decisions of Young Scientists

In Austria as in other European contexts, recent years have seen a considerable rise in science communication activities targeted at young people. Against the background of a dense policy discourse lamenting the fact that too few pupils choose lines of study and careers in sciences and engineering, there is a flurry of classical media coverage, brochures, videos and initiatives aimed at drawing young people into science. In the preceding section, we have briefly commented on how "open labs" and other related activities convey a misleading idea of actual scientific practice. Similarly, the glossy brochures, videos and websites produced in the framework of public communication activities may be criticized as favouring some elements of research perceived as attractive, while other issues that may be just as crucial for deciding about future careers are omitted. In our focus groups and interviews, young life science researchers retrospectively comment that they had not been aware of some crucial aspects of a career in the life sciences until well into their postgraduate work. These brochures and activities do not talk about the ever-increasing temporalization of academic employment and careers, about the intense competition for the few more long-term positions, or about the downsides of the mobility required until well into the post-doc phase.[11] In our fieldwork, particularly researchers at the post-doc level were very critical about the fact that even most undergraduate students are not aware of these issues when choosing a PhD, let alone pupils who choose a certain line of study. Consider the following quote from a post-doc focus group:

---

[11] See Felt et al. (2010a) for a more detailed discussion of young life scientists' perceptions of academic careers.

> I know a lot of people, students, [. . .], that say, they had a totally different perception of what it's like to be a scientist. And if they had known, they would have done something else (post-doc, female).

However, medialized representations of science, scientists and scientific careers do not only play a key role in tacitly governing young people's decisions about a career in science. They also play an important role in determining how PhD-students and early post-docs think and talk about academic careers and their own potential future lives in the life sciences. The reason for this is a strong uncertainty about the rules which govern current academic careers, and in particular, which career pathways (if any) exist beyond the "excellence career", which they know will be possible only for very few. In talking about careers and lives in science, PhD-students and post-docs refer to the medialized representations of scientists available to them, which are nicely characterized by a senior scientist commenting on how these representations have changed from the time when he was a junior scientist:

> Now there are glossy brochures in which scientists are portrayed. And they are portrayed in a completely different way, they are like pop stars partially, so they have a completely different character than the role models I saw in my youth (lab leader, male).

Media representations of scientific careers, both in classical media as well as in the career brochures issued by scientific institutions or funding agencies, tend to depict those "stars" who "have made it", who have won some prestigious prize or achieved a professorship in their mid-thirties. Hence, the biographies told in these contexts are always successful ones, and they are often told to stress particular values perceived as crucial, such as mobility, early independence, or a high publication output. While younger students in particular may see these figures as role models, comparing one's own track record with these accounts becomes a more and more frightening exercise for researchers moving into the post-doc phase. For many in this career stage, these medialized representations embody values and expectations they can or do not want to fulfil anymore.

Our analytic point in writing about this is that the general homogeneity of the pictures of successful researchers serves a normative function by significantly curtailing how junior researchers talk and think about their careers. Certain desires, such as long term positions that do not first require winning a Nobel Prize, or professional biographies which may be better reconciled with the wish of having a stable relationship or even a family, or to continue working in the lab instead of becoming the grant-writing group leader, are perceived as illegitimate from the start under these discursive conditions – and hence our interviewees at times hardly dared to voice them, or implicitly thought that doing so would actually label them as potential drop-outs not worthy of doing good science. Other role models and stories beyond the excellence discourse are mostly missing, and hence junior scientists quite frequently talked about asking themselves, "Where do the post-docs go, then? So, not everybody can be a professor, but what happens to all the post-docs?" (PhD student, female). Medialized representations both in the classical media as well as from within science thus play a crucial role in, and hence tacitly also govern how junior scientists think and decide about their careers. Beyond this, an issue of high analytic salience is that these medialized representations may only achieve tacit governing powers because junior researchers employ them as a form of observing science and its dynamics, which they obviously cannot make sense of from within and in the context of their everyday experiences. The lack of alternative accounts lends the "standard excellence biography" a particular kind of normative power.

7.5 Discussion and Conclusion

The main aim of this article has been to analyze how medialization affects research cultures and practices, specifically in the life sciences in Austria. Here, we have understood medialization in the context of an increasing co-evolution of science and society, and thus as a set of processes in which scientific and societal rationales are increasingly intertwined. Hence, medialization refers to an ever-increasing coverage of science in the media, as well as to the growing number of contexts in which scientists themselves (or their PR managers) present and re-present their work to the public. Our empirical argument shows that the practices of both the media and scientists share the same logics and forms of representing science, because scientists often tacitly reproduce their perception of media logics as they do "media work", be it in writing press releases or in doing public engagement in an "open lab".

One of our main theoretical arguments in this paper is that an analysis of medialization which takes the idea of co-evolution seriously should not only consider these new forms of representing science at the science/society interface, but also needs to pay attention to how medialization actually affects science and society as they co-evolve. In merely studying changes in the public images of science, most work on science in the media focuses exclusively on one side of this co-evolutionary process. Tacitly this approach assumes a linear communication model, in which science is re-presented to the public, but in which scientists' basic self-understanding as well as the epistemic core of knowledge production remain largely untouched.

Using the concept of epistemic living spaces (Felt 2009), we have attempted to show that the feedback of media representations into science, as well as the very practices researchers engage in to medialize their research, influence and change research cultures and practices, along with how the researchers understand what living in and doing research means to them. Studying these processes is of particular importance, as we assume that the ways in which researchers perceive and inhabit their epistemic living spaces is deeply intertwined with the kinds of knowledge they will produce. Hence, the progressive medialization of science can be expected to shift both the social and symbolic orders in which research is done, as well as its epistemic orientation. Concretely, we would like to highlight three processes in which this re-ordering of epistemic living spaces takes place.

First, medialization touches upon the symbolic dimension of researchers' epistemic living spaces. Our results point to a convergence of the ways researchers describe and contextualize their own work and how it is framed by the media. Here, being able to make promises about the future societal relevance of one's own research plays a crucial role in doing boundary work. In an "economy of promises" these arguments become a central currency in competing for attention in the media, but also in research funding and within science itself. This creates new kinds of images of what doing science and being a scientist means, both in terms of the aims of scientists' epistemic pursuits, as well as in terms of the skills and virtues expected from the scientist as a person. Whether they embrace or reject these new images, researchers have to position themselves with respect to them in their orientation work and when developing their own professional identity. This may have a particularly strong impact on junior researchers, for whom this game of "press-packaging promises" has become part and parcel of their socialization into science as they struggle to survive in the project-centred and temporalized logics of current careers in research.

Second, relating again to the symbolic as well as the social dimension of epistemic living spaces, the researchers themselves, and junior researchers in particular, increasingly use media representations of science in orientation work, as a method for observing and making sense of science itself. The junior researchers we talked to in our fieldwork found it particularly hard to make sense of the rationales and dynamics of science, be they connected to careers or epistemic developments, beyond their own very personal and narrow experiences. Also, they lacked any institutionalized spaces of reflexivity within science, which might foster discussions around these questions or in which they may discuss them with their peers. Hence for them, media was a resource for making the meaning of research visible and understandable. In doing so, they of course ran the risk of taking the representation as a substitute for reality. For example, the fact that the media mostly portrays highly successful scientists, whose story complies with all of the normative requirements seen as crucial for scientific careers today, was often taken as an indicator that only these kinds of scientists could successfully remain in science, and hence could trigger processes of self-selected exclusion. The way media is used as means of observing science thus transforms what being in research means; it interferes with established value systems, forces their reinterpretation, and also creates anxiety. Yet there is also an interesting absence: The media is seen as merely re-presenting science, not as a possible space for re-imaging it in creative and dynamic ways (Felt 2000).

Third, and in close relation to the first two points, medialization influences how researchers perceive and act towards the temporal dimension of their epistemic living spaces. As we have described, medialized ways of re-presenting and thinking about one's own research lead to a shrunken perception of its temporalities. On the one hand, processes and aims which may take very long in actuality are portrayed as imminent and short in media logics. As we have shown, this creates considerable tensions in researchers' self-perception of their practices. On the other hand, medialization also may impinge on researchers' choice of topics, as epistemic puzzles that might deliver on relevant promises in short time frames could receive preference over more long-term topics.

All arguments we have made so far point to the conclusion that processes of medialization are an important force tacitly governing contemporary research. Doing research on medialization not only means analysing the expansion in the number and places where accounts of research are delivered, but it means, most of all, that we need to consider the effects of these expansions on science itself and its relations to society. Change caused by medialization needs to be conceptualized as both triggered from the outside by the increasing media representations of science as well as performed from within through the regular rehearsal of public expectations of both scientists and science, its products and its relations to society. Hence medialization takes place as much from within science as it is imposed by external actors.

However, the governance effects of medialization practices are hardly ever considered by the central actors involved. In effect, medialized processes and events play an important role in the governance of science. However, the respective actors associated with them, be they journalists, university PR-professionals or scientists neither assume nor are asked to take any responsibility for the consequences of their actions. In talking about science/society relations, Nowotny et al. (2001: 259–260) stress that "in an age of intense contextualisation, images of science need to have a strong 'reality content', that is, be closer to actual practices and their rapid changes than the traditional and timeless images." Thus "the gap between images of science and the actual practices should not become too wide" (ibid.). Our account of the effects of medialization suggests that this normative statement is not only true for societal perceptions of science, but all the more relevant for the epistemic and social development of science itself.

Acknowledgements This paper is based on empirical work done in several research projects – for details please see the section on material and methods. The authors acknowledge the contribution of all colleagues involved in these projects, and would like to thank the editors and reviewers of this volume as well as the participants of the yearbook conference in Bielefeld for their comments and suggestions. We are also grateful to Martha Kenney for the final language editing.

References

Etzkowitz, H. and L. Leydesdorff (2000). The dynamics of innovation: From national systems and "mode 2" to a triple helix of university-industry-government relations. *Research Policy*, 29(2), 109–123.

Felt, U. (2000). Why should the public "understand" science? A historical perspective on aspects of the public understanding of science. In M. Dierkes and C. V. Grote (eds.), *Between understanding and trust. The public, science and technology*. Amsterdam: Harwood Academic Publishers, pp. 7–38.

Felt, U. (2009). Knowing and living in academic research. In U. Felt (ed.), *Knowing and living in academic research: Convergence and heterogeneities in European research cultures*. Prague: Institute of Sociology of the Academy of Sciences of the Czech Republic, pp. 17–39.

Felt, U. and M. Fochler (2010). Machineries for making publics. Inscribing and de-scribing publics in public engagement. *Minerva*, 48(3), 219–238.

Felt, U. and T. Stöckelová (2009). Modes of ordering and boundaries that matter in academic knowledge production. In U. Felt (ed.), *Knowing and living in academic research: Convergence and heterogeneities in European research cultures*. Prague: Institute of Sociology of the Academy of Sciences of the Czech Republic, pp. 41–126.

Felt, U., M. Fochler, and R. Müller (2010a). *Career and/or biography? Essential tensions in knowing and living in contemporary academic research*. Manuscript.

Felt, U., L. Sigl, and V. Wöhrer (2010b). *Multiple ways of being together alone – A comparative analysis of collective and individual dimensions of academic research in two epistemic fields.* Paper submitted to "Science and Public Policy".

Felt, U., et al. (2007). Taking European knowledge society seriously. Report to the Expert Group on Science and Governance to the Science, Economy and Society Directorate, Directorate-General for Research. European Commission, Brussels.

Garforth, L. and A. Cervinková (2009). Times and trajectories in academic knowledge production. In U. Felt (ed.), *Knowing and living in academic research: Convergence and heterogeneities in European research cultures*. Prague: Institute of Sociology of the Academy of Sciences of the Czech Republic, pp. 129–225.

Hakala, J. (2009). Socialization of junior researchers in new academic research environments: Two case studies from Finland. *Studies in Higher Education*, 34(5), 501–516.

Jasanoff, S. (2004). *States of knowledge. The Co-Production of science and the social order*. London: Routledge.

Knorr-Cetina, K. (1999). *Epistemic cultures. How the sciences make knowledge*. Cambridge/London: Harvard University Press.

LaFollette, M. C. (1990). *Making science our own. Public images of science 1910–1955*. Chicago/London: University of Chicago Press.

Nelkin, D. (1994). *Selling science: How the press covers science and technology*. New York: W. H. Freeman.

Nowotny, H., P. Scott, and M. Gibbons (2001). *Re-thinking science. Knowledge and the public in an age of uncertainty*. Cambridge: Polity Press.

Peters, H. P., et al. (2008). Science Communication: Interactions with the mass media. *Science*, 321(5886), 204–205.

Power, M. (1997). *The audit society. Rituals of verification*. Oxford: Oxford University Press.

Rödder, S. (2009). Reassessing the concept of a medialization of science. A story from the 'book of life'. *Public Understanding of Science*, 18(4), 452–463.

Shapin, S. (2008). *The scientific life. A moral history of a late modern vocation*. Chicago: University of Chicago Press.

Strathern, M. (2000). *Audit cultures: Anthropological studies in audit, ethics and the academy*. London: Routledge.

Weingart, P. (1998). Science and the media. *Research Policy*, 27(9), 869–879.

Ylijoki, O.-H. (2005). Academic nostalgia. A narrative approach to academic work. *Human Relations*, 58(5), 555–576.

Ylijoki, O.-H. and H. Mäntylä (2003). Conflicting time perspectives in academic work. *Time & Society*, 12(1), 55–78.

Chapter 8
The Ambivalence of Visible Scientists

Simone Rödder

8.1 Introduction

In 1953, James D. Watson and Francis Crick discovered the double helix structure of the salt of deoxyribose nucleic acid (DNA). Fifteen years and a Nobel Prize for this research later, the general public discovered the "truth" about scientists in Watson's popular book *The Double Helix*: "they can be boastful, jealous, garrulous, violent, stupid" commented the St. Louis Post-Dispatch (as quoted in Merton 1973: 325). The sociologist of science Robert K. Merton saw the media debate, and even more the outrage amongst scientists, as revealing of the fact that scientific behaviour is governed by a certain ethos.[1] The relevance of social norms that any scientist takes in just as he acquires technical skills was put forward in a seminal paper for the sociology of science (Merton [1942] 1973: 268f). It is only if we *expect* scientists to behave universalistic, disinterested and solely geared towards common scientific progress that we can be disappointed to read the first sentence of the book "I have never seen Francis Crick in a modest mood" (Watson [1968] 2001: 7); or to hear Watson's reflection on his professional prospects: "It was certainly better to imagine myself becoming famous than maturing into a stifled academic who never risked a thought" (ibid.: 35). Outrage directed towards contraventions of the norms – namely universalism, communism, disinterestedness and organised scepticism – is but one demonstration of their structural importance. That scientific practice is governed by a normative structure is furthermore apparent in the occurrence of contradictions between different norms. Studying priority struggles in the history of science, Merton found a widespread ambivalence towards priority issues that he explained by conflicting normative expectations (1976). On the one hand, the value placed on the

[1] For reviews on the reception of the book in science and in the media, see Stent (1968) and Yoxen (1985).

S. Rödder (✉)
Bielefeld University
e-mail: sir@uni-bielefeld.de

S. Rödder et al. (eds.), *The Sciences' Media Connection – Public Communication and its Repercussions*, Sociology of the Sciences Yearbook 28,
DOI 10.1007/978-94-007-2085-5_8, © Springer Science+Business Media B.V. 2012

originality of scientific contributions demands the pursuit of knowledge production. This includes seeking and getting credit for the priority of discovery. On the other hand, the institution supports the ideal of the intrinsically motivated researcher who is driven by curiosity, not by a quest for recognition. Merton concludes that the juxtaposition of the reward system and the normative structure results in ambivalence in which "scientists are contemptuous of the very attitudes which they have acquired from the institutions to which they subscribe" (Merton 1973: 285). This "ambivalence of scientists" is structurally induced; it is a conflict that can be attributed to a social role, not to an individual role occupant (see Merton and Barber 1976: 6ff).[2]

Whether the Mertonian ethos is an appropriate description of science has since been called into question. The norms have been critically reassessed as "vocabularies" for ideological constructions of science (Mulkay 1976: 646; Gieryn 1983), and a review of the field states "a widespread sense in science studies that the Mertonian paradigm has been vanquished and relegated to the museum if not the attic of science studies" (Restivo 1995: 97). Other reviews, in contrast, leave no doubt that scientific practice is governed by social norms as well as by technical standards and emphasise that the "ethos of science" is a valuable description thereof (Stehr 1978: 184ff; likewise Weingart 2001: esp. 68ff). Weingart understands the ethos as an analytic subsumption of the social organisation of science which allows distinguishing short-term surface phenomena from fundamental change (Weingart 2001: 86). Transformations in the normative structure of science are thus indicative of changes in the relationship of science and other social systems such as politics and the mass media.[3] To draw on the Mertonian description thus has its merits in particular for analyses of those conflicts in the relationship of science and its environment which seem to be rooted in normative expectations of scientific and other institutions that directly contradict one another (ibid.: 75).

This paper argues that there is scope for such conflicts in the context of what has been called "medialization of science". The medialization concept postulates a tendency to orient science towards the interests of the media because their attention for issues, scientists, research institutes and scientific journals is perceived as crucial for public support (see Chapter 1). To investigate the relevance of media attention and demands for the professional role of scientists as one aspect of medialization is the intention of this chapter. To tackle the science-media interface at the individual level yields two complementary research questions: How do scientists perceive

[2] This sets the sociological concept of ambivalence apart from its use in psychology where it refers to ambivalent feelings of an individual. A psychological ambivalence, however, may be rooted in a structurally induced conflict. A role occupant's mixed feelings can thus be revealing of a tension built into a particular social role (for an example, see Section 8.4.1).

[3] Merton himself develops his analysis with regard to influences of the political system on science in the 1930s.

themselves as visible scientists[4]? And how does the peer community evaluate the media presence of an individual scientist?

In the following, expectations linked to the social status of the scientist are discussed in detail, and the concept of *boundary role* is proposed to describe the role of the visible scientist (Section 8.2). This allows complementing the analysis of individual role preferences with the question of how the role performance of a visible scientist is integrated into the scientific community. For the high profile field of human genome research (Section 8.3) it will be shown that scientists react to expectations to be visible. For the visible scientist as well as for her colleagues, this induces ambivalence that is grounded in conflicting expectations (Section 8.4). The genome case provides evidence for a role ambiguity of visible scientists and furthermore suggests that the view that visibility is now firmly built into the role expectations of scientists is policy talk rather than empirical fact (Section 8.5).

8.2 "Shun the Limelight" or "Thou Shalt Communicate"?

As its point of departure, this paper understands science as a social system in its own right. The production of knowledge evolved into what is now modern science when the question of the reliability of scientific knowledge was disconnected from its accordance with religious beliefs, its political correctness or its news value.[5] The therewith established cognitive and social imperatives distinguish science from other spheres such as the political system or the mass media as the arena of public communication. Scientists, on the contrary, closed their communication by using technical jargon and developed the scientific community as their primary audience. It is for this reason that science differs in its relation to its publics from most of the major social spheres in our society. Other than the political or the law system, where everybody is eligible to vote or endued with rights enforceable by law, science is not socially inclusive (see Stichweh 2005).

> Ordinary people may readily observe carpenters, lawyers, or nurses at work; but even in a technologically advanced country such as the United States, few people ever watch scientists working in any stage of the research process (LaFollette 1990: 18).

---

[4] The term "visible scientist" was introduced by Rae Goodell (1977). In this paper, the analytical concept of visibility as public communication, i.e., communication with audiences other than the scientific community, is operationalised as visibility in the mass media, in line with the argument that these media form an important resource for the public legitimacy of science in modern democracies (see Chapter 1). The role of the visible scientist thus encompasses intellectuals and media experts. This distinction is not relevant for my concern; it would be, however, if one wanted to compare the media visibility of life scientists with that of social scientists, for instance.

[5] "If you nowadays try to hold against biologists that one of their discoveries is politically right or left, catholic or not catholic, you will provoke open exhilaration, but this has not always been the case" (Bourdieu 1998: 20, my translation). It is obvious that scientific disciplines differ in the degree of exhilaration when facing such allegations.

Science is not genuinely a "visible" occupation and it is for this reason that the scientific community has been found to be "as uncomfortable about the democratization of science communication as the rest of us are about some of the other effects of technology" (Goodell 1977: 8). Describing the attitudes of the scientific community towards visible scientists, Goodell concludes from qualitative interviews conducted in 1977 that scientists lose their colleagues' respect if they voluntarily seek or accept media prominence (ibid.: 90ff.). There is furthermore ample anecdotal evidence that scientists frown on colleagues on magazine covers, but so far the issue has not been examined from the perspective of the sociology of science. If a scientist's primary audience is the scientific community, public communication is to shun because it disregards the audience of the peers and thus jeopardises the quality control that scientific knowledge owes its reliability to. The norm of the community protects the peer's communication system and reads: "*Shun the limelight*!" In line with that norm, the traditional relationship of scientists and journalists has been characterised by misunderstandings and tensions: "The tension tends to decrease communication between the two groups, which decreases the quality of science news, which in turn increases the tension" (Goodell 1977: 132).

These tensions have long been noticed and viewed as a problem with regard to science communication efforts (e.g., Chapter 13). There is a vast interest of science policymakers in communication activities that complements the self-interest of the media in science news to be used (see Chapter 1). With regard to our concern here, the role of scientists in this process, it is worth mentioning that policy statements and reports address the scientific community as their primary audience: "Our most direct and urgent message must be to the scientists themselves: Learn to communicate with the public, be willing to do so and consider it your duty to do so" (The Royal Society 1985: 36). Criticised is above all the lack of respective modules in the training to be a scientist; junior researchers are typically identified as the first target group for the teaching of media skills. In the form of "Criterion 2 – broader impacts", outreach duties are part of the funding policy guidelines of the world's largest funding agency, the US National Science Foundation. Alongside research and teaching, dissemination is anchored as the "third task" of scientists in university acts, for instance, in Denmark (Act on Universities 2003; for a related survey of Danish scientists, see Nielsen et al. 2007). German science policy goes as far as to postulate that engagement in outreach activities should be eligible for reputation (Stifterverband für die Deutsche Wissenschaft 1999: 60; Streier 2006). Academic institutions consider their organisational visibility as a prerequisite to be competitive for funds and students. All over the world, they include the participation in communication activities as a professional duty (see for Germany, Peters et al. 2008; for Norway, Kyvik 2005; for France, de Cheveigné 1997). The expectations have been found to be formally integrated especially in organisational leadership roles (Peters et al. 2008). It is a feasible option to entrust a few organisational leaders with "representation" duties of public communication but the mainstream view on science communication policy reads differently: Every scientist is requested to actively participate in the visibility of science by engaging in communication with its diverse publics. That science operates in a society where it is unavoidably linked to the political, economic and media system implies that scientists have the choice to address diverse publics that dispose of resources such as funding, prominence or prestige. In line with the science policy goals and demands from funding agencies and research organisations, recent studies state that scientists "are at the very least aware of a push toward public communication" (Davies 2008: 414). In editorials and science policy pieces, science policymakers and scientists themselves express the need to engage every scientist:

> It would be convenient to leave this task in the hands of a few representatives selected especially for their communication skills, but that won't work. Given the breadth of issues and the intensity of the effort required, we need as many ambassadors as we can muster (Leshner 2007: 161).

Obviously, this is wishful talk. But one does not need to share the policy assumptions to ask sociologically whether there are any implications of what is now broadly considered as a "new commandment": "*Thou shalt communicate!*" (Gregory and Miller 1998: 1, original emphasis). What does this imply for the personal role preferences of a scientist and how visible scientists are seen by their colleagues? Before this question can be empirically addressed, we need to look at the professional role of the scientist in more theoretical detail.

A professional role such as scientist is characterised by a bundle of associated roles, or a role-set (Merton 1957). Traditionally, the role-set of a major category of scientists, academics, is seen to be composed mainly of three different roles: research, teaching and academic self-administration (see Stichweh 1984: 70ff; Klima 1969; and for a normative conception that focuses on organisations, Kalleberg 2000).[6] Other than research, teaching and administration are not vital elements of knowledge production but role expectations towards scientists as professors or as staff members of scientific organisations. These roles thus refer to interrelations of science and other social systems, such as the educational system in the case of teaching or to an organisation in the case of academic self-administration. To study such interfaces, one can speak of "boundary-spanning activities" (Thompson 1967) or "boundary roles", occupied by "boundary role persons" (Adams 1976). The boundary role concept focuses on the individual who provides the linking mechanism across organisational or other system boundaries and addresses the internal and external set of expectations that a boundary role person is exposed to in the course of boundary activities[7] (see Tacke 1997). It brings to attention how boundary spanning activities are integrated into the original system (Luhmann 1999 [1964]: 220ff). In organisations, boundary spanning

---

[6] For my interest in individual role preferences the focus on the academic role-set is promising because their role-bundle includes teaching and self-administration and is thus inherently more complex than that of a scientist who researches in a private firm.

[7] With regard to science, the concept of "boundary work" has been mainly used as a critique of the "professional ideologies of scientists" (Gieryn 1983, 1999).

units ("Grenzstellen" in Luhmann's terminology) are formalised to deal with a defined part of the environment. Examples are university PR offices to handle media requests or human resources to minister staff membership in an organisation.[8] Boundary spanning units can be thought of as contact points; they have privileged access to the environment, and vice versa: it facilitates a journalist's job if a university clearly indicates who is responsible for media contacts. The functioning of boundary spanning activities can be illustrated using the example of the human eye. Visual information reaches the brain by way of the eyes only, which allows for a focused, unrivalled processing in the brain. Boundary spanning units thus function as channels: they canalise a particular type of information from the environment and relieve the strain of dealing with that environment from the rest of the system. Because the PR office observes the media by way of clippings of the coverage on the respective institution and because it handles media requests, the remaining employees normally do not need to wrack their brains with the university's presentation in the local press. The establishment of boundary spanning units is thus efficient in dealing proactively and reactively with external expectations and constraints and allows the organisation or any other social system to be aware of relevant changes in its environment (Luhmann 1999 [1964]: 224). Interfaces with important environments such as the media are institutionalised on the organisational level, but they also occur in the form of face-to-face interaction: a scientist is interviewed by a science journalist. The interview situation interface develops as a social system in its own right. This implies that boundary spanning activities are characterised by a split loyalty, because a boundary role person is simultaneously integrated in the original social system and in the boundary system. In the interview example, the scientist acts as a member of a scientific community and as a part of a face-to-face interaction. Split loyalty typically results in adaptations to the communication at the interface. In the course of a journalistic interview, a scientist is much more likely to communicate without using jargon than in a discussion with her peers, i.e., a communication internal to the science system. At the interface, these adaptations enhance the chance of mutual understanding and are thus both functional and legitimate; split loyalty allows mobilising problem solutions that would otherwise not be available (ibid.: 228). But the question arises in how far the public presentation of science can be separated from its production without disintegrating the science system.

To complement the analysis of individual role behaviour with a perspective that focuses on the consequences of visibility for the scientific community, I propose to conceptualise visible scientists as occupants of a boundary role at the science-media-interface – just as the institutionalisation of science PR is the formalisation of this interface at the organisational level. The claim that will be developed in this paper is that it is not possible to establish the role of the visible scientist without

---

[8] The use of the concept has so far been confined to organisational boundary activities (see Luhmann 1999 [1964]; Tacke 1997). Parsons (1955: 11f) was the first to analyse the role of the husband-father as the paradigm of a boundary role outside an organisational context.

introducing a systematic tension to other role expectations in the role-set of scientists. One can therefore assume that the average scientist's role performance is not very much influenced by the mass media. But given that some individuals choose to act as visible scientists, i.e., in the boundary role, leads to the question how this role behaviour is integrated into the science system. The boundary role concept assumes that its in-built split loyalty is noticed within the internal system and leads to tolerance for boundary activities (see Luhmann 1999 [1964]: 228). How much tolerance can a visible scientist expect in the peer community? And does the tolerance increase in times of medialization? To answer these questions requires investigating a case that is characterised by a high public profile. Such a case is human genome research, a field that in the course of the last decade has been repeatedly exposed to high media attention.

8.3 Material and Methods

The human genome project (HGP) was the first "big science" project in the history of biology (Hilgartner 1995). Starting in 1990, an international consortium mapped, sequenced and functionally analysed a reference sequence of the human genome and made the data available to the scientific community free of charge. In May 1998, however, the launch of the private firm Celera Genomics was announced whose goal was to sequence the human genome as a money-making venture. Triggered by the competition between publicly and privately funded scientists to finish draft versions of the genome sequence, the field became highly visible, as witnessed by the staging of media events and the regular appearance of prominent scientists in the press (Nerlich et al. 2002; Gerhards and Schäfer 2009; Rödder 2009a). To assess the consequences of this visibility, the author conducted 55 in-depth interviews with researchers from the publicly and privately funded human genome projects in France (6), Germany (17), the United Kingdom (14) and the United States of America (18).[9] The visible high-profile heads of the projects as well as junior and senior researchers without media experience were interviewed. The scientists' positions in the authority structure of science ranged from Ph.D. students to presidents and directors of research institutions, their reputation in the reward system from junior researchers to Nobel laureates. Each interview was transcribed in full. The interviews were analysed using a combination of individual case-based and comparative steps (see Witzel 2000; Glaser and Strauss 1967; Strauss and Corbin 1990). In representing relevant categories, the quotes that are used in the presentation of the case ensure the empirical grounding of hypotheses and conclusions.[10]

---

[9] For the empirical study, a scientist was a priori classified as "visible" if he had repeated media prominence in more than one context (a context being a journal publication, etc.). The sample was composed of visible and not visible scientists from each of the four countries; the classification was based on a literature review and verified by a triangulation with quantitative data (Rödder 2009b).

[10] The expert interview situation is itself a kind of boundary system rather than peer communication. This methodical weakness was dealt with both on the level of data acquisition and data

8.4 Results

8.4.1 Normative Expectations Towards Visible Scientists

The genome researchers' self-descriptions are revealing of the traditional paradigm: "The old school was that if you had your name in the paper then you were doing something wrong" (10:25).[11] But what does it mean when they characteristically speak of an "old school" (31:57)? Do its rules still apply? A visible German scientist has experienced that "some colleagues are extremely uncomfortable if someone gets a high public profile" (22:43); a not visible US researcher confirms: "Highly visible scientists are viewed as slightly suspect by their colleagues" (11:158) Media prominence "smells" (23:67) and therefore public communication is fundamentally in need of legitimation, first of all for one's own professional self:

> To be willing to communicate in public needs to be justified. [. . .] I have been doing that for a while [. . .] and I have always looked for an excuse to myself, why do I do that? (21:89)

How a visible scientist is perceived by a colleague is revealing of this scientist's genuine expectations towards his peer:

> He is the most modest man and he is a true scientist *although* he does some media presentation and you see him on the BBC (28:189, my emphasis).

A junior researcher who was chiefly involved in the UK genome project and who describes himself as "someone who usually is perfectly happy to chat away to journalists" mentions that he turned down a science journalist's request in the final phase of the draft sequencing. The journalist wanted to write a series on human genome research focused on the young scientist but the junior scientist felt "that a series was gonna be quite destructive to my career" (26:76). In the light of his colleagues' attitudes, his worries to repeatedly appear in the media seem to not be without cause:

> The names of [scientist x] or [scientist y] appear in the newspapers again and again and this poses the question, are they really aiming at publishing high quality research in prestigious journals or are they geared towards sound bite production to get media interest? (20:20)

All of the genome researchers see the general public as a relevant audience in several respects, as taxpayers, as patients, as users, some also as a valuable voice in science policy debates. They thus welcome institutionalised interfaces such as PR offices and outreach activities and stress the importance of being in the public eye to secure the resources for big genome science. But they strictly differentiate the visibility of the institution from the media prominence in the boundary role of the visible scientist:

---

analysis. The interview was set up as research communication on the issue of visible scientists, and reconstructive analysis was applied in looking for coherence in the interview data (for details on the methodology and a validity assessment, see Rödder 2009b).

[11] The first digit represents the number of the interview, the second digit the coded sequence. Quotes in German and French were translated by the author.

> Scientists are glad that science is out there in the public eye; that is a good thing. [...] At the same time, people who appear to be a little too comfortable with the camera and the microphone are often sort of looked down upon by their a bit more professional colleagues (1:157).

For a visible scientist, a tension arises:

> You have to sell your work and sell yourself. But you have to always keep in mind to maintain a certain kind of seriousness (44:14).

This tension is prevalent in reputable and less reputable researchers and on all levels of organisational hierarchies. The ambivalence is intensified by the central role of the media for public debates and their functioning as a system in its own right: "One should not exaggerate. But only sensational claims hit the headlines" (32:72). As assumed, it is only a small number of scientists who become visible. They remain a minority even in the high profile genome case, where this fact cannot be attributed to a general lack of media interest. For the average scientist, the relevance of the media and of being visible remains marginal:

> I have many colleagues whose work will never attract any media attention and there is no need for them to have it and they never even think about it (5:221).

The evaluation that – referring to all scientists in all fields – very few scientists act as visible scientists is reflected in a general scepticism towards the necessity of media training: "I don't know if it would be a right cost versus gain to train scientists like this" (34:23).

But while both visible and not visible scientists in all four countries express unease towards visibility, they also describe some recent changes to the "old school view":

> It's getting less of a hindrance for promotion, it's getting some funding from the agencies, it's encouraged in some places and less frowned upon in some others (13:64).

The diffuse expectation "thou shalt communicate" is now integrated in leading organisational roles:

> When journalists call and say, 'Look the human genome is gonna be published next week, I need to talk to a scientist to have an opinion', they will ask Genopole and Genopole will often redirect that to [director x] as the public figure of Genoscope (34:84).

The head of the institute who is described here as its "public figure" by a member of his team perceives his own media presence as "probably useful to collect funds but I have mixed feelings about that" (17:145). His words express ambivalence that he can neither dispose of for the purpose of resource generation, nor as an expectation that is built into his role as head of an institute. That visibility is part of an organisational role becomes apparent in feelings of guilt of not living up to this expectation, as is exemplified in the self-description of a leading scientist in the UK public genome project:

> There are people who are a bit nervous, or for them it's gonna be a big effort [so they] avoid it. I put myself in that category though I have some guilt about it having been involved in the human genome and large scale things (43:102).

This quote is revealing of another fact. The researchers characteristically attribute their reservation to their personality: they are "shy" or maybe "not used to the public eye". In the following, however, I will draw on the data to show that this psychological ambivalence is rooted in a systematic tension between external expectations and the internal expectations of the peers. As a starting point, however, the reasoning of the genome researchers themselves is examined.

8.4.2 Are Scientists "Media Shy"?

The researchers characteristically locate their unease in their personality structure: "Media and visibility isn't my thing" (7:102). The choice of the primary audience is attributed to a personal preference that apparently is prevalent amongst scientists: "Most scientists are very shy, we want to talk about our science to peers" (33:5). A junior researcher who describes himself as "media shy" (4:7) was asked what this actually meant. He replied that he is "media shy" as a *postdoc* but that he would be willing to communicate if he were a *research group* or *project leader*:

> To be media shy is absolutely relative, I mean, in my rank, with my scientific achievements, I would be absolutely out of place in the media (4:262).

That the postdoctoral researcher points to his current "rank" and a lack of "achievement" indicates that being "media shy" is not his individual trait but a normative expectation towards a junior scientist. A second personality trait that the researchers typically name as a source of ambivalence is jealousy. They consider jealousy as a human condition that is ubiquitous in social situations. Media prominence hence evokes unease because it produces inequality: "Everybody wants to be equal with all their colleagues and that creates unease in any community" (9:39). This explanation is in marked contrast to the meritocratic reward system of science that (ideally) attributes recognition on the basis of scientific achievement. A hierarchy of scientists that is based on their contributions is the ordering principle of peer communities. It is hence unlikely that scientists generally perceive recognition differences as inappropriate. For the attribution of a rank, however, evaluation criteria are needed. Ambivalence towards visible scientists reveals worries that the recognition of a visible scientist is not rooted in scientific evaluation, and that media prominence therefore endangers the reputation autonomy of the peers. Indicative of these worries is the case of the astronomer and visible scientist Carl Sagan:

> There was a famous episode in which he was blocked from being elected to the National Academy of Sciences, and the basic argument of his opponents was 'What has he done besides making all these nice TV programs?' There was a feeling amongst some of his defenders that the bar was actually set higher for him in terms of his papers and so forth than it would have been if he weren't as suspect for having had such a high profile (11:177).

Sagan's case shows that the bar for scientific achievement "in terms of his papers" is set higher *because* he is prominent. As a high profile is based on media criteria, his peers are particularly attentive in rewarding him scientifically with the election to an academy. All of the interviewees who mention the Sagan case assume that his prominence has damaged his scientific reputation. The "Carl-Sagan-effect" points to structurally induced ambivalence; the attribution to individual personalities that has here been discussed with regard to being "media shy" or "jealous" falls short of an adequate explanation.

I rather argue to understand the ambivalence in the light of the scientific community's attempt to safeguard the integration of a scientific discipline, i.e., the allocation of rewards according to scientific criteria. This claim can be substantiated with empirical data that reveal systematic tensions between the requirements of public communication on the one hand, and the normative structure of science on the other. In the data, one can distinguish structural issues with communicating to the media and other publics in several respects.

Firstly, this communication requires *media skills*, which future scientists do not necessarily acquire in the course of their socialisation. The interviewees struggle to speak "free of jargon" (50:170) to bridge the gap to public communication without losing their scientific credibility. In the routine relationship of science and the media, the genome researchers also typically perceive a lack of media interest:

> People are always saying that we scientists are supposed to talk to the media, but they are not going to be interested in us as we ring them up (41:109).

A second point is that communication is *time demanding*. The small number of visible scientists is explained by the need of resources: "It is a tiny fraction of scientists who have the means and can spare the time to establish a media image" (2:215). To become more visible would imply to "let science slide" (2:345); the production of knowledge would suffer in support of its presentation. It does not come as a surprise that the genome researchers perceive a tension between the role of the researcher and the role of the visible scientist:

> If you spend a lot of time communicating, you must be spending less time in the lab doing science and therefore you can't be a good scientist (34:78).

The third and main point, then, are *conflicting normative expectations* towards visible scientists. Such norms govern behaviour without that we usually call them into question or reflect on them. A certain perplexity in the face of ambivalence towards individual visibility can thus be regarded as a first indicator that normative expectations are involved: "Scientists tend to lose respect for scientists who communicate very much and it is not very clear to me why" (34:79). The interviewees typically found it "bloody difficult" (19:32) to name why there is an issue with visible scientists.

Contradicting expectations that are built into the role of the visible scientist can be specified in a material, a social and a temporal respect.[12] A first tension arises with regard to the material respect in scientific versus media communication:

[12] In all three dimensions of meaning, the interviewees focus on the point emphasised by the differentiation theory perspective that we propose in this volume: the difference in meaning between scientific and mass media communication, i.e., between an orientation towards "truth" and an orientation towards news values (see Chapter 1).

> What makes a nice story? That is the question. It may be scientifically unsound, but it reads well, it writes well. It may not be of scientific interest, not as much as other results, results that deserve a Nobel Prize but that are not easy to understand and to report on (2:709).

The criteria for a nice story differ from the criteria for the Nobel Prize; the media success of a research result is linked to its news value rather than its scientific substance. All of the genome researchers observe, on the one hand, a lack of media interest for most of the areas that scientists are interested in and, on the other, media hype of a few other issues.

> If it is good, it's fantastic; if it is bad, it is like the world is coming to an end. Everything has to be large and that, for a scientist, is completely counter to the honesty that you are supposed to have in all communications about what you are doing (1:31).

Tensions arise between the media's *search for sensations* and the norm of *intellectual honesty* that is binding for visible scientists as well. Even though the scientists acknowledge that they themselves overemphasise the societal relevance of their work in most grant proposals and publications, they view the exaggerations and polarisations in the media's presentation of their work as a major problem.

> For the scientists, the thing to be proud of is the fact that yes, we thought it was gonna work out this way, and after four years of very hard work indeed we can distinguish two types of breast cancer and cure one type and not the other type. And yet the media would like to portray that as 'scientists fail to find cure to breast cancer', there would be this very negative, shocking headline. I read one recently, actually, where it said 'Darwin's theory of life disproved'. As I was reading that I was thinking, 'Bloody hell, that's a strong statement, scientists disprove Darwin's theory of life'. But Darwin originally stated that he thought that hot springs were where life originated, and somebody had gone round and figured out that, really, organic molecules don't last long in hot springs, it's all a bit of a disaster, and it is more likely to be temperate springs. But the headline was almost deliberately misleading and caught your eye. So that's why I think there is a conflict between what the media wants to report and what science wants to state, and that's because the media is looking for more eye-catching statements and these are usually the more negative ones. So that inhibits scientists from going to the media which I think is unfortunate because I think there is quite a lot to be proud of in a lot of science, and it's a bit unfortunate really that we are always countering some negative impressions (26:47).

In the temporal respect, there is a tension between the *time demands of knowledge production* and the *media's time frame*:

> They've been spending 50 years to get the crystal structures. It makes a good publication but not very good press (28:143).

A second conflict in this respect is a temporal restriction for the complexity of statements, i.e., a systematic tension between *sound bites* and *scientific arguments*:

> Even if what you say has no basis at all, it doesn't take very long to make highly damaging statements about someone. To try to defend against those takes a lot of explanation, so you lose dynamic with this sound bite statement (11:444).

A third tension concerns the point in time at which scientific knowledge is published in the popular media. The media release of a research result prior to its scientific publication is marked as deviant behaviour: "What is frowned on is rushing results out before they are verified" (24:65). The expectation that scientific knowledge has been *scrutinised by peer review* conflicts with *science by press release.* An example for the genome case is the announcement of sequence drafts in June 2000, much prior to their publications in *Science* (Venter et al. 2001) and *Nature* (International Human Genome Sequencing Consortium 2001) respectively. With regard to the timing of scientific publication and media release, the formalisation of PR in scientific organisations poses problems:

> Frequently what happens is that the institution says, 'We have a fundraising dinner coming up next week and Bill and Linda Gates are coming and if we have a cure for cancer they may give us a lot of money'. Do we think this [not yet published research] is important, let's have a press conference (24:41).

A visible scientist critically assesses this formalisation:

> Now the norm is that you issue a press release if you got something exciting before you publish, and this causes a lot of trouble. I don't know what we are going to do about this because, at the same time, if things really are important, the public has a right to be informed (50:227).

In a social respect, a tension builds up between the visible scientist as a boundary role *person* and as a *representative* of the scientific community, i.e., between the personality orientation of the media and the imperatives of universalism and communism in science. Ambivalence that is rooted in that juxtaposition can be found in the self-description (statement 1) of a scientist who got some media prominence as speaker of a consortium, as well as in his perception by a member of the consortium (statement 2):

> (1) On the one hand, it would be great, if everybody who was involved would get some attention but, on the other hand, this is not possible (44:135).
>
> (2) It wasn't even mentioned that this is a consortium of seven different labs and institutions. On the one hand, of course, this is understandable and he was the coordinator but the way it was presented was not correct (18:15).

This junior researcher struggles to accommodate the boundary role behaviour with her expectations and while she takes into account the circumstances of the situation, she evaluates the performance as deviance from scientific norms. Some interviewees propose to solve this issue by having a different person represent the community each time; some have even tried to direct journalists to colleagues they consider more knowledgeable in the respective field – with modest success, as they report. More media experienced scientists link this failure to the interest of the media in the same person over and over again, to "rent a mouth" (50:313). Known to the media as experts in a field, they are subsequently asked to comment on very different issues.

The second tension in the social dimension concerns the criteria that are used to evaluate achievements in science and the media, the attribution of *prominence* and *reputation* respectively. Visible and not visible genome researchers unanimously note that visible scientists are judged by their "performance" (50:150). "It entirely depends on the way we would perform. Not at all on the content. This is the problem" (15:33). A tension builds up between external *recognition of presentation skills* and internal *credit for the production of original knowledge*. The media interest in entertaining individuals is contradictory to any of the norms. As I have shown for the case of Carl Sagan, the ambivalence of the genome researcher can first and foremost be rooted in the fact that media prominence is not attributed for scientific achievement and should therefore not be relevant for a scientist's reputation. An individual researcher's primary audience is the peer community and the aim of addressing an audience scrutiny. Peer review cannot be outsourced to the general public: "They have no way of deciding" (34:140). The genome researchers therefore do not regard the public as an audience relevant for what is at the core of their role expectation: "When I think about whether this work is publishable, is it interesting, I don't think about the public. I think about peers" (8:9). As the general public cannot scrutinise scientific claims, it is not the appropriate addressee of any communication that is oriented towards the production of reliable knowledge. This induces the *ambivalence of visible scientists* as the sum of the conflicts that have just been discussed. It is not possible to establish the role of the visible scientist without introducing a systematic tension to the reputation autonomy of the scientific community. Because conflicting expectations cannot be met concomitantly, one has to look at the conditions under which internal and external expectations take the lead in governing a scientist's behaviour. The question whether visible scientists are exposed to conflicting normative expectations can be rephrased: Under what conditions does a scientist become visible *although* public communication induces ambivalence?

8.4.3 Ambivalence Management

Other than the role of the scientist-researcher, the role of the visible scientist is socially inclusive in terms of its publics. It is thus noteworthy that one of the most important mechanisms of ambivalence management in a role-set, the "insulation from full observability" (Merton 1957: 115), is not available. Scientists' proverbial worries about how their colleagues will judge their public appearance illustrate this point: it is not possible to shield the role performance as media visible scientist from anyone's insight.

Boundary role activities address different publics at once and conditions need to be specified under which they are perceived as legitimate: "You wanna do it for the right reasons" (33:115). From the interview data, three conditions can be derived: a scientist's professional merits, whether he is proactive in the media contact, and whether he has an organisational position or goal. Ambivalence management strategies in the boundary role of the visible scientist are thus sound scientific work, the reference to an institutional context and reacting to being asked by the media. While the first condition is unanimously emphasised by visible and not visible scientists in all four countries, the second and third conditions are more controversial.

8.4.3.1 Condition 1: Sound Scientific Work

The quality of a scientist's work is a basic precondition of visibility: "That is where the difference is, whether or not your science is good" (27:31). Grounding media prominence in scientific achievement is a first strategy of ambivalence management: "He has very high media presence but he has got extremely high reputation scientifically" (43:63). Provided that this criterion is met, the actual media format plays a rather minor role. As author of a sound publication, a scientist can happily accept an invitation to a TV talk show:

> Oprah Winfrey, that's about as mass market as you're gonna get. So she was invited on this show, and this is just typical because she wrote a very respectable scientific paper (11:88).

As is marked here as "just typical", prominence as a consequence of sound scientific work is the usual direction of influence at the science-media interface. Subsequent media prominence links visibility directly to the scientific achievement. This kind of visibility can therefore be reconciled with the peers' recognition; it amplifies public as well as scientific attention and may have positive consequences for the career path and reputation of a scientist.

By contrast, the genome researchers identify the visibility of scientists without credibility and the pre-publication of not peer reviewed scientific claims as major visibility problems. Cases in point in the life sciences are claims to have cloned human life by the Raelian sect and the debates on intelligent design. All of the genome researchers emphasise the disintegrating effect of science by press release and stress the importance of the norm "no claims": If knowledge is to be regarded as scientific knowledge, it needs to be certified by peer review. Visible and not visible scientists in all four countries thus unanimously safeguard organised scepticism. This points towards the relevance of linking credit to contribution for the social order in science.

8.4.3.2 Condition 2: Reference to an Institutional Role

A second criterion is linked to the institutional context of a scientist. Visibility is now built into certain organisational roles such as the head of an institute or the speaker of a consortium. The ambivalence management strategy that derives from this condition is the reference to a role expectation: "It's part of my role, my position" (43:106). This condition therefore safeguards the norms of universalism and communism. It is closely linked to lobbying: "There is nothing against promoting the discipline you're in" (7:129). Lobbying is seen as a task that secures resources for the field and is thus an important duty of leading positions, but not restricted to these roles exclusively. Lobbying serves a valuable purpose and, as a boundary activity, can be accommodated with the peers expectations because a lobbyist addresses external publics to gain resources these publics dispose of and, in that respect, these publics are utterly relevant. In the perception of visible scientists this is acknowledged as "Part of [scientist x's] job is to sell the policy". Lobbying is set apart from the not scrutinised-claims-mode of visibility that is problematic.

> He didn't predict any science in that. He just said, 'I got 2.5 million dollars from some foundation to do this.' That's a press release. That's fine (33:138).

In line with this condition, it is consensus that visibility without any reference to an institutional position or goal is not appropriate, i.e., media presence that appears to be motivated by the quest for "self-aggrandisement" (11:47): "If their message is that they are important people, that isn't much of a message" (24:29). Self-promotion is criticised but regarded as human: "Scientists like to impress their mother-in-laws" (9:391). Ego-trips are much rather forgiven than deviance from the expectation "no claims". To address a "mother-in-law-audience" is apparently considered as without much consequence for the integration of the scientific community; in such cases, the individual respect for the person may suffer but not the scientific recognition.

8.4.3.3 Condition 3: No Proactive Media Contact

As a third condition it is relevant how a scientist's visibility came about. Visibility appears appropriate if the initiative at the science-media interface is on the side of the journalist. Again, visibility in this case is usually restricted to the authors of recent publications or to scientists who are asked as experts, which is, in most cases at least, indirectly related to their scientific merits. Visible scientists thus can justify their public appearance by stating that they did not deliberately seek media attention: "I have been asked" (2:681). This strategy safeguards the imperative of disinterestedness. At times, a 350-page popular book develops unintentionally, as its author recalls:

> In a certain sense it was not my choice, or at least not a deliberate choice that's what I should say. [. . .] It was something that just emerged (50:16/19).

Altogether, the strategies of ambivalence management that have been derived from the data show that the researchers distinguish "kinds of visibility" (1: 177), and that it is the kind of visibility that determines whether a particular boundary communication can be accommodated with the science system, as indicated by the very specific reactions of the peers towards the different kinds. Preconditions of legitimate boundary activity are sound scientific work, being asked by the media or the reference to an institutional context. Tolerance for boundary roles appears bound to these conditions.

But how stable is the integrative authority of the scientific community? Can circumstances be identified in which the tolerance for boundary role activities increases? It has been argued elsewhere that the human genome project can be understood as a phase of medialization (Rödder 2009a; Schäfer 2008; Nerlich et al. 2002). With regard to our concern here, the visibility of individual scientists, the data show that in this phase boundary role activities gain importance:

> Some of the more charismatic gentlemen, shall we say, went out and tried to make the arguments in the popular press (33:129).

Facing this situation, some visible scientists adapt their communication to the sound bite style of the media which, in turn, is noticed by their peers as a change in their mode of communication. Not visible scientists accuse visible scientists of self-promotion in a political style:

> The closer the scientists are to the policy level, the more excited they are about showing off as well [as politicians] (26:43).

Tolerance for the split loyalty of a boundary role person increases:

> I am not gonna be quick to criticise people like [scientist x] who were heavily exposed on this issue and trying to improvise as they went along" (11:486).

But how far can the peers' tolerance increase without disintegrating the system?

> That was a difficult position to be in and I think he managed it fairly well by political standards. But he said a lot of things along the way that don't withstand close scrutiny and didn't at the time (11:495).

The quote suggests that the performance as visible scientist is closely observed and linked to scientific criteria in an attempt to integrate the role behaviour into the science system. How far, then, goes the tolerance in times of medialization? To assess the question whether increasing tolerance for boundary role communication – and thus system disintegration – is a trend in ambivalence management, I will finally examine the relation of the role of the visible scientist to the other roles in the scientific role-set.[13]

8.4.4 Visibility as Part of the Role-Set of Scientists?

What are the conditions that determine the preference of a certain role over others in the role-set? Is it personal taste or individual skills that decide whether a scientist devotes time and energy to prepare an experiment, a lecture, or a media occurrence? Sociologically, one would rather assume that the investment in a particular role is weighted against the recognition that can be expected in return. Whether a scientist spends his time writing an article for a local paper or whether he conducts an additional experiment that the reviewer of a publication asked for therefore depends on how the respective effort is rewarded. One thus needs to look at the incentives to act in the different roles.

The data show that the genome researchers' main incentive is the recognition of their peers rather than disinterested quest for knowledge and much rather than media spotlight. This holds true for visible and not visible scientists alike. Indicative of the potential of a role to govern behaviour is therefore whether it is eligible for recognition. The genome researchers express strong expectations that this is the case for the production and presentation of research results exclusively. The interviewees' talk is focused on the production of peer reviewed knowledge as their professional goal.

[13] For a discussion of hierarchies in the role-set, see also Merton (1957: 113f).

As the core of their job, they talk about original and innovative ideas, experiments, publications in scientific journals, discussions at meetings and conferences and the review of their colleagues' proposals and papers. A career in genome research, however, cannot be based on research only. The interviewees describe activities that are not essential to the process of knowledge production itself but that are prerequisites thereof and therefore part of any scientist's job. Most importantly, the acquisition of resources is perceived as a task that is vital for a successful career and a mandatory quality for a scientist. Additionally, the management of a larger group is an important element of the work in genome research; this job is complicated by the interdisciplinary nature of the field. With the rise in organisational hierarchies, time demands for membership duties increase. It is precisely the recognition in terms of a large research group, etc. that turns successful researchers into managers and withdraws them from the work in the lab (see Luhmann 1992: 678). Characteristically, scientists speak of this paradoxical development as "frustrating" (42:141). This perception reflects the primacy of the reward system over the authority structure of scientific organisations (see Goodstein and Woodward 1999) and points towards the hierarchy of the roles in the scientific role-set. The emphasis on the functional primacy of knowledge production suggests that the role of the researchers is at the core of the role-set; the pursuit of recognition by way of scientific contributions the top priority. The role of academic teaching is unanimously separated from the core in its relevance for the scientist – it is not eligible for reputation. Teaching as well as self-administration can thus be conceptualised as secondary roles, likewise the boundary role of the visible scientist (Table. 8.1).

The self-descriptions of all of the genome researchers suggest that the expectation to be visible is an expectation that is not perceived as an essential part of their job duties. Visibility is "somehow optional" (43:24), "tertiaire de la recherche" (13:54) and "rather irrelevant for the real thing" ["für die Wirklichkeit relativ unwichtig"] (22:44); in all four countries acceptable as a boundary activity, but not eligible for reputation. The autonomy to allocate reputation remains for peers only.[14]

Table 8.1 The role-set of scientists

Professional role		Scientist	
Core (eligible for reputation)		Researcher	
Secondary/boundary roles	Teacher	Manager	Visible scientist

[14] The reputation autonomy of the scientific community is safeguarded by all of the interviewees, but beyond this common ground, types of scientists can be differentiated on the basis of their construction of the public and their willingness to adapt their communication: the geek, the missionary, the advocate of knowledge and the public scientist (Rödder 2009b).

8.5 Conclusions

For the case of human genome research, the data show a striking ambivalence towards the visibility of individual scientists. The ambivalence is prevalent in the talk of visible and not visible scientists in four countries that work in a high profile field, indicating that every scientist is confronted with conflicting expectations with regard to the boundary activities of some colleagues. The role ambiguity is revealing of the fact that contradictory normative expectations are involved in addition to the well known time constraints and lack of presentation skills.

Drawing on Merton's analysis of ambivalences of scientists,[15] this paper extends the sociological study of conflicting expectations in the role-set of scientists to public communication. That the expectations induce ambivalence verifies, first of all, that the normative structure, which has been described as the ethos of science, is firmly in place in the researchers' talk. At the same time, a context in which science is increasingly asked to justify tax dollars or private investment brings about the demand for public communication. For science as an institution, the public becomes relevant in several respects: as tax payer and venture capitalist they fund research, as patient and user they apply products and technologies; and a positive or negative public opinion towards a certain field cannot be ignored in science policymaking. With regard to the institution, these publics can allocate or withdraw resources; to strive for these resources by way of societal relevance is a sine qua none for the institution. The human genome researchers acknowledge that by broadly accepting the visibility of science as an institution. The interviewees in all four countries unanimously welcome the institutionalisation of PR and outreach activities at the level of scientific organisations. At the same time, they express a striking unease *as* and *towards* visible scientists. Scientists who become media visible are confronted with expectations that suggest a different practice in a material, temporal and social respect. In a material respect, there is a tension between the reliability of scientific knowledge and between the media's interest in news value. In a temporal respect, a tension arises between the time frame of knowledge production and peer review and the issue attention cycles and presentation formats of the media. In a social respect, finally, the role of visible scientists as members of a scientific community and as media stars is conflicting with regard to the relation of scientific and media recognition.

---

[15] Merton's ambivalence was found in the interview data as well: "Well, I am not really after prizes, but, on the other hand, to get one is not bad at all" (42:110). Further ambivalences are also present: for example, the tension between different role-sets of a person such as being torn between the professional role and the role as a family member. The relation of the professional and the family role shows that a structural dominance of the professional role is rather widespread amongst scientists in general and, in particular, amongst geek-type scientists. This allows to speak of a "hyperinclusion" in science, a phenomenon first noticed with regard to professional athletes (Göbel and Schmidt 1998). This tension, however, is not a sociological ambivalence in the narrow sense of the term, but a "derivative type" (Merton and Barber 1976: 9).

Facing this situation, the *ambivalence of visible scientists* arises: as a reaction to the demands for visibility but as unease towards visible scientists. Visibility is in need of justification because publics other than scientific peers are not the audience to judge scientific claims and thus not the addressees of a scientist's primary role as researcher. The role of the visible scientist can be integrated as a boundary role in the role-set, occupied by a minority of scientists and with the main incentive to secure public and private research funding. But the genome case provides evidence that it is not a regular job duty. On the basis of the analysis of this high profile case, one can conclude that the view that visibility is now built into the role expectation of every professional scientist, including being eligible for reputation, is policy talk rather than an empirical fact.

Because the demands which boundary role occupants are expected to fulfil are inconsistent, they cannot be fully formalised, i.e., illegitimate visibility cannot be precluded categorically. Other than institutional boundary activities, that can be formalised in a mission statement for the PR office or an outreach activity, the purpose of individual visibility is much less determined. The split loyalty in the boundary role allows adapting scientific communication much more freely at the interface. This is useful in face-to face interaction between scientists and journalists and may account for the prevalent finding that – despite conflicting expectations – the routine interactions between scientists and science journalists are mutually perceived as satisfying (e.g., Chapter 11). This functional solution, however, has consequential costs. It matters with regard to individual scientists, not with regard to scientific organisations, that, in the boundary system, the presentation of scientific knowledge is not systematically linked to its production. This offers an explanation why scientists strictly differentiate institutional and individual visibility. Individual visibility is legitimate as a boundary role, but there is a deep ambivalence towards it. The widespread occurrence in the self-descriptions of both visible and not visible scientists indicates that the ambivalence of visible scientists is not a psychological problem of individuals but structurally induced by trying to live up to conflicting expectations. To be visible – meeting the expectations to communicate with extra-scientific publics – and to be a scientist – bound to the normative structure of science – is an ambiguity that is built into the role of visible scientists. The ambivalence is intensified because the science-media coupling links visibility to the rules of the mass media.

At the science-media interface, the institutionalisation of science PR (Peters et al. 2008), more science in the news (Elmer et al. 2008) and a number of hypes and media events have been witnessed in recent years, but there is much less evidence that public science communication is by now seen by scientists as part of their job. If the issue was solely one of lack of skills or time demands, one would expect that two decades of media and communication training should have started to eliminate it. If the reservations towards visibility were rooted mainly in the absence of presentation skills, such training should enhance the number of visible scientists, at the least in times of high media attention, where a lack of public interest cannot account for the fact that visible scientists remain a minority.

I have argued instead that, in addition to time demands and the gap between scientific and public communication, the unease towards visible scientists can be explained sociologically by an ambivalence that is built into the role of the visible scientist. Most work in science and technology studies sees the difficulties in communication at the science media-interface as stemming from an urge of the scientific community to protect its autonomy, and therefore does not recognise this role ambiguity. For the theory and practice of science communication, where the time constraints and media competences are now frequently discussed, the study allows concluding that the issue that needs to be addressed is beyond the scope of current media and communication trainings because it is not merely a lack of skills or time. If visibility is to become a professional duty, media prominence needs to be established as eligible for recognition. The price is yet another ambivalence of scientists.

Acknowledgements I want to thank André Kieserling and Veronika Tacke for valuable comments on a draft of this chapter.

References

Act on Universities (2003). *Rep. No. Act no. 403 of 28 May 2003, English translation*. Kopenhagen.

Adams, J. S. (1976). The structure and dynamics of behavior in organizational boundary roles. In M. D. Dunnette (ed.), *Handbook of industrial and organizational psychology*. Chicago: Rand McNally, pp. 1175–1199.

Bourdieu, P. (1998). *Vom Gebrauch der Wissenschaft. Für eine klinische Soziologie des wissenschaftlichen Feldes*. Konstanz: UVK.

Davies, S. R. (2008). Constructing communication: Talking to scientists about talking to the public. *Science Communication*, 29(4), 413–434.

de Cheveigné, S. (1997). La science médiatisée: Les contradictions des scientifiques. In S. de Cheveigné (ed.), *Sciences et médias*. Paris: CNRS Éditions, pp. 121–133.

Elmer, C., F. Badenschier, and H. Wormer (2008). Science for everybody? How the coverage of research issues in German newspapers has increased dramatically. *J & MC Quarterly*, 85(4), 878–893.

Gerhards, J. and M. S. Schäfer (2009). Two normative models of science in public. Human genome sequencing in German and US print media. *Public Understanding of Science*, 18(4), 437–451.

Gieryn, T. F. (1983). Boundary-work and the demarcation of science from non-science: Strains and interests in professional ideologies of scientists. *American Sociological Review*, 48(6), 781–795.

Gieryn, T. F. (1999). *Cultural boundaries of science: Credibility on the line*. Chicago, London: University of Chicago Press.

Glaser, B. G. and A. L. Strauss (1967). *The discovery of grounded theory. Strategies for qualitative research*. Chicago: Aldine.

Göbel, M. and J. F. K. Schmidt (1998). Inklusion/Exklusion: Karriere, Probleme und Differenzierungen eines systemtheoretischen Begriffspaars. *Soziale Systeme*, 1, 87–117.

Goodell, R. (1977). *The visible scientists*. Boston: Little, Brown and Co.

Goodstein, D. and J. Woodward (1999). Inside science. *The American Scholar*. Magazine of the Phi Beta Kappa Society.

Gregory, J. and S. Miller (1998). *Science in public. Communication, culture and credibility*. New York: Plenum Press.

Hilgartner, S. (1995). The human genome project. In S. Jasanoff et al. (eds.), *Handbook of science and technology studies*. Thousand Oaks: Sage, pp. 302–315.
International Human Genome Sequencing Consortium (2001). Initial sequencing and analysis of the human genome. *Nature*, 409(6822), 860–921.
Kalleberg, R. (2000). Universities: Complex bundle institutions and the projects of enlightenment. *Comparative Social Research*, 19, 219–255.
Klima, R. (1969). Einige Widersprüche im Rollen-Set der Soziologen. In B. Schäfers (ed.), *Thesen zur Kritik der Soziologie*. Frankfurt/M.: Suhrkamp, pp. 80–95.
Kyvik, S. (2005). Popular science publishing and contributions to public discourse among university faculty. *Science Communication*, 26(3), 288–311.
LaFollette, M. C. (1990). *Making science our own. Public images of science 1910–1955*. Chicago, London: The University of Chicago Press.
Leshner, A. I. (2007). Outreach training needed. *Science*, 315(5809), 161.
Luhmann, N. (1992). *Die Wissenschaft der Gesellschaft*. Frankfurt/M.: Suhrkamp.
Luhmann, N. (1999 [1964]). *Funktionen und Folgen formaler Organisation*. Berlin: Duncker & Humblot.
Merton, R. K. (1957). The role-set: Problems in sociological theory. *British Journal of Sociology*, 8, 106–120.
Merton, R. K. (1973). *The sociology of science: Theoretical and empirical investigations*. Chicago: University of Chicago Press.
Merton, R. K. ([1942] 1973). The normative structure of science. In R. K. Merton (ed.), *The sociology of science: Theoretical and empirical investigations*. Chicago: University of Chicago Press, pp. 267–278.
Merton, R. K. (1976). The ambivalence of scientists. In R. K. Merton (ed.), *Sociological ambivalence and other essays*. New York: The Free Press, pp. 32–55.
Merton, R. K. and E. Barber (1976). Sociological ambivalence. In R. K. Merton (ed.), *Sociological ambivalence and other essays*. New York: The Free Press, pp. 3–31.
Mulkay, M. J. (1976). Norms and ideology in science. *Social Science Information*, 15(4–5), 637–656.
Nerlich, B., R. Dingwall, and D. D. Clarke (2002). The book of life: How the completion of the Human Genome Project was revealed to the public. *Health*, 6(4), 445–469.
Nielsen, K. H., C. R. Kjaer, and J. Dahlgaard (2007). Scientists and science communication: A Danish survey. *Journal of Science Communication*, 6(1), 1–12.
Parsons, T. (1955). *Family, socialization and interaction process*. New York: The Free Press.
Peters, H. P., et al. (2008). The medialization of science as a prerequisite of its legitimation and political relevance. In B. Schiele et al. (eds.), *Communicating science in social contexts: New models, new practices*. Dordrecht: Springer, pp. 71–92.
Restivo, S. (1995). The theory landscape in science studies: Sociological traditions. In S. Jasanoff et al. (eds.), *Handbook of science and technology studies*. Thousand Oaks: Sage, pp. 95–110.
Rödder, S. (2009a). Reassessing the concept of a medialization of science: A story from the "book of life". *Public Understanding of Science*, 18(4), 452–463.
Rödder, S. (2009b). *Wahrhaft sichtbar. Humangenomforscher in der Öffentlichkeit*. Baden-Baden: Nomos.
Schäfer, M. S. (2008). From public understanding to public engagement. An empirical assessment of changes in science coverage. *Science Communication*, 30(4), 475–505.
Stehr, N. (1978). The ethos of science revisited. Social and cognitive norms. In J. Gaston (ed.), *Sociology of science*. San Francisco, CA: Jossey-Bass, pp. 172–196.
Stent, G. S. (1968). What they are saying about Honest Jim. *Quarterly Review of Biology*, 43(2), 179–184.
Stichweh, R. (1984). *Zur Entstehung des modernen Systems wissenschaftlicher Disziplinen*. Frankfurt/M.: Suhrkamp.
Stichweh, R. (2005). Inklusion in Funktionssysteme der modernen Gesellschaft. In R. Stichweh (ed.), *Inklusion und Exklusion*. Bielefeld: transcript Verlag, pp. 13–44.

Stifterverband für die Deutsche Wissenschaft (1999). *Dialog Wissenschaft und Gesellschaft*. PUSH Symposium. Essen.
Strauss, A. L. and J. Corbin (1990). *Basics of qualitative research: Grounded theory procedures and techniques*. London: Sage.
Streier, E.-M. (2006). Wissenschaft in einer Pressestelle I: Vermittler, Vermarkter und manchmal Verkäufer. In H. Wormer (ed.), *Die Wissensmacher. Profile und Arbeitsfelder von Wissenschaftsredaktionen in Deutschland*. Wiesbaden: VS Verlag für Sozialwissenschaften, pp. 270–283.
Tacke, V. (1997). Systemrationalisierung an ihren Grenzen – Organisationsgrenzen und Funktionen von Grenzstellen in Wirtschaftsorganisationen. In G. Schreyögg and J. Sydow (eds.), *Managementforschung 7*. Berlin, New York: de Gruyter, pp. 1–44.
The Royal Society (1985). *The public understanding of science*. London: The Royal Society.
Thompson, J. D. (1967). *Organizations in action. Social science bases of administrative theory*. New York: McGraw-Hill.
Venter, J. C., et al. (2001). The sequence of the human genome. *Science*, 291(5507), 1304–1351.
Watson, J. D. ([1968] 2001). *The Double Helix*. New York: Simon & Schuster.
Weingart, P. (2001). *Die Stunde der Wahrheit? Zum Verhältnis der Wissenschaft zu Politik, Wirtschaft und Medien in der Wissensgesellschaft*. Velbrück: Weilerswist.
Witzel, A. (2000). Das problemzentrierte Interview. *Forum: Qualitative Sozialforschung*, 1(1).
Yoxen, E. (1985). Speaking out about competition. In T. Shinn and R. Whitley (eds.), *Expository science. Forms and functions of popularisation*. Sociology of the Sciences Yearbook IX, Dordrecht: Kluwer, pp. 163–181.

Stifterverband für die Deutsche Wissenschaft (2000). *Dialog Wissenschaft und Gesellschaft*. PUSH-Symposium, Essen.

Strauss, A. L. and J. Corbin (1990). *Basics of qualitative research. Grounded theory procedures and techniques*. London: Sage.

[illegible]

Chapter 9
Practitioner's Perspective: Medialization and Scholarship: A Historian's Point of View

Paul Nolte

9.1 Getting Started

When I decided to study history, with a minor in sociology, after finishing high school in Germany in 1982, it certainly was not for the prospect of future media attention. I had developed strong political interests, as was quite common for academic youth during the time of the "Second Cold War" and peace movements, but somehow political science did not seem as differentiated and intellectually fascinating to me. Through my Master's and Ph.D. studies, mostly at the University of Bielefeld, with one year of Graduate Study at Johns Hopkins University, the field of concentration was nineteenth century German history instead of more contemporary times that presumably would have been better suited for media coverage. Only in the mid-1990s, when working on the "Habilitationsschrift," did I move into the twentieth century, with a focus on post-1945 societies. In an attempt at combining the "Bielefeld school" social history with approaches in conceptual history also present in Bielefeld through Reinhart Koselleck, plus some sociology of knowledge, the resulting book discussed perceptions of social order in Germany from around 1900 into the 1960s: hopes and visions, anxieties and apprehensions of unity and disunity; from the Kaiserreich and Weimar class divisions to the Nazi Volksgemeinschaft and the Federal Republic's emerging middle class society.

With this project largely finished in 1999, several strands came together that, step by step and certainly not following a "master plan," drew me into more essayistic writing and public commentary in the following 5 years. Without a doubt, there was a major influence of my main academic adviser, Hans-Ulrich Wehler, a highly productive and original historian and public intellectual, whom I had observed for a long time switching back and forth from scholarly rigor to sharp political intervention. In both roles he had been a model for the public responsibility of the historian's work to me. In addition, I had always enjoyed writing and,

P. Nolte (✉)
Free University Berlin
e-mail: paul.nolte@fu-berlin.de

S. Rödder et al. (eds.), *The Sciences' Media Connection – Public Communication and its Repercussions*, Sociology of the Sciences Yearbook 28,
DOI 10.1007/978-94-007-2085-5_9, © Springer Science+Business Media B.V. 2012

frankly, "being published," and I wanted to test out publishing possibilities beyond the purely academic sphere and readership. (I assume, though, that many scholars would admit to that.) During a fellowship at the Wissenschaftskolleg Berlin in 1998–1999, my ambition was fostered in a community of high-ranking academics, many of whom drew media attention and fashioned themselves as public intellectuals. At the same time, these were years of new social conflict in reunified Germany, as the expansive prosperity of the post-war Federal Republic gave way to stagnation and new poverty, and the welfare state was being reshuffled even under the leftist Red-Green government.

So I started out experimenting with essay pieces on current issues in German society and politics that were published in the liberal weekly *Die Zeit*. In one particular essay, I claimed the return of class divisions and the emergence of a new underclass ("neue Unterschichten") in Germany characterized by certain cultural modes of collective self-stylization, e.g., a favor for trash TV for which I coined the term "Unterschichtenfernsehen." This term developed a sort of career of its own, furthered by a collection of my essays that were published in 2004 by C.H. Beck under the title *Generation Reform*. When talk show host Harald Schmidt presented the book later that year in his first appearance on German public television ARD, media attention for the book and for my propositions rose to dangerous heights, with daily demands for texts and interviews, for radio and even some TV news or talk shows.

It took me some time to sort out and understand what was going on here. Part of the attention had to do with the fact that, at roughly age 40, I was considered still very young and a rather rare species among public intellectuals mostly in the "wise old man" category; and new faces are always interesting to the media. Media interest also was increased by the fact that with concepts such as "Unterschichtenfernsehen," I was actually speaking about themselves; a fact highlighted by the way Harald Schmidt had recommended my book as an ironical statement of his own no longer working for commercial trash TV. Most of all, I learnt about the insatiable hunger of the media, and that you do not have to offer yourself to them, but would hardly be able to escape them chasing you. On the other hand, I was well aware of my own contribution – not so much, I hope, in "prostituting" myself to the media (as accusations from some colleagues soon went) – but in offering clear theses and concepts, such as the term "neue Klassengesellschaft," that public discussion and the media need for the staging of controversy.

9.2 The Roles of "Public Intellectual" and "Expert"

From the point of view of scholarly work and its never-ending differentiation and qualification of any problem, this may sound ambivalent at best. However, clear-cut language and even some simplification of academic subject matter is not just a demand of modern *media* society, but is indispensable for processing information in a democratic *expert* society in which the wider public can claim a right to understanding and participating in expert knowledge. But there is more to this than just the tension between experts and a lay public. Political controversy that is essential to the functioning of democratic societies, and their possibilities for dynamic development, also needs aggregation of opinion instead of endless differentiation and individualization – that is why we have political parties. And it is dependent upon people, be they politicians or public intellectuals, who at times overemphasize one side of the matter for the sake of an argument and a lively culture of conflict. Why would, to take one recent example from a transatlantic public controversy, high-ranking European intellectuals such as Jacques Derrida and Jürgen Habermas so aggressively denounce the United States for their belligerent anti-terrorist efforts? Would they not know, as extremely smart people, that this is a matter of innumerable pros and cons, of almost inextricable complications? Of course they would. However, the way they argued and staged themselves has not only rewarded them with more media attention than they would otherwise have received. It has also been extremely helpful in stimulating discussion and clarifying substantial points.

Overall, it is not surprising to find historians participating in current affairs and drawing attention by the media. Especially in the German cultural tradition, this can be regarded a well-established pattern, dating back to such eminent founders of the discipline such as Leopold von Ranke, adviser to kings, or Heinrich von Treitschke, who gained an (in)famous reputation in the so-called "Antisemitismusstreit" around 1880. The second half of the twentieth century saw the rise of a particular intellectual type within this pattern: critical historians who acted as transmitters between scholarly work and public interest in the wake of the Nazi dictatorship and the Holocaust – topics that continue to thrive in political discourse, memory culture, and media treatment. They mostly belonged to a typical age cohort: Born roughly around 1930–1935, they had experienced the Nazi regime as youths and turned into ardent fighters for Western-style democracy and liberal society after 1945, and especially since the 1960s, when the general reformist climate of that time met with their professional securing of university positions.

Among historians, Hans-Ulrich Wehler, Heinrich August Winkler, and Hans Mommsen are prime examples. Generally, sociologists Jürgen Habermas and the late Ralf Dahrendorf are probably the best-known figures in that game, as for many decades they have skillfully, and with enormous influence, commuted between deep scholarly work and their roles as public intellectuals, or even, in the case of Dahrendorf, as politicians. The particular connection of historical circumstances and biography that justifies labelling those scholar-intellectuals as "Forty-fivers" of course cannot be reproduced. However, the larger idea of scholars in the humanities and social sciences taking public responsibility in any given historical situation still seems appealing to me, and still is a necessary task in the current "post-post-Nazi" configuration.

Historians may be particularly well prepared to take on that role for several reasons. For one, history – notwithstanding all specialization and "scienticization" over the past decades – still retains its general interest appeal and in the public meets with many people who have "always wanted to study history." With an often-made distinction, historians are somewhat more likely to be "lumpers" than "splitters." They are capable of answering – or at least, addressing – broad and general questions, and they are still being trained (although not always successful) to speak and write in a language that is close to general usage: largely free of a scientific and technical jargon, with hardly any mathematical formulas involved, etc. Some would lament the lack of precision and database in what historians are saying. But overall, these features make them very attractive for a media world in which the ability to express oneself clearly and understandably, and to talk about broad issues instead of just presenting a small piece in a jigsaw puzzle, is highly appreciated. History is always in demand; the more so since we seem to live in the age of a new historicism, in which questions of memory and remembrance, of preservation and cultural heritage, haunt the general public. In Germany, the legacy of the Nazi past continues to influence not just the culture sections of the quality newspapers, but politics; and with every Auschwitz comparison being made by a politician, phone calls from newspapers and radio stations reach the history departments.

As explained earlier, my own public appearance is not primarily in this field, but rather in more contemporary issues of conflicts in modern societies, with an emphasis on social structure and on problems of party politics and democracy. Although I received an interdisciplinary training and could hardly imagine doing my job – both the more scholarly and the more public aspects of it – without what I learned from sociology and political science, I sincerely profess to the specific value of a historian's knowledge and perspective on those issues. Interestingly, the media sometimes have difficulties with my disciplinary identity. Although a historian commenting on current affairs is a well-established, time-honored model, journalists seem distrustful of selling a historian as commentator on, say, the results of the latest federal elections; perhaps because they fear their audience would wonder about legitimacy or credibility of my statements or, even worse, their journalistic work. Therefore, depending on the issue, I often find myself cited as a political scientist or a sociologist. I remember one conversation with a radio station many years ago in preparation for an interview: "How should we introduce you?" – "Well, I'm a historian, a professor of contemporary history." – "Now, can't we better say you're a political scientist?" In the meantime, I have mostly given up on trying to correct them, and it might be argued that from an outside perspective, disciplinary differentiation rightfully follows other criteria than within the scholarly field.

A further complication arises from two different, and at times almost contradictory, roles of scholars in the public and media arena, and I have had to learn to deal with that. On the one hand, the academic world is in demand as provider of specialized knowledge – that is the role of the "expert." The advance of modern science usually comes in small bits and pieces of a highly complicated nature, and experts are expected to work at this forefront of research, to know the most intricate details, and to be able to explain them in a semi-technical language that is accessible to a wider audience. Public intellectuals share some of this translating function with experts; both have to bring about intellectual and cultural transfer between distinct spheres of communication and meaning. However, public intellectuals are in some ways anti-experts. Their focus is not so much on tiny bits of empirical evidence, but on the larger webs of meaning and interpretation. Experts strictly stick to their particular field of legitimate academic competence, whereas public intellectuals have often, as in a prominent article by German sociologist M. Rainer Lepsius, been defined as deliberately, and necessarily, transgressing the boundaries not just of disciplinary knowledge, but indeed of their own scholarly legitimation (Lepsius 1964). To put it bluntly, they talk about things of which they do not understand much.

Of course, in reality this is somewhat more complicated. Being an expert requires more than specialized knowledge, namely broad experience and the capacity for contextualization; otherwise, any graduate student would be an expert. Other aspects add on to this, especially from a media perspective, not least reputation and to some extent Merton's "Matthew effect." But still, the contrasting roles of expert and public intellectual fit the aforementioned distinction between "splitters" and "lumpers" very well. It was first introduced by biologist George G. Simpson in 1945 and framed a controversy between historians Jack H. Hexter and Christopher Hill in the "Times Literary Supplement" three decades later. One is also reminded of Isaiah Berlin's famed essay of 1953, which played with the metaphors of the hedgehog and the fox taken from an ancient Greek fable by Archilochus: "The fox knows many things, but the hedgehog knows one great thing." Experts tend to be splitter-foxes, while public intellectuals prefer the lumper-hedgehog mode of dealing with knowledge: both inside their academic community (e.g., in a preference for writing syntheses or research papers that assess the current state-of-the-art) and in "selling" it to an outside public.

From my own experience, I should say that the role of a public intellectual, on the other hand, is much more than grand design and free-style opinion. It may be different with novelists and artists whose claim to being a public intellectual rests on their fame, their independence, and their moral authority – take Günter Grass's role in German public debates as an example. The legitimacy of "scholar-intellectuals" and their credibility with the media, however, is grounded in a particular linkage between expertise and dilettantism, between objectivity and opinion. Moreover, there is hardly a clear boundary between one and the other. On the contrary: it may very well be argued that the role of a public intellectual actually requires a blurred zone of transition between scholarly competence and opinionated transgression. Journalists, however, sometimes get it mixed up, or they just do not know in what role they should address a particular scholar. It often happens that they call me as an "expert" for some factual bit of historical information, more or less accidentally as they were looking for just some historian of contemporary Germany, not being aware that I have – for better or worse – a public reputation in that particular area that is strongly connected to my opinionated judgments, e.g., on the welfare state and the German lower class.

Then again, similar things may happen to pure "expert scholars" as well, because journalists sometimes do their research in a lazy and superficial manner. Thus, I can well imagine a famous physicist, perhaps a Nobel laureate, who is approached for explaining some rather basic issue in particle physics, with the journalists unaware that they are talking to the greatest capacity in the field. If journalists call me to get an introductory history lesson on some topic, which would spare them the effort of reading, I politely ask them to get the basic information first, and then return to me for an evaluation of the state-of-the-art, or ongoing controversies in the field. Luckily, though, these are exceptions, and I mostly find the media well prepared and doing their job in a very professional manner.

9.3 Into the Mass Media – Transformations and Tensions

To speak about "the media" in a summarizing way is, of course, highly problematic for a number of reasons. From the point of view of the academic world, and again from my personal experience, the different forms and technologies of media communication make a great difference. Writing for a newspaper is different from a radio broadcast, and the visual appearance on television again changes the rules of the game. Academics are used to writing and they focus on publishing, be it articles in peer-reviewed journals or, as with the humanities and not least with history, old-fashioned books. Small wonder, then, that writing an article for a magazine or newspaper is relatively secure territory for scholars, and after many years, I still feel most comfortable with this. The transformation of one's identity is a gradual one in this genre. The most important characteristic of media contact through writing is probably that one retains basic sovereignty over one's own intellectual product: the text. Comments and criticism by editors or peer-reviewers is well known in the world of academic publishing; and in that sphere, too, certain external limitations apply, such as a maximum number of words or pages in a journal's guidelines or in a book contract.

Still, the transformation happens, with ambivalence: It is enjoyable and rewarding to find a piece published quickly, in a matter of weeks or, in the case of a newspaper article, within 24 h, instead of waiting months or even a year, as it happens with journal articles or book chapter publications in the humanities. But then you will find your original manuscript changed, sentences rewritten, entire passages cut, sometimes without explicit consultation with you, the author. And the title that you have given your text has disappeared in favor of a headline that is not your own and may, at least from the perspective of scholarly purity, slightly misrepresent your intentions and nuances in exchange for a more dramatic appeal to the readership. Another type of written statement in the media is the interview that appears in print. I find interviews extremely difficult, perhaps because they do not fully belong to the written world, and not entirely to the spoken world either. You talk to a journalist on the phone or in person in your office for a taped conversation that is truly in-depth and may take an hour or more of everyone's time. A day or two later, the edited version of the interview will be in your mailbox, and there are only few things that are more painful in my work than going through those texts for corrections and final approval, because I often hardly recognize my own words, or at least there are subtle changes in mood and expression that seem to shift the meaning of what I had intended to say. Of course, that is exactly the famous moment when politicians (or their staff) entirely rewrite interviews, regardless of whether something has actually been said or not. But somehow, as a scholar, I would not want to do this. However, most media are entirely co-operative and understanding during that ordeal.

So why not have the spoken word transmitted in the first place? For the scholarly community, radio is an immensely important bridge of communication to the wider public, both in the roles as "experts" and as "public intellectuals." At least in Germany that is the case, with its public and federal broadcasting system that allows for the existence of two major national channels for "highbrow" culture and science broadcasts, the Cologne-based "Deutschlandfunk" and the "Deutschlandradio Kultur" in Berlin. All of the broadcast stations of the Länder maintain some program with an emphasis on high culture and science, on serious politics and social debates that heavily depends on the contributions of academics. These programs have come under economic pressure, but overall they have proven resilient, not least because of the many excellent journalists, usually with an academic background, whom I have often come to know as congenial partners in a common project of making academic problems and intellectual viewpoints accessible to a wider audience. That partnership may take the form of a live debate on a current political issue or a recent book release; or it could be a radio essay, i.e., a written text that is spoken by the scholar himself.

Although one might guess that all print journalism belongs to one group and all audiovisual media to a second one, I have experienced the gap between radio and television to be extremely wide and difficult to cross for scholars and scientists. The visual element of being on screen changes almost everything; the message retreats for image and appearance; and any differentiation or complication is hard to convey. That is especially true for talk show formats, which often feature a bunch of politicians plus one outside person who may be a scholar, and who has to play the role of an expert – then he is reduced to delivering facts which the politicians try to make sense of – or the role of a maverick intellectual in which case his job is to throw in some provocation that the politicians can chew on. Generally, on television the pressure on what you say, and how you say it, is much greater than on radio. Again, this is not the rule, but on several occasions during taped TV interviews I was pushed to rephrase some statements over and over ("That was nice, but now could you say it again, making clearer that ..."), so as to make it fit with certain editorial intentions of that particular news format. It is probably not accidental that many public intellectuals have started blogging in recent years, as it provides modern mass communication while reclaiming autonomy and sovereignty over one's words. But I have not tried this format yet.

9.4 The Organizational and Social Context

Many other issues deserve discussion. There is, of course, the economic aspect: Appearing on the media can provide an additional income. A few hundred Euros for an extended newspaper article or a radio essay are certainly nice, and sometimes they give an additional incentive when the decision has to be made to write a text under heavy time constraints. But for tenured university professors that is not a major rationale, let alone a strategy towards riches.

Then there is the role of the university administration. Mostly, a scholar appears as an individual vis-à-vis the media, not so much as the member of an organization. For the media, "branding" their content is largely about persons and names. But a name alone would not work, especially when scientific legitimacy is at stake – that is why the name of the institution accompanies the scholar's name: "We talk about this with Prof. Paul Nolte from the Free University in Berlin." Reversely, this is about reputation for an academic institution, and therefore universities are developing an awareness for the public appearance of their professors and, indeed, mechanisms for promoting their media presence. My own university has established an "Expertendienst," an "expert service," which is a data-bank and tool for interested journalists to find experts in topics of their current investigation. Hence, I may receive an email, or a phone call, referring to the "Expertendienst" in which my name is connected to certain issues or areas of specialization such as "history of reunified Germany" or "political parties and democratic development in Germany." Although this can at times be annoying, it is of crucial importance in reassuring me that my media visibility is not a strange hobby outside of what I really should be doing, but is an integral part of academic work and appreciated by my employer.

Not everyone is appreciative, of course. Scholars who tend to be attractive for the media do experience some criticism by colleagues in their own field who happen to concentrate on research, "splitters," as it were, and on strictly academic publishing. The allegation is that media contact corrupts scholarship and damages one's academic integrity. I have come across that attitude quite often, although it is mostly expressed indirectly: Colleagues who are generally sympathetic to me would express their concern by saying: "I find it fascinating, but I have heard others wonder about your media business." In other cases, critique and rejection is a mere disguise for envy, or just for not understanding, for not being able to conceive of such a role for oneself. Nevertheless, I do not disregard those critical-friendly comments, and they remind me that it is important, of course, to retain and build a reputation in other fields of academic activity, be it research and academic publishing, or teaching and advising graduate students, or even committee work and service to the academic community. With this in mind, I hope to continue in the coming years what I see as an essential duty of scholars, and of historians in particular: to stay in touch with a wider public, and to contribute to lively public debates and an engaging democratic culture.

Reference

Lepsius, M. R. (1964). Kritik als Beruf. Zur Soziologie des Intellektuellen. *Kölner Zeitschrift für Soziologie und Sozialpsychologie*, 16, 75–91.

Part V
Organizational Responses to Media Expectations

Chapter 10
Staging High-Visibility Science: Media Orientation in Genome Research

Stephen Hilgartner

10.1 Introduction

Scientists and scientific organizations in certain times and places orient some of their activities toward the media – a phenomenon that Weingart (1998) in his analysis of medialization glosses as science exhibiting a "media orientation."[1] The medialization concept holds that an intensification of science-media coupling in contemporary societies is producing changes in science. One line of evidence for this view is the increasing coverage of science by the media; the other is evidence that science is increasingly oriented toward media. Rödder (2009) describes these as two dimensions of medialization. The main strategy for documenting the media orientation of science has been to look at several indicators of the phenomenon, such as the creation of media events, the practice of "publication by press conference," the existence of scientists who become media stars, and the coupling of scientific, political, and media discourses. The medialization concept was defined and has been applied analytically at the level of systems (Weingart 1998), using differentiation theory to consider coupling and system change. In contrast to a system-level approach, this paper develops a complementary perspective for considering medialization that focuses on media orientation at the level of actors and interaction. How do scientists, scientific organizations, and others associated with high-technology research interact with the media? How do actors manifest an orientation to the media? In what ways, and how intensely, does the media fit into their activities and practices?

To address these questions, this paper develops a framework that conceptualizes media orientation at the level of the actor (e.g., the individual scientist or science-intensive organization), defining media orientation as a specific form of "theatrical

[1] In some communication theory, the term media orientation is used to refer to such things as people's patterns and proclivities in media consumption, a usage that differs from that of Weingart (1998).

S. Hilgartner (✉)
Cornell University
e-mail: shh6@cornell.edu

S. Rödder et al. (eds.), *The Sciences' Media Connection – Public Communication and its Repercussions*, Sociology of the Sciences Yearbook 28,
DOI 10.1007/978-94-007-2085-5_10,

self-consciousness" (Goffman 1959) that attends to the media. This formulation, which brings the tools of dramaturgical analysis to the staging of science (Hilgartner 2000), provides a vocabulary for exploring science-media coupling not as connections abstract systems but as strategic interaction. The analysis is grounded in data gathered in an ethnographic study of genome research conducted during the Human Genome Project (HGP), which was officially launched in 1990 and officially completed in 2003, and focuses on an episode that generated a burst of media coverage. This ethnographic approach provides a look at the specific and varied ways in which a particular research community related to the media.

Genome research is a good field for examining how scientists orient to the media. As a large research enterprise for biology, the HGP won the ongoing attention of the scientific, science policy, biotechnology, venture capital, and bioethics communities. From its inception, genome researchers imagined the HGP as a highly visible undertaking. Indeed, according to one influential account of the launching of the genome project, the first meeting about the possibility of sequencing the human genome was itself born in part in the pursuit of visibility, spurred on by a desire to put the University of California at Santa Cruz "on the map" (Cook-Deegan 1994: 79). Genome researchers also framed the significance of mapping and sequencing the human genome using a variety of metaphors, comparing it to the Manhattan project and the Apollo Program. As these examples suggest, genome researchers approached this enterprise with historical self-consciousness, conceiving of their effort as something that would rank among the greatest of scientific achievements.[2]

The HGP also attracted substantial media coverage. To be sure, the volume of coverage fluctuated (e.g., Schäfer 2009), but the project was discussed in major newspapers and magazines; on television; in specialized media such as the news pages of *Science*, *Nature*, *The Scientist*, and *Genetic Engineering News*; in investor-oriented publications; and beginning in the late 1990s in an increasing variety of Internet-based "new media." The HGP also became something of a "lightning rod" for bioethical issues connected to genetics and human biotechnology, inspiring the funding of research programs on the ethical, legal, and social aspects of genomics and generating ongoing analysis and commentary in both professional and popular publications.

This list of "media" admittedly exhibits some lack of clarity at the margins about what should be included in that category, but rather than regarding that as a deficiency to be rectified through more precise conceptualization, this mixture – or better, *web* of communication genres (Lewenstein 1995) – indexes the visible, Mode 2 (Nowotny et al. 2001) environment that genome researchers inhabited. The HGP was discussed publicly in so many diverse forums that genome scientists developed a sense that their field was on stage. Media became part of their professional

[2] For an ethnographer studying a group of scientists, this historical self-understanding proved helpful in gathering data; the scientists involved often seemed to regard the fact that someone would want to study them as natural.

world in a significant way, especially for leaders of the project, who had to concern themselves with its legitimation.[3] Many of the genome scientists I talked with seemed to regard the media as an active agent that produces potentially consequential representations (and misrepresentations). They often expressed concerns consistent with the "deficit model" of the public understanding of science (Irwin and Wynne 1996), and they also used the notion of distorted popularizations of science to do boundary work (Hilgartner 1990). In addition, they sometimes expressed negative attitudes toward "hype," attributing this discursive overbidding (Weingart 1998) not to the media alone but also to actors (such as individual scientists or companies) that they charged were over-promoting their wares.[4]

Below, this paper examines more specifically how actors involved in genome research manifested an orientation toward media. I begin by elaborating the conceptualization of media orientation as a form of theatrical self-consciousness. Next, the paper illustrates the framework by considering a specific episode: the intense science/media interaction surrounding the announcement in 1998 that a private company, later named Celera Genomics, planned to sequence the human genome before the publicly-funded genome project could complete the work. The empirical material focuses on a 10-day period immediately preceding and following the announcement. Examining this episode offers an opportunity to consider interaction with the media in ethnographic context and to examine how orientation toward the media is connected to concerns about a variety of purposes and audiences. Given the limited scope of this study, conclusions must be tentative. For one thing, to the extent that medialization involves system-level changes that develop over time as science-media coupling increases, a single case study focused on events occurring within 10 short days cannot address change because it lacks a baseline for comparison. In addition, the Human Genome Project – a highly-visible "big biology" project forged amid a historically unique constellation of scientific, political, and commercial circumstances – was unusual in many respects, raising questions about whether the findings are applicable to other areas of science. Even so, the analysis suggests that examining how a community of researchers interacts with the media makes it possible to analytically differentiate four facets of media orientation, posing new questions for research. In addition, the analysis may raise methodological questions for macro-level studies of medialization.

[3] Peters et al. (2008) found that a willingness to serve as a media contact has become an institutionalized expectation for scientists in leadership roles.

[4] Fortun (2001, 2008) argues that hype cannot be eradicated from genomics because promisory rhetoric and promises are constitutive elements of genomics. Nelkin (1987) argues that much media coverage is promotional in character and that scientists and scientific institutions encourage and generate such coverage.

10.1.1 Theatrical Self-Consciousness and Media Orientation

To develop a framework for examining media orientation in interaction, this paper builds on the dramaturgical sociology of Erving Goffman (1959, 1963) and its concept of theatrical self-consciousness. In his analysis of interactions among individuals, Goffman examines how participants, who possess the capacity for mutual monitoring and self-reflection, experience themselves as performers who aim to present themselves strategically to others in ways that foster desirable impressions. Through "information control" – the selective revelation and concealment of information – actors highlight some information and present it on the front stage, while relegating other information to the backstage, invisible to the audience. Goffman stresses that strategic impression management does not imply dishonesty (although dishonest performers certainly exist); on the contrary, the honest performer must also artfully manage impressions to avoid unintentional misunderstandings. The performer, thus, must attend to the audience, imagine its reactions, monitor his or her self-presentation, and make adjustments as needed. Audiences, for their part, are far from being passive recipients of information; they often monitor carefully the acts presented to them to avoid being taken in, attending both to the impressions that performers deliberately advance and also those that they unintentionally communicate. In face-to-face interaction, playing the role of the audience is itself a kind of performance. For example, audiences must at times tactfully overlook flaws in a performance, feigning an appropriate reaction to it. Goffman's analysis applies at various levels of social organization. Most of the time, his focus is on individuals, but the analysis also applies to organizations, which clearly must collectively manage impressions.

Goffman's dramaturgical perspective was designed to examine face-to-face interaction and it has mainly been used in that context. Since so much scientific communication uses written texts or visual displays, conducting a dramaturgical analysis of science entails adapting the perspective to apply to written documents and other texts (Hilgartner 2000). This involves focusing on the existence of audiences who are not present in the same physical space as the performer but nonetheless are considered during the creation and delivery of the performance. When preparing a science advisory report, for example, a committee of experts "interacts" – albeit in a virtual manner – with its imagined future audiences, attempting to anticipate and address questions or objections in advance (ibid.; see also Bijker et al. 2009). The stage management of a written document often involves a complex dialectic of revelation and concealment: performers present some information directly and visibly in the document; they also keep some information "backstage" by omitting it from the text; and they erect temporal, social, and spatial boundaries that control access to the document. Through such techniques, performers use stage management to constitute audiences (and nonaudiences) with specific capacities and incapacities of perception and speech (Hilgartner 2000).

In bureaucracies, the information control techniques surrounding written documents are often quite elaborate (Smith 1974; Hilgartner 2000; Riles 2006), with theatrical self-consciousness embodied not only at the individual level but also in procedures and material practices that instantiate it as a kind of distributed cognition. Of course, just as the audience of a performance presented in a face-to-face encounter need not simply accept the show uncritically, so too can the audience of a written document interrogate the text (and what is omitted from it) for clues about its subject matter or about the character of its creators. This is not to say that face-to-face and written communication are identical; on the contrary, the interaction between performer and audience takes diverse shapes depending on countless details about the specific communications tools employed. A press conference offers its audience opportunities for participation that little resemble those provided by a press release, for example. Nevertheless, the central concepts of dramaturgy – such as theatrical self-consciousness – are useful for examining performances packaged in many means of communication and for examining a variety of actors, including individuals, organizations, and "teams" engaged in dramaturgical cooperation.[5]

As the above discussion suggests, theatrical self-consciousness is a phenomenon experienced by performers and audiences alike, although in somewhat different ways. The differences, and even the distinction between the roles, vary with the mode of communication. In a two-person face-to-face interaction, each participant may simultaneously occupy the roles of both performer and audience, whereas the readers of a written text may experience themselves primarily in an audience role (unless they are expected to comment on it). Members of media audiences also bring an understanding of the theatricality of social life to their encounters with media content, which they recognize to have a staged quality. When the media are involved, theatrical self-consciousness is complex; for the media can be fruitfully understood in several ways: as audience, because journalists are a primary audience for certain communication genres, such as press releases; as performer, because journalists actively create news accounts and present them to audiences; and as stage, because the media offer performers a platform for communicating with audiences, such as newspaper readers or television viewers.

These observations provide a starting point for analyzing in dramaturgical terms how scientists and other actors interact with the media. At the level of the actor, media orientation can be conceptualized as a species of theatrical self-consciousness that attends to the significant, and arguably increasing, role of the media in contemporary societies. It is not surprising that many scientists and scientific institutions recognize that the media offer dangers and opportunities. Beyond simply believing that media coverage matters, many scientists and science-intensive organizations, including the genome scientists discussed in this paper, sometimes engage with the media in direct ways, approaching media organizations and media content with a sense of the theatricality of situations that involve the press.

To provide an empirical look at genome scientists interacting with the media, we now turn to the dramatic announcement on May 10, 1998 that a private company was

---

[5] Goffman (1959: 77–105) analyzes performances by "teams" – groups of people who work together to sustain a definition of the situation. The membership of a team will often cross the boundaries of formal organizations.

being founded to sequence the human genome in three years, sooner and at lower cost than was then planned by the publicly funded HGP. After briefly introducing the context in which this announcement was made, I turn to the media-oriented activities of genome researchers and others engaged in human genome research in the period from May 8 to May 17, immediately before and immediately after the formation of the New Company was announced. My account is based on a variety of sources, including press releases; selected news coverage; participant observation at the Cold Spring Harbor Meeting on Genome Mapping, Sequencing, & Biology, the main annual scientific meeting of the HGP community where roughly 400 scientists gathered from May 13 to 17; three audiorecorded interviews and many informal conversations with genome scientists conducted during the week following the announcement; and books later published by some of the most visible scientists (Sulston and Ferry 2002; Collins 2006; Venter 2007) and by science journalists (Davis 2001; Shreeve 2004).

Because events took place in geographically distributed locations and occurred simultaneously, and because many of them took place in backstage spaces that left no publicly available records, my account makes no pretensions of completeness. However, this ethnographic approach provides a look at a research community in the process of learning about and trying to understand the significance of an unexpected and potentially major development in its field. Media were deeply entangled in the events, so this case study offers an opportunity for considering how media and science interact in practice and provides an occasion for illustrating various aspects of media orientation in action.

10.2 The Announcement and Mediated Reception of a Commercial Human Genome Project

Let us begin with the context in which the announcement took place. In the United States, the HGP was officially launched in 1990 under the leadership of the National Institutes of Health (NIH) and the Department of Energy.[6] The goals of the project – to map and sequence the human genome and the genomes of several model organisms, to develop mapping and sequencing technology, and to anticipate and address the ethical, legal, and social issues associated with genomic knowledge – were to be accomplished by 2005 at a cost of $3 billion. The plan was to house the sequence data in public databases, such as Genbank, which would make this information freely available to any and all (Hilgartner 1995). During the early years of the project, work focused on creating basic maps of the human genome, developing technology, and conducting some small pilot sequencing projects, with the expectation that large-scale sequencing would take place only after significant

[6] The word official here signals the staged nature of the project, whose "start date" was contrived for presentational purposes with bureaucratic, Congressional, and media audiences in mind.

improvements in technology had occurred.[7] By the mid 1990s, some genome project leaders believed that the technology had matured to the point that large-scale sequencing of the human genome should commence (e.g., Olson 1995). DNA sequencing machines, especially those manufactured by the Applied Biosystems division of Perkin-Elmer (PE), had proven reliable, and success in completing genetic maps, demonstrating physical mapping technology, creating effective laboratory informatics systems, and finishing a number of pilot sequencing projects, made obtaining the complete human sequence seem achievable. The NIH-funded genome centers began moving toward scaling up sequencing, as did the Sanger Centre in the United Kingdom, with financial backing of the largest funder of biomedical research in Britain, the Wellcome Trust.

Orchestrating this multi-laboratory and international collaboration required establishing policy agreements about such matters as data quality and data access. The details of data access policies – which specify what types of data should be provided to whom, when, and under what terms and conditions – had long been a topic of debate in genome research (Hilgartner 1998). A variety of kinds of "public release" of data are possible, depending on the precise details of policies. The international collaborators developed standards for data release at a series of meetings held in Bermuda. The so-called Bermuda principles required *daily* submission of all newly-generated sequence data to publicly accessible biological databases. This unprecedented policy was intended to prevent the hoarding of data for commercial purposes or to gain a competitive edge in biological science. The U.S. and U.K. programs also committed themselves to producing a "complete" sequence of the human genome, not simply "skimming" the genome for genes or other entities of particular interest. The task of producing "finished" sequence, high quality data with no gaps in coverage, was considered difficult but worthwhile given the importance of the data.

During the 1990s, commercial versions of genome research also became prominent. No private firm prior to 1998 had launched an effort to sequence the entire human genome, although several tentative schemes along those lines had been explored and failed to materialize. However, several companies, notably Incyte and Human Genome Sciences, had developed businesses based on sequencing lots of tiny fragments of the genome – believed to be parts of genes. J. Craig Venter was the first scientist to undertake a project to sequence large numbers of these fragments – which he called Expressed Sequence Tags (ESTs). Venter formed an alliance for a time with Human Genome Sciences, which provided funding for a non-profit organization that Venter founded, The Institute for Genome Research (TIGR). Patents filed on ESTs inspired international controversy and opposition from many HGP scientists and the international Human Genome Organization. A few years later, Venter led the group that completed the sequence of the first free-standing (non-viral) organism, using what became known as the "whole genome

[7] See Gaudillière and Rheinberger (2004), esp. the papers by Bostanci (2004), Hilgartner (2004), and Kaufman (2004).

shotgun" technique. By the mid 1990s, he had also become something of a visible scientist, at least for aficionados of genome research. After the formation of Celera, his public profile grew (Rödder 2009) and it has continued to do so with his involvement in synthetic genomics.

The May 1998 announcement that a new company was being founded to sequence the human genome altered the course of the HGP. From the start, the news was taken very seriously – by the genome research community, the media, and by investors – in part owing to the fact that the team who advanced the plan consisted of Venter and Michael Hunkapiller, head of Perkin-Elmer's Applied Biosystems group, the world's leading manufacturer of DNA sequencing machines. Venter enjoyed much credibility as a genome scientist, as well as some notoriety for his association with the EST patent applications. Many HGP researchers worried that the company would try to capture large parts of the human genome as intellectual property. Intense competition developed between the new company, later named Celera Genomics, and the public HGP, despite repeated attempts to stage détentes. The most dramatic of these occurred in June 2000 at a White House press conference, where U.S. President Bill Clinton and (by satellite) U.K. Prime Minister Tony Blair helped celebrate the completion of a "draft" of the human sequence, a feat that was presented as the joint achievement of the public HGP and Celera.[8]

In late 1997, Hunkapiller began making overtures to Venter to see if he was interested in discussing the prospects for sequencing the human genome with a new automated sequencer that the company was developing. In early 1998, Venter went to California to see the instrument and began talks. PE indicated that it was willing to invest $300 million in an effort to sequence the genome if Venter believed it would work and would lead the project (Venter 2007). Venter agreed and the parties developed a plan for PE to form a new company, with Venter as CEO. Because the company did not yet have a name (the name was announced in a July 1998 news release), people availed themselves of such descriptors as "the New Company" or "the Venter/Hunkapiller proposal." In keeping with the situation during the 10-day period focused on here, I will avoid the name Celera, generally referring to the firm and its principals with the generic name the New Company or as the Venter/Hunkapiller proposal.

10.2.1 The Announcement

Until May 10, 1998 when the story appeared on the front page of the Sunday *New York Times* and in a press release issued by the New Company, there was no media coverage of the plan. However, once the decision to form the company was made, the question of how to announce the plan – to whom, when, and how – became salient. According to Venter's autobiography, there was "considerable

[8] See Rödder (2009) for a discussion of news coverage of this event in the German and British press.

debate" about how to unveil the news. "Some wanted to do it by press release," he wrote, "but I preferred to touch base with key leaders in the genome community first to see if we could have real cooperation in one effort" (Venter 2007: 237). Initially, the New Company told a small number of selected people; for example, a handful of well-known scientists were invited to join the company's advisory board. Shortly before the announcement, a few HGP scientists learned that Venter was planning to launch a commercial sequencing project. These included such genome project leaders as James Watson, John Sulston of the Sanger Centre, the NIH's Francis Collins of the National Human Genome Research Institute (NHGRI), and Michael Morgan of the Wellcome Trust. On Friday, May 8, Venter and Hunkapiller personally met with Harold Varmus, the director of the NIH, and Francis Collins, head of the NHGRI; it was then that NIH officials first learned any details of the plan or that PerkinElmer was involved (Sulston 2002: 149–152; Venter 2007: 238–243). Two days later, the news hit the press. In the week following the announcement, articles appeared in leading newspapers, including the *New York Times*, the *Washington Post*, and the *Independent* (London) as well as such magazines as *The Economist*. In light of the startling news and the incomplete and sometimes confusing media coverage, HGP researchers experienced some of the extreme uncertainty described by Lewenstein (1995) in his analysis of the cold fusion case.

The New Company carefully orchestrated the announcement in a manner designed to produce well-placed news coverage. Venter (2007: 241) describes how "we decided to offer the story on a plate to a reporter who we thought could get the space to do it justice," namely, Nicholas Wade of the *New York Times*. The plan was for the New Company to "issue a press release before the markets opened on Monday, May 11, 1998" while Wade's story would appear on the front page of the Sunday *New York Times* on May 10 (ibid.). The press release, dated May 9, was loaded into the *Business Wire* press release distribution service on Sunday, May 10. It announced the plan for the new genome sequencing company and also told reporters and investors that they could phone into a conference call on Monday, May 11 at 10:00 am. Under the headline "Perkin-Elmer, Dr. J. Craig Venter, and TIGR Announce Formation of New Genomics Company; Plan to Sequence Human Genome Within Three Years" the press release began:

> The Perkin-Elmer Corporation (NYSE:PKN), Dr. J. Craig Venter, and The Institute for Genomic Research (TIGR) announced today that they have signed letters of intent relating to the formation by Perkin-Elmer and Dr. Venter of a new genomics company. Its strategy will be centered on a plan to substantially complete the sequencing of the human genome in three years (Perkin-Elmer and TIGR 1998).

It went on to suggest that the company's business plan would be based on providing genome data.

> The new company's goal is to become the definitive source of genomic and associated medical information that will be used by scientists to develop a better understanding of the biological processes in humans and deliver improved healthcare in the future. Using breakthrough DNA analysis technology being developed by Perkin-Elmer's Applied Biosystems Division, applied to sequencing strategies pioneered by Dr. Venter and others at TIGR, the

> company will operate a genomics sequencing facility with an expected capacity greater than that of the current combined world output.
>
> Concurrently, the new company also intends to build the scientific expertise and informatics tools necessary to extract valuable biological knowledge from genomic data, including the discovery of new genes, development of polymorphism assay systems, and databases for the scientific community. Perkin-Elmer and Dr. Venter believe that this information has significant commercial value and that the new company can provide this information more rapidly and more accurately than is currently possible (ibid.).

The press release presented a number of narrative elements that could easily be repackaged into news stories with themes of progress, market value, and private sector efficiency versus government inefficiency. A second press release by PE also announced its new sequencing machine (Perkin-Elmer 1998). As is typical of the genre, neither press release contained much technical information; for example, they said nothing about the proposed method (known as a whole genome shotgun) or the ongoing internal discussions of possible technological strategies for making that method more robust. Nor did the announcement provide much detail on the business plan through which the company planned to make money. Importantly, however, the press release announcing the New Company said that it "plans to make sequencing data publicly available to ensure that as many researchers as possible are examining it and that applications, such as the development of diagnostic tests and new drug discovery, are as broad as possible" – a formulation that made the plan seem, at least potentially, to be consistent with the HGP goal of making sequence data public – although the terms and conditions of access were not discussed (Perkin-Elmer and TIGR 1998).

In staging the announcement, the New Company managed to keep its developing plans backstage, preventing all but a few people from learning what was happening until it was ready to present a neatly packaged account for front-page placement in a leading U.S. newspaper with a notable role in agenda setting for the U.S. media. The company clearly treated the media as an important audience, but its attention to the media did not entail viewing the media as the only or the ultimate audience. The press release was also aimed at investors, and it conformed as a text to requirements relevant to both audiences. Thus, the press release provided a differentiated list of contacts for people who sought more information:

> Contact: Perkin-Elmer (Investors)
> Charles Poole, 203/761-5400
> or
> Noonan/Russo Communications (Media)
> Timothy Sipols, 212/696-4455 Ext. 272
> or
> Hill & Knowlton (For TIGR and Dr. Venter)
> Jim Jennings, 202/944-5102 (ibid.)

When speaking to investors, Perkin-Elmer – a company with shares traded on the New York Stock Exchange – was legally obligated to disclose major developments pertaining to its operations and to ensure that investors are informed of risks. Accordingly, the press release ended with a lengthy standard disclosure paragraph, indicating that "certain statements in this press release and its attachments are forward-looking" and stating that in order to comply with the "safe harbor" provisions of The Private Securities Litigation Reform Act, PE "notes that a variety of factors could cause actual results and experience to differ materially from the anticipated results or other expectations expressed in such forward-looking statements" (ibid.). In other words, the document not only complied with the conventions of the journalistic genre of the press release, but also with a set of legal requirements from securities law. At the same time, the disclosure paragraph underlined in a conventional way (much commented on in recent STS work on genomics) the promissory and speculative nature of such ventures (see Fortun 2001, 2008; Rabinow and Dan-Cohen 2005; Sunder Rajan 2006; see also Hedgecoe 2004).

Other salient audiences for the New Company included the biological research community and the scientists associated with the HGP. Indeed, the company took into account a variety of audiences as it planned and performed its presentations to the press. For example, according to a scientist with knowledge of company decision making, the New Company's reached a decision about precisely when to unveil its plan to the media for "lots of reasons": several investment firm meetings were coming up; the Cold Spring Harbor meeting offered a chance to explain the plan to the genome community; rumors were starting to circulate; and the company needed to announce its new sequencing machine because word of the new machine was getting out and starting to hurt sales. Orientation to the media, thus, was intertwined with orientation to a variety of other audiences, including investors, scientists, and customers.

As planned, the *New York Times* published Nicholas Wade's exclusive article on page 1 on the same day that the press release was distributed. The article, headlined "Scientist's Plan: Map All DNA Within 3 Years," made the challenge to the publicly funded HGP a main theme.

> A pioneer in genetic sequencing and a private company are joining forces with the aim of deciphering the entire DNA, or genome, of humans within three years, far faster and cheaper than the Federal Government is planning. If successful, the venture would outstrip and to some extent make redundant the Government's $3 billion program to sequence the human genome by 2005 (Wade 1998a).

The article said that Hunkapiller believed that the new sequencing machines were "so fast" that they could sequence the human genome "far sooner and 10 times more cheaply than envisioned by the National Institutes of Health." Wade also presented a picture of the HGP leadership's immediate response:

> The director of the Federal human genome project at the National Institutes of Health, Dr. Francis Collins, first heard of the new company's plan on Friday, as did the director of the N.I.H., Dr Harold Varmus. Both said that the plan, if successful, would enable them to reach a desired goal sooner. Dr. Collins said he planned to integrate his program with the new company's initiative. The Government would adjust by focusing on the many projects that are needed to interpret the human DNA sequence, such as sequencing the genomes of mice and other animals (ibid.).

The article also seemed to imply that the HGP leadership planned to move on to new projects:

> Both Dr. Varmus and Dr. Collins expressed confidence that they could persuade Congress to accept the need for this change in focus, noting that the sequencing of mouse and other genomes has always been included as a necessary part of the human genome project (ibid.).

10.2.2 The NHGRI's Response

The New Company's plan clearly put the leadership of the HGP in delicate position. The first article in the *New York Times* raised the question of whether a publicly funded HGP was still needed. Yet the details of the plan remained sketchy, the technical feasibility remained open to question, and no one outside of PE had worked with the new sequencing machine. In light of doubts about both the technical issues and the availability of data, the HGP leadership did not want to allow the New Company to take over the sequencing. The HGP had already discussed the possibility of using a whole genome shotgun method, and it had concluded that this technique was risky and unlikely to yield a complete sequence of the quality that it sought to produce. Yet simply dismissing the plan, which was backed by some serious scientists and considerable financial and managerial resources, might make the HGP appear rigid and defensive. Genome project leaders worried that in the United States, with its enthusiasm for privatization and small government, people might perceive the HGP as an inefficient government program. Not surprisingly given this context, the NHGRI chose to respond by "welcoming" the New Company's plan as a positive development that might speed the sequencing effort while simultaneously raising questions about whether the proposed method would work and, even if successful, whether the company's product would meet the HGP's goals for data quality and access.

In analyzing the situation, genome scientists imagined how salient observers would perceive things given their lack of sophistication and exposure only to a mass-mediated version of events. As one scientist put it in a May 15 interview:

> I think even my mother would look at this and say, 'oh, wait, somebody's saying they're going to do the genome.' I'm not sure – My mother may not be savvy enough to know the difference, and the Congressmen may not be either, [between] what's an intermediate product versus the final product. 'I think it sounds like a good shortcut.' 'We don't need to be spending all this extra money.' And so, you just have to be concerned about that, and we need to keep enthusiasm constantly up for continued good, healthy funding of the genome project. And if somebody can run out and get a shortcut, it may not dawn on them until several years later that, oh my God, that wasn't really quite what we wanted.[9]

---

[9] Audiorecorded interview. Interviewees were told their names would not be revealed in publications.

For the NHGRI, the U.S. Congress, which at the time was controlled by Republicans with a strong anti-government ideology, was an especially salient audience. On Monday, May 11, the NIH held a press conference at 3:00 pm in which Varmus, Collins, Ari Patrinos of the DOE, Venter, and Hunkapiller all participated. The director of the Sanger Centre, John Sulston, later called this joint press conference "the first of what was to become a series of bizarrely staged shows of unity" (Sulston 2002: 152). I was not in Washington and did not attend the press conference, but post hoc accounts from people who did indicated that the message the NIH tried to convey was that the New Company's initiative was something that "should be embraced" if it plays out as portrayed. *If* was the operative word; so the message was that its initiative could be a helpful supplement to the HGP *if* the method in fact produces high-quality data and *if* the data are in fact made truly publicly available. The main point was that the public program should take no hasty steps given that the instrument was new and the whole genome shotgun method remained untested. Whether this message would be effectively delivered via the media was a source of worry because superficially the plan looked like a sensible cost-saver.

Media coverage generated by the NIH press conference presented a somewhat confusing picture. Immediately after the press conference, an Associated Press wire story, headlined "Gene Technology Must Be Proven Before Government Will Use It," led with the idea that the new technological strategy was unproven:

> A new technology that promises a faster and cheaper way of piecing together the human genetic pattern will have to prove itself before a National Institutes of Health agency will adopt it, federal health officials said Monday. Dr. Francis Collins, head of the National Human Genome Institute, said it was premature to change the plan now being followed by government-supported laboratories to sequence the human genome, which is the pattern of genes in each human cell (Recer 1998).

In contrast, the *New York Times* published an article the next day that used the past tense to suggest that the "takeover of the human genome project" was a fait accompli:

> The sequencing of the human genome, a historic goal in biomedical research, was snatched away last Friday from its Government sponsor, the National Institutes of Health, by a private venture that says it can get the job done faster. Now Government officials are scrambling to adjust to the stunning turn of events, saying that the task of interpreting the genome may begin much sooner now, and that there is every reason for Congress to continue to fund the project (Wade 1998b).

It also presented criticism of the management of the HGP:

> There have been serious problems of organization and management both at the Department of Energy and at N.I.H., together with internal dissension among the senior scientists involved, said Dr. William A. Haseltine, chief of Human Genome Sciences, a genome sequencing company in Rockville, Md (ibid.).

And reported that NIH officials

> are preparing to persuade Congress to continue funding the genome project but to switch the focus from getting the sequence to the enormous tasking of interpreting it. Dr. Venter plans to enter his findings in a public database. One essential aid to understanding the human

> genome is to sequence the surprisingly similar genome of the mouse. Though all biologists recognise the need for such a project, it may not be immediately clear to members of Congress that, having forfeited the grand prize of human genome sequence, they should now be equally happy with the glory of paying for similar research on mice (ibid.).

A contradictory message appeared that same day in the *Washington Post*, which reported that NIH officials expected no immediate change in course:

> Federal officials said the accelerating government effort to find and decode all 60,000 or more genes in the human body would remain on its current course for the next 12 to 18 months, by which time it will be clearer whether the project should change its approach to accommodate the new players in the field. 'It would be vastly premature to go out and [...] change the plan of our genome centers,' said Francis Collins [...] (Gillis and Weiss 1998).

On May 12, the day before the annual international meeting of genome scientists was to begin, the heads of the largest American genome laboratories gathered in Cold Spring Harbor. Venter and Hunkapiller attended part of this pre-meeting gathering and presented their plan to the genome center directors. By all accounts the proposal was not well received, and Venter and Hunkapiller left Cold Spring Harbor (Sulston 2002: 153–158; Venter 2007: 243–245).

10.2.3 The Wellcome Trust's Response

The news of the New Company's plan provoked intense discussions among HGP scientists about how to respond. Some of the most consequential discussions took place among genome scientists in the U.K. and the Wellcome Trust, funder of what was then the largest genome laboratory, the Sanger Centre. In his account of events, Sulston describes how he and other genome scientists worried that the American program was in jeopardy. What was needed, he believed, was a powerful statement that could counter the mounting perception that the public HGP was clumsy, inefficient, and no longer needed.

> Trying to get reporters to print the admittedly more complex analyses that we felt were being ignored was going to be an uphill struggle. We were learning fast that we would have to play the public relations game if we were to survive. But that didn't mean indulging in empty hype. What we needed was a big vote of confidence in the public project as a counter to Craig's [Venter's] hints that it was an expensive white elephant (Sulston and Ferry 2002: 162).

On Wednesday, May 13, Sulston, along with Michael Morgan of the Wellcome Trust, made a presentation to a previously scheduled meeting of the Trust's Board of Governors that persuaded them to make such a statement in the form of a serious financial commitment (ibid.). That same day, the Trust issued a press release saying that it would increase its funding of human genome sequencing by "£110 million over seven years, bringing the total Trust investment in the Human Genome Project to £205 million" (Wellcome Trust 1998). The press release also said that

> The Trust is concerned that commercial entities might file opportunistic patents on DNA sequence. The Trust is conducting an urgent review of the credibility and scope of patents based solely on DNA sequence. It is prepared to challenge such patents (ibid.).

Over the next few days, a new frame focusing on two competing groups took hold. The *New York Times* published an article called "International Gene Project Gets Lift." After saying that the $3 billion project had been "upstaged" by a private company that was aiming to complete the human genome sequence at a fraction of the cost, the article continued:

> Now the Wellcome Trust of London, the world's largest medical philanthropy, has stepped into the fray in an effort to maintain the impetus of the publicly financed program and to prevent the human genome sequence from falling under the control of a private company [...] 'To leave this to a private company, which has to make money, seems to me completely and utterly stupid,' said Dr. Michael J. Morgan, program director for the Wellcome Trust. Asked if the Trust was prepared to finance the sequencing of the entire human genome, Dr. Morgan said, 'If we had to and if we wanted to, we could do it.' The Wellcome Trust, he noted, has assets of $19 billion (Wade 1998c).

Some of the media accounts used war metaphors; for example, *The Gazette* (Montreal) published an article saying Venter had "secured $300 million of private funding to launch his assault," saying his "chief weapon" was speed, and describing the Wellcome Trust funding as "retaliation" (Irwin 1998).

The Wellcome Trust press release was not only distributed to the media; copies of it were also provided with the registration materials to the attendees of the Cold Spring Harbor meeting, where it helped boost sagging morale. During the meeting, the New Company's plans were discussed extensively, both in side meetings and informal settings and in a special session hastily squeezed into the meeting agenda.

10.2.4 Extracting Information, Criticizing Coverage

For most genome scientists, the media (or press releases) was the only source of information about the New Company's plans during the first days after the announcement. Only a few HGP scientists learned anything about the New Company before the May 10 *New York Times* article, although a few leaders picked up bits of news prior to the announcement. A consistent theme in discussions among genome scientists, however, was how sketchy the news coverage was – about technical details, about the company's business model, even about the NHGRI's response. One scientist who was in Washington at the time of the announcement explained in an interview that shortly before the news broke he learned "something was cooking" from people in NHGRI.

> I started to get pieces of it but with incredibly vague scientific details. [...] I did know that they all met on Friday, but I still didn't get a whole lot. And the press was already calling me actually on that Monday. And they were asking me questions and I didn't know if it was

> a BAC-by-BAC shotgun, I didn't know if it was a whole genome shotgun.[10] I didn't really – I didn't know anything about these instruments. So I really learned about a lot of the details pretty much by, you know, the press advisory that took place with NIH on Monday at 3:00.

Another scientist described how he was so busy that he initially missed the news, even though someone had tried to call it to his attention.

> I was in [name of city] trying to get some grant out, so I didn't see it. Although someone gave me, Monday, a webpage announcement and got it in my hands, and I looked at it and I just thought this is a press release from a company. There's a lot of these things. Someone said do you know about this? I looked at it and I just thought press release so I didn't take it seriously. It's only when I heard about it on Tuesday. Someone told me about it and I picked [the *New York Times*] up at the airport [on the way to Cold Spring Harbor].

He also said he had been baffled by the *New York Times* accounts, and described trying to fill in the blanks.

> I guessed they meant whole [genome] shotgun. I guessed that they meant looking for lots of polymorphisms quickly. I guessed that they weren't serious about closing the genome [that is, completing it without gaps]. What shocked me was that NHGRI, from reading the newspaper, I got the impression that the leaders had said we're going to move to mouse and we're going to move to other things, and I said how can you retreat? There's so much information that you don't know here that we shouldn't be changing our plan. I met [genome project leader] Tuesday night when I arrived here the day before for a pre-meeting. I confirmed the shotgun and learned NHGRI hadn't said these things about retreating.

By combining his knowledge of genome science with the fragments of information in the news, he formed some guesses about both the sequencing method the company was planning to use and about the business plan, which he imagined to be based on quickly skimming the genome for commercially valuable bits (e.g., polymorphisms). He went on to explain that as he learned more, he concluded that the response of the HGP leadership had not been misguided. He had assumed from the way the news stories were written that the leaders were retreating, but he was very pleased to realize that they had been merely taking a few days to get more information.

As these examples suggest, genome scientists attempted to extract information relevant to technical questions from the media, as in the cold fusion episode (Lewenstein 1995), combining the sparse details provided with their own knowledge and analysis to make educated guesses. In addition to narrowly technical information, the missing details that they sought included questions about the company's business model and precise plans for making data public. Among the genome researchers gathered in Cold Spring Harbor there was much speculation about how the company planned to make money; for example, by patenting genes, selling access to sequence data, building databases of polymorphisms for use in research and diagnostics. While speculating, some genome scientists exhibited skepticism about certain claims presented in the news coverage, which they clearly approached

[10] BAC stands for bacterial artificial chromosome. A BAC-by-BAC shotgun method would be much closer to the technical strategy then taken by the HGP than the whole genome shotgun that Venter had planned.

as a performance that in principle could be misleading. For example, an Associated Press story reported on May 11 that Venter said the company "would release its results every three months and would not patent any genes discovered by his group. However, the new partnership will extract promising genetic sequences from the mapping effort and use that information to develop gene products" (Recer 1998). Many scientists at Cold Spring Harbor regarded these reassuring words as unconvincing; for one thing, plans could easily be changed. Similarly, the New Company's assertion that it would make the data "publicly available" seemed to many to omit information on the crucial issue; namely, the terms governing availability.

Genome scientists pursued a number of strategies to try to acquire additional information. One strategy was to piggyback on the New Company's efforts to reach the media, such as the conference call that it held on Monday, May 11 "to discuss this press release" (Perkin-Elmer and TIGR 1998). One participant in the call told me that when the teleconference began, the company read a U.S. Security and Exchange Commission disclaimer. The teleconference was audiorecorded for legal reasons. She said that the investment banking community was calling up and asking questions like "How come you are introducing a new instrument when you already lead in this instrument area. Aren't you going to cut into your own sales?" The investment people were often friendly and congratulated people or said something nice, she said, whereas the reporters just called in and coldly asked their questions. During the call, she had said something like "congratulations" to Craig Venter because, she said, I know him. Later, at Cold Spring Harbor, someone gave her shit, as she put it, for having congratulated him. That person was someone who was listening in, not someone who had asked questions. She could not tell how many people were doing that, but she concluded that a number of people associated with genome mapping and sequencing community had listened in.

Beyond attempting to extract information from the media coverage, many of the scientists at Cold Spring Harbor also commented on its quality. During the meeting, many HGP scientists complained about the media coverage in conversations and informal settings. Some genome researchers were particularly incensed by news coverage reporting that the HGP was half way through its projected 15 years but only 3 percent of the genome had been sequenced (e.g., Wade 1998a). The following excerpt from an audiorecorded interview is worth quoting at length:

Scientist: I mean, a lot of this is a heavy amount of propaganda coming and publicity coming from them. And on top of it, it's a very harsh treatment by the press, and particularly the *New York Times*, of this whole thing. So it's not only that you have this scientific agenda that could be perceived as undermining the public [project], but immediately, instead of, you know, instead of having time to scientifically evaluate and discuss this, immediately there's been some major darts thrown at the organization in the execution of the public genome project. [...] 'How come we're half way through the genome project and we only have 3% of the genome sequence?' I mean, this is a really stupid argument. [...] And it is just absolutely ridiculous. It was a planned program from the beginning.

[. . .] And it's measured in milestones and we [. . .] didn't even think we'd be this far along in sequencing now. And forget the fact that we've built physical maps. Forget the fact that we've built genetic maps. Forget the fact we've built infrastructure for sequencing and just focus on the 3%. This is just really naïve and it's just a stupid way to look at it. So it's a lot of publicity and it's that type of publicity. [. . .]

SH: Have you seen, coming out of the New Company or the group, any kind of documents? Any kinds of papers?

Scientist: [disdainful] Press releases. Only thing I've seen.

In this comment on the news coverage, this scientist argues that the media accounts are misleading and aims to set the record straight. His statement about "propaganda coming [. . .] from them" seems to implicate people associated with the New Company. He also hints at a strong distaste for press releases as a genre of scientific communication and stresses the importance of a proper scientific evaluation of the New Company's plan.

10.2.5 Unavoidable Visibility

During the Cold Spring Harbor meeting there was a great deal of informal discussion of the Venter/Hunkapiller proposal. The events also occasioned a departure from the announced schedule of scientific talks – a move that was rather unusual, though not unprecedented, at the Cold Spring Harbor genome meetings. On the morning of May 15, the chair of the session titled "Large-Scale Sequencing Symposium" announced that owing to the excitement of the last week, the allotted time would be divided into two parts: the first would follow the pre-organized session, and the second, would address the Venter/Hunkapiller proposal and the Wellcome Trust/Sanger Centre plans. No official representative of the New Company venture was present, so the meeting organizers asked a well-known sequencer and leader in the HGP, who I will call Matthews, to summarize what they then knew about the proposal, filling in some of the gaps in the news reports with information from the meeting that Venter and Hunkapiller had held with genome center directors on May 12.

For the leadership of the HGP, this discussion offered an opportunity to build morale and challenge the idea that the project had been snatched away from them. The gathering of so many HGP scientists offered an opportunity to begin to build a collective response based in shared framings of the situation. But there were also issues that they sought to leave backstage. Even before the Cold Spring Harbor meeting had officially started on May 13, genome project leaders had begun vigorous closed-door discussions about how to respond – both in public statements and in their scientific strategy. The issues were complex because an effective change of course would have to maintain the HGP's commitment to its original goals while speeding up sequencing and addressing multiple challenges, including the possibility of a political campaign to cut funding of a "wasteful" public program and the potential threat of large-scale patenting of features of the genome that the shotgun sequencing might reveal. Final decisions were far from being made, however, and the leadership was not ready to announce a change in course. Anything "officially said" during the discussion had a decent chance of appearing in the press. The registrants for the meeting included several science journalists, as well as a handful of employees from PE Applied Biosystems or TIGR. In this context, the scientists leading the HGP – and especially the representatives of funding agencies such as NIH – could not avoid performing in an official role, for their words would surely be interpreted that way. Unlike a press conference, this session did not address journalists directly; in effect, however, the visibility of events and the nature of the audience made it into something different from an ordinary discussion at the typical scientific meeting.

Matthews prefaced his remarks by framing his role in the session as "unusual," noting that his charge placed him in a "somewhat awkward" situation that offered "some temptations," but nonetheless, he would try to be "dispassionate" and present the proposal fairly. With the audience thus notified that he had adopted a self-consciously neutral persona, distinct from his everyday self, he proceeded in a just-the-facts mode, describing some technical details about the whole genome shotgun approach and the new PE instrument. Some of the information in Matthews' description had previously appeared in news accounts and press releases announcing the New Company, but other information had not. The PE press release on the new instrument, for example, merely referred to it as "ultra high throughput," whereas Matthews had sufficient quantitative details to shed light on its theoretical throughput (96 capillaries/run $\times$ 500 bp/capillary $\times$ 3 runs/day $=$ 144,000 bp/day) – very fast by 1998 standards. Such information, while outlining some features of the sequencing plans, the capabilities of the instrument, and the vision for the company, raised many additional questions. For experts in the field, knowing a few details (e.g., that the strategy for assembling the sequence was designed to proceed incrementally and make use of BAC ends and ESTs to help with anchoring and closure) was informative, but raised additional questions about precisely what the sequencing strategy was and how well it would work. The question of how closely the instrument – which was still under development – would approximate its theoretical capacity remained open, as did issues of its accuracy, the technician time required to operate it, and so forth. Matthews also addressed what he called "other important issues," such as the data release policy, noting that current plans call for data to be released quarterly and that the company had said it planned to patent 100 to 300 genes. At one point commenting on the scale of the $300 million investment, Matthews remarked "given that amount of money, there has to be a business plan," then, apparently concluding he had stepped outside his neutral role, added "sorry, that was an editorial comment."

Matthews was immediately followed by a representative of the Wellcome Trust who said that the Trust had already been discussing options for its continuing support of sequencing that included the one announced in its May 13 press release. He added that the decision to announce the increased commitment on that day publicly was a response to the New Company. The press release was issued with the "enthusiastic support" of the "entire" Wellcome Trust Board of Governors. He then said that he would like to read the press release "into the record of this meeting." When he finished reading, extremely enthusiastic applause followed.

In the open discussion following these formal presentations, genome project leaders from both the United Kingdom and the United States reiterated what had become the official line of "welcoming" the new initiative while also stressing the need to complete the entire sequence and not be derailed from that goal. An American HGP leader stressed that the company's sequence data would be useful but emphasized that we cannot be satisfied by disconnected paragraphs in a bag without order and connection. The noble goal of decoding the "book of life" cannot be satisfied with an incomplete view. He also said that the press accounts saying either that the HGP will pay no attention to this development or that it will move on to the mouse are not true. We should be energized by this announcement, he said. An additional $300 million! Let's see if we can use it to speed up the project.

One questioner sought more information than was available from the Wellcome Trust's press release about its plans to challenge the patentability of DNA. The representative of the Trust replied that it would make a serious effort to understand the law and to consider challenging the patents, adding that the Trust has the resources needed. Some questions stayed close to technical issues. For example, one member of the audience asked whether it might be possible for HGP scientists to use simulation to understand what this proposed project would be able to do. He was answered from the floor by a well-known mathematician turned genome scientist, who argued that such a simulation would be hard to do, but that the whole genome shotgun was likely to produce an incomplete sequence with many gaps.

Late in the session, the discussion turned to the details of data release and patenting, with a stream of people asking questions: Would the company patent single nucleotide polymorphisms? Mike Hunkapiller and Craig Venter said they would not, right? Can we rely on their quarterly data release policy? They said they would not release traces, raw data, right? As these issues came up in rapid succession, orderly turn taking began to break down. A scientist with connections to the New Company jumped up from the audience, and was given the floor with the words "Let's hear from someone who knows what's going on." The scientist said that many details remain to be decided: the people who are launching the New Company do not have a really concrete plan of exactly what to do. It is a company in process, not in concrete. Part of the business plan includes early release for a price with later release to all for free. Be really careful making judgments, he stressed, because they are still working out the details themselves. When he was done speaking, the session reached the end of its scheduled time and was adjourned.

The Cold Spring Harbor meeting ended two days later, with the stage set for ongoing competition between the public HGP and the New Company, which later that summer was named Celera. In the fall, the HGP announced a change in scientific strategy: it would produce a rough draft of the human genome and subsequently complete the entire finished sequence. This change resulted from many considerations, including: ensuring that the sequence ended up in the public domain rather than being encumbered by intellectual property claims; shoring up Congressional support for the public program; making a valuable "intermediate product" available to the scientific community; and defining a new milestone so that observers would not inappropriately equate the incomplete sequence expected to result from Celera's whole genome shotgun with the "finished" sequence that the HGP was committed to eventually produce. This shift in the research program stemmed from too many considerations to be simply attributed to media orientation, but there is no doubt that concerns about audience perceptions and legitimating continued governmental expenditures for a public domain project figured prominently in shaping this response.

10.3 Discussion

Above, this paper has examined how genome researchers interacted with the media during a salient episode in the history of the HGP; namely, the announcement of a well-funded and credible private effort to sequence the human genome. What remains at this point is to consider what this study suggests about how media orientation is manifested in action. Because this study takes an action orientation that complements the medialization perspective, it does not address precisely the same questions. Nevertheless, it is useful to ask what implications the perspective developed here might have for research that takes a more macro-level approach to medialization. Conclusions must be tentative owing to the focus on a single episode with many unusual features.

This paper has argued that the orientation of scientific actors to the media can be understood as a specific form of theatrical self-consciousness expressed in interaction. The genome scientists involved in this episode paid significant attention to the media and approached the media and media content with intense theatrical self-consciousness. The events described suggest that, media orientation is an extremely complex phenomenon – thoroughly embedded in the particularities of specific situations and the performers and audiences who inhabit them. But even given situational particularities, this account suggests some tentative conclusions about how media orientation is manifested in action.

First, for analytical purposes, one can discern in this episode four facets of media orientation – the actor as performer, as audience, as commentator, or as builder of media relations infrastructure – each of which expresses a different aspect of theatrical self-consciousness. In practice, these facets are often intertwined and are sometimes experienced simultaneously, but it is useful to consider them separately.

(1) *Actor as Performer*. One facet of media orientation is the theatrical self-consciousness of the scientists and other actors who present themselves, their organizations, and their science to the media, aiming to shape the amount, location, timing, and content of media coverage. In the episode described above, salient performers included officials from the various organizations (e.g., Venter and Collins) as well as the organizations themselves (e.g., PE, NIH, and the Wellcome Trust). Whether they interacted interpersonally with journalists or via the circulation of texts, such as press releases, they addressed the media in

ways that involved extensive information control. The New Company, for example, was careful to prevent news of its emerging plan from seeing the light of day until it was ready, and when it issued its announcement, it projected confidence while backstaging the fact that some major questions remained unsettled, such as the details of the business plan and the continuing discussion about how to tweak the whole genome shotgun strategy. Similarly, the NHGRI worked to combat certain kinds of narratives, such as the inefficient government program frame, while keeping its internal discussions of how to respond private. In addition, as they addressed the media, these actors did not simply perform for journalists; they also spoke to additional audiences. The New Company and the HGP leadership alike aimed their performances not only *at* the media but also *through* the media, hoping that news coverage would convey their version of the story to some very salient imagined audiences. These imagined "reach-through audiences" included not only generic categories defined in terms of the media (such as "readers of the *New York Times*"), but also generic categories defined in other ways (such as "members of the U.S. Congress" or "investors") and perhaps even in some cases specific individuals (such as Collins or Venter).[11] Other salient reach-through audiences included the broader biological research community.

(2) *Actor as Audience Member*. A second aspect of media orientation is the perspective of the audience member who self-consciously recognizes that he or she is on the receiving end of a performance. As they scrutinized the news accounts surrounding the announcement of the New Company, the genome scientists described above clearly understood newspaper readers (including themselves) as part of an audience for staged performances that journalists or their "sources" constructed, possibly in order to foster certain impressions. In this audience mode, actors may engage with the performance to extract information. The term *information* here encompasses not only scientific information, such as the shape of an experimental apparatus that one might discern from the news coverage (Lewenstein 1995), but also information about the character and intentions of the performers. Genome scientists approached performances with a variety of purposes and with varying degrees of skepticism, sometimes treating media accounts as containing useful information and at other (and often the same) times regarding them as thin on detail and prone to distortion arising from the constraints of the genre, the lack of sophistication of journalists, or deliberate efforts to "hype" developments or "spin" stories. Actors operating in this critical mode sometimes attribute the distortion to the media itself, but they may also "reach back" to the sources used or quoted and speculate about their strategies, credibility, competence, and motivations.

[11] Just as the authors of journal publications may envision them being read not only by "the scientific community" but also by specific, often-namable colleagues in the face-to-face worlds (Shapin 2008) or core-sets (Collins 1995) of contemporary science, so too do actors sometimes imagine the reactions of specific people to their media performances.

(3) *Actor as Commentator*. A third way that scientists interacted with media was as commentators who actively presented evaluations of media coverage to an audience. This mode of interaction is closely related to the performer and the audience modes, in that it includes aspects of each. These commentaries are instances of a distinctive kind of performance that focuses on the media coverage itself as a *topic*. These commentaries are distinguishable from the audience mode because they are less about attempting to acquire information from news coverage than about attempting to evaluate – and often discredit – the coverage before an audience. These commentaries sometimes "reach back" via the media to evaluate the sources quoted in or displayed in the news coverage, often negatively; for example, when scientists complain about "propaganda," publicity seeking, or publication by press conference. Actors perform such commentaries in various settings, and for various purposes. Most of the comments that I observed were not made to the press or before large groups, but to colleagues in informal conversations or, under the veil of confidentiality, to me as an ethnographer. A notable exception was the statement that an American leader of the HGP made at the special session on the New Company's plans contending that the press accounts had misrepresented the HGP's response to the announcement of the New Company.[12] The comments seemed to be aimed at dispelling rumors, reassuring worried scientists, and rallying the troops for action. Even when made before small, targeted audiences of colleagues and observers, these commentaries can contribute to framing situations collectively and building responses to them. On one level, attacks on news coverage are complaints about how the media and its sources represent science and science-intensive activities; but they can also be a means through which controversies among scientists are conducted.

(4) *Actor as Builder of Media Relations Infrastructure*. The account above also provides indirect evidence of another aspect of media orientation: self-conscious efforts to build or enhance organizational infrastructure or personal skill aimed at interacting with the media more effectively. Examples include creating media relations offices in large organizations, hiring publicity firms, taking media training seminars, and so forth. Other examples include contracting with services that monitor television coverage or having staff regularly produce summaries of recent media coverage – things that some organizations involved in the HGP did do, at least during the final years of the project. One can infer the existence of this facet of media orientation from the ethnographic and media data presented above, although I did not observe people deciding to invest money or time in media relations infrastructure during the ten days discussed above. Nevertheless, such investments clearly had previously been made, and the major organizational players – the NIH and the Wellcome Trust as well as

[12] In addition, in my fieldwork I learned of instances in which genome scientists sometimes complained directly about coverage to reporters or editors following an article that they did not like. During certain episodes in genome research complaints about publicity seeking were made on the record, appearing, on occasion, in the press.

PE, TIGR, and the New Company – made use of media relations infrastructure and professionals. The hiring of communications firms by PE and TIGR is a case in point. Such efforts express theatrical self-consciousness because they indicate that actors are making concerted efforts to improve future performances directed at the media and to monitor how they themselves are appearing in it.

Described at this level of generality, these forms of theatrical self-consciousness no doubt appear in varied organizational settings, some far removed from science. In the case of the episode investigated above, the behavior of the main actors seems at an abstract level to match that which we might expect from any organization operating under a media spotlight when stakes are high and events are rapidly-unfolding. PE – like any company launching a potentially controversial new initiative – formulated a communications strategy and made use of public relations professionals. HGP leaders, for their part, arguably did what the managers of any enterprise would: they tried to react strategically to emerging events, seeking to acquire information, develop a short- and longer-term plans for responding, and tailor media messages that would defend their legitimacy. These observations suggest an important question for further research: what, if anything, is distinctive about how theatrical self-consciousness regarding media operates in the everyday action of scientists and science-intensive organizations, as opposed to actors of other kinds? To address this question, research might compare events such as the New Company announcement with structurally similar episodes involving rapidly-unfolding events, ranging from hostile corporate takeovers to technological disasters, that feature mass-mediated battles for the hearts and minds of audiences. Similarly, examining how patterns in the four facets of media orientation vary across cases would be useful. In particular, international comparisons (Jasanoff 2005) and change over time might offer especially promising lines of investigation.

Finally, the focus on theatrically self-conscious actors may raise some methodological questions for more macro-level studies of science-media interaction. Most studies of science and the media begin with the media (or relatively close to the media) and select a set of science-related media phenomena for detailed examination. This study proceeded in something approaching the opposite manner, beginning with a research community and examining its interactions that involved the media or took place via the media. In the social sciences, as in other fields of inquiry, the sources of data employed and the results obtained are often highly coupled, and studies of science in the media are by no means an exception. Content analysis of published news stories, for example, offers an aggregated picture of topics addressed, issue classification, emotional valences, metaphors and symbols, and other aspects of media content, such as tracking the volume of coverage over time. But studies of published news stories, unless supplemented by other methods, shed only a limited light on how scientists and others use media to acquire information or how they "reach back" to evaluate sources and look ahead to imagined "reach through" audiences. Such studies also tell us little about how scientists and others work to control the backstage/front-stage boundary.

Conceiving of media orientation as a form of theatrical self-consciousness casts a spotlight on questions about precisely what actors seek to make visible to whom and when. The dramaturgical framework suggests that the volume of media coverage about a topic (especially in the short term) is not necessarily closely related to the intensity with which actors are oriented toward media and various reach-through audiences in their day-to-day action. For example, the New Company's interest in shaping news coverage clearly began long before it handed the story to *New York Times* reporter Nicholas Wade "on a plate." Efforts to keep certain things out of the press – often one of the central goals of public relations strategy – express an important form of media orientation, used both to control the timing of public performances and to prevent some things from ever being made public. Performers who aim to keep imagined audiences from learning their secrets may do a great deal of work to construct sealed backstage spaces, to enclose information, to fight leaks and unauthorized disclosures, to battle against rumors. On the one hand, avoiding certain types of news coverage is a feature of many areas of high-technology research, especially in the commercial and military sectors. On the other hand, performers who attempt to attract media attention often fail to win it. For these reasons, the volume of media coverage may at times be an imperfect proxy for the degree of media orientation found among actors in a particular domain; for volume alone misses both failed efforts to win coverage and successful efforts to operate in an unobserved backstage.

References

Bijker, W. E., R. Bal, and R. Hendriks (2009). *The paradox of scientific authority: The role of scientific advice in democracies*. Cambridge, MA: MIT Press.

Bostanci, A. (2004). Sequencing human genomes. In J.-P. Gaudillière and H.-J. Rheinberger (eds.), *From molecular genetics to genomics: The mapping cultures of twentieth-century genetics*. New York: Routledge, pp. 158–179.

Collins, H. M. (1995). *Changing order: Replication and induction in scientific practice*. Chicago: University of Chicago Press.

Collins, F. S. (2006). *The language of god*. New York: Free Press.

Cook-Deegan, R. (1994). *The gene wars: Science, politics, and the human genome*. New York: Norton.

Davis, K. (2001). *Cracking the genome: Inside the race to unlock human DNA*. New York: Free Press.

Fortun, M. (2001) Mediated speculations in the genomics future markets. *New Genetics and Society*, 20(2), 139–156.

Fortun, M. (2008). *Promising genomics: Iceland and deCODE genetics in a world of speculation*. Berkeley, CA: University of California Press.

Gaudillière, J.-P. and H.-J. Rheinberger (2004). *From molecular genetics to genomics: The mapping cultures of twentieth-century genetics*. New York: Routledge.

Gillis, J. and R. Weiss (1998). Private firm aims to beat government to gene map. *Washington Post*, 12 May, A1.

Goffman, E. (1959). *The presentation of self in everyday life*. New York: Anchor Books.

Goffman, E. (1963). *Stigma: Notes on the management of spoiled identity*. New York: Prentice-Hall.

Hedgecoe, A. (2004). *The politics of personalised medicine: Pharmacogenetics in the clinic*. Cambridge, UK: Cambridge University Press.

Hilgartner, S. (1990). The dominant view of popularization: Conceptual problems, political uses. *Social Studies of Science*, 20(3), 519–539.

Hilgartner, S. (1995). Biomolecular databases: New communication regimes for biology? *Science Communication*, 17, 240–263.

Hilgartner, S. (1998). *Data access policy in genome research*. In A. Thackray (ed.), *Private science: Biotechnology and the rise of the molecular sciences*. Philadelphia, PA: University of Pennsylvania Press, pp. 202–218.

Hilgartner, S. (2000). *Science on stage. Expert advice as public drama*. Stanford, CA: Stanford University Press.

Hilgartner, S. (2004). Making maps and making social order: Governing American genome centers, 1988–1993. In J.-P. Gaudillière and H.-J. Rheinberger (eds.), *From molecular genetics to genomics: The mapping cultures of twentieth-century genetics*. New York: Routledge, pp. 113–128.

Irwin, A. (1998). Gene scientist races to grab a quick billion. *The Gazette* (Montreal), 15 May, B6.

Irwin, A. and B. Wynne (1996). *Misunderstanding science? The public reconstruction of science and technology*. Cambridge, UK: Cambridge University Press.

Jasanoff, S. (2005). *Designs on nature: Science and democracy in Europe and the United States*. Princeton, NJ: Princeton University Press.

Kaufman, A. (2004). Mapping the human genome at Généthon: The French Muscular Dystrophy Association and the politics of the gene. In J.-P. Gaudillière and H.-J. Rheinberger (eds.), *From molecular genetics to genomics: The mapping cultures of twentieth-century genetics*. New York: Routledge, pp. 129–157.

Lewenstein, B. V. (1995). From fax to facts: Science communication in the cold fusion saga. *Social Studies of Science*, 25(3), 403–436.

Nelkin, D. (1987). *Selling science: How the press covers science and technology*. New York: Freeman.

Nowotny, H., P. Scott, and M. Gibbons (2001). *Re-thinking science: Knowledge and the public in an age of uncertainty*. Cambridge, UK: Polity Press.

Olson, M. V. (1995, October 20). A time to sequence. *Science*, 270, 395–396.

Perkin-Elmer (1998). *Perkin-Elmer to introduce new instrument based on breakthrough DNA analysis*. Press release, 9 May 1998.

Perkin-Elmer and The Institute for Genome Research (TIGR) (1998). *Perkin-Elmer, Dr. J. Craig Venter, and TIGR announce formation of new genomics company; Plan to sequence human genome within three years*. Press release, 9 May 1998.

Peters, H. P., H. Heinrich, A. Jung, et al. (2008). Medialization of science as a prerequisite of its legitimization and political relevance. In D. Chen et al. (eds.), *Communicating science in social contexts: New models, new practices*. New York: Springer, pp. 71–92.

Rabinow, P. and T. Dan-Cohen (2005). *A machine to make a future: Biotech chronicles*. Princeton, NJ: Princeton University Press.

Recer, P. (1998). Gene technology must be proven before government will use it. *Associated Press*, May 11.

Riles, A. (2006). *Documents: Artifacts of modern knowledge*. Ann Arbor, MI: University of Michigan Press.

Rödder, S. (2009). Reassessing the concept of a medialization of science: A story from the "book of life." *Public Understanding of Science*, 18, 452–463.

Schäfer, M. S. (2009). From public understanding to public engagement: An empirical assessment of changes in science coverage. *Science Communication*, 30(4), 475–505.

Shapin, S. (2008). *The scientific life: A moral history of a late modern vocation*. Chicago: University of Chicago Press.

Shreeve, J. (2004). *The genome war: How Craig Venter tried to capture the code of life and save the world*. New York: Alfred Knopf.

Smith, D. E. (1974). The social construction of documentary reality. *Sociological Inquiry*, 44, 257–268.

Sulston, J. and G. Ferry (2002). *The common thread: A Story of science, politics, ethics, and the human genome*. Washington, DC: Joseph Henry Press.

Sunder Rajan, K. (2006). *Biocapital: The constitution of postgenomic life*. Durham, NC: Duke University Press.

Venter, J. C. (2007). *A life decoded: My genome, my life*. New York: Viking.

Wade, N. (1998a). Scientist's plan: Map all DNA within 3 years. *New York Times*, 10 May, A1.

Wade, N. (1998b). Beyond sequencing of human DNA. *New York Times*, 12 May, F3.

Wade, N. (1998c). International gene project gets lift. *New York Times*, 17 May, 20.

Weingart, P. (1998). Science and the media. *Research Policy*, 27, 869–879.

Wellcome Trust (1998). *Wellcome Trust announces major increase in human genome sequencing*. Press release, 13 May 1998.

Chapter 11
Scientific Sources and the Mass Media: Forms and Consequences of Medialization

Hans Peter Peters

11.1 Introduction

The "medialization of science" concept comprises two complementary observations: first, increasing media attention for science, and, second, adaptation to or even anticipation of media criteria within science as a response to the increasing necessity of legitimating science by means of public communication (Weingart 2001: 244–253). This chapter is mainly concerned with the latter aspect of how science adapts to the conditions of media communication and the resulting consequences, the empirical evidence quoted coming from two biomedical research fields that undoubtedly attract media attention: stem cell research and epidemiology, however. The analysis in this chapter will show the forms of medialization in the routine interactions between two types of scientific communicators serving as media sources – individual scientists and PR departments of research organizations – and journalism. Furthermore, possible consequences of medialization for the public constructs of science, for the research process, and for the science-policy relationship will be addressed.

The main theses are, first, that as media sources individual researchers and research organizations have institutionalized a communication approach towards the mass media that implies strong anticipation of media criteria, and, second, that anticipation of media criteria, strategic orientation and organizational involvement systematically affect the public construct of science, have repercussions on the research process, contribute to the governance of science and also increase the political impact of scientific expertise.

There has been a long discussion about the relationship between science and journalism which has generally been analyzed as relationship between individual

H.P. Peters (✉)
Research Center Jülich
e-mail: h.p.peters@fz-juelich.de

S. Rödder et al. (eds.), *The Sciences' Media Connection – Public Communication and its Repercussions*, Sociology of the Sciences Yearbook 28,
DOI 10.1007/978-94-007-2085-5_11,

scientists and journalists. Many commentators and communication researchers diagnosed "gaps," "tensions," "barriers" or miscommunication such as inaccuracies, and lack of motivation and communication skills on the part of scientists (e.g., Willems 1976; Dunwoody and Ryan 1985; Singer 1990). If the social context was included in the analysis it was mainly the scientific communities and their norms that were taken into account – and mostly seen as a factor impeding the involvement of scientists in public communication. However, not only the phenomenon of "media stars" among scientists (see Weingart 2001: 262–272) but also the prevalence of routine interactions of scientific sources and the widespread satisfaction of scientific sources with the media coverage of their own research (Peters et al. 2008a) suggest the need for a reassessment of the science-media relationship. In this reassessment, the significance of contexts for the science-media relationship other than the still important scientific communities (see Chapter 8) has to be considered, namely research organizations as employers of scientists and high-ranking journals as publishers (see Chapter 17).

The empirical evidence used to support the arguments of this chapter is mainly drawn from an international mail survey of 1,354 biomedical researchers in France, Germany, Japan, the United Kingdom and USA regarding their relations with the mass media, and 45 semi-structured face-to-face interviews with public information officers from research organizations in France, Germany and the United Kingdom. A detailed description of the methodology of the mail survey, including an analysis of response rate and sampling bias, can be found in Peters et al. (2008a, supporting online material). The overall response rate was 43 percent; we found no indication of a strong sampling bias. The methodology of the survey of public information officers is explained in Kallfass (2009: 103–110).[1]

11.2 Media Relations of Scientists and Research Organizations

There are many reasons for media interest in science, which can be stimulated by events inside and outside science: a new scientific discovery or finding, an innovative technical or medical application, a problem requiring scientific explanation, or an issue of science governance. The media actively monitor science by linking

[1] Both surveys are part of the project "Integration of scientific expertise into media-based public discourses (INWEDIS)" (see Peters et al. 2008c; Peters 2009) which was supported by a grant from the German Federal Ministry of Education and Research (BMBF) in the research program *Knowledge for Decision-making Processes – Research on the Relationship between Science, Politics and Society*. The survey of scientists was conducted in collaboration with Dominique Brossard, Suzanne de Cheveigné, Sharon Dunwoody, Monika Kallfass, Steve Miller and Shoji Tsuchida. The semi-structured interviews with public information officers were conducted by Monika Kallfass (see Kallfass 2009). The conclusions in this chapter are based on the author's secondary analysis of the interview transcripts and do not necessarily match the interpretations of the primary researcher.

into inner-scientific communication, regularly reading *Science* or *Nature*, for example, or they look for scientists with specific expertise by turning to PR departments of relevant research organizations. The media are also the target of communication initiatives that originate in science itself: researchers chatting with journalists about their latest research during scientific conferences, or universities issuing press releases about scientific "successes," for example. In one way or another, individual researchers are usually involved in all these public communication processes, but the initiative for making something public may come from very different actors inside and outside science: the media, research organizations' PR departments, sponsors, publishers, or science critics. This section looks at the interdependencies of science and the media focusing on the "science side" of that relationship. Throughout this chapter the term "media" is used as an abbreviation meaning "journalistic news media," irrespective of the technical dissemination channel such as print, broadcasting or Internet.

11.2.1 Institutionalization

In all five countries and both research fields studied in the INWEDIS survey, biomedical researchers consider the impact of their media contacts on their career generally to be mostly positive, neutral or balanced, but hardly ever negative (Peters et al. 2008a). Furthermore, when they receive feedback on specific media appearances by colleagues, other journalists and superiors, this feedback is mostly positive with little variation between countries and research fields. Less than one percent of the respondents reported "mostly negative" feedback. This is even true when focusing on responses from colleagues. One may speculate, of course, that "visible" scientists are not directly criticized because of their media visibility but that such criticism is spread like rumors behind their backs. However, it is unlikely that researchers grossly misjudge the impact of factors supporting or damaging their own career. We thus have to acknowledge that actors and mechanisms influencing the allocation of career-relevant resources – jobs, promotions, institute budgets, research grants, acceptance of publications in high-impact journals – in part respond positively to the media visibility of scientists and thereby effectively express a normative expectation that scientists should be prepared to interact with the media.

This finding seems rather strange given the classical and still widespread assumption that public visibility sets a scientist's reputation at risk because it conflicts with inner-scientific norms (e.g., Boltanski and Maldidier 1970; Goodfield 1981). If our thesis is true that media contacts by researchers are now institutionalized, we have to challenge the assumption that these contacts violate (effective) norms of the scientific community. A question on the possible motives of researchers for engaging in or refusing contacts with the mass media in the INWEDIS survey included three statements regarding the relevance of scientific culture and peer responses to media visibility as motivators for or against interactions with the mass media: compatibility of media contacts with scientific norms, concerns because of critical responses from peers, and expectations of enhanced personal reputation among peers (see Peters et al. 2008a, supporting online material, table 8). The distribution of answers shows that about one third of the respondents consider “incompatibility with the scientific culture” to be a very or somewhat important concern that increases scientists’ reluctance to agree to contact with the media; two thirds consider this concern to be not or not very important. For about an equal number of respondents (42 percent and 39 percent, respectively) “possible critical reactions from peers” and “enhanced personal reputation among peers” are very or somewhat important motives for rejecting or accepting media contacts. The impact of scientific norms on readiness to become involved in media interactions can best be described as ambivalent. This ambivalence is not only obvious on the aggregate level (i.e., coexistence of researchers with different perceptions of the encouraging or discouraging character of scientific norms in the same scientific community) but also exists on the individual level (i.e., respondents being concerned about possible critical reactions from peers and at the same time being motivated by the possibility of enhanced personal reputation among peers). Rather strangely, answers to both these items are not – as one would expect – negatively but (mildly) positively statistically associated ($\tau_b = 0.14$, $p \leq 0.001$) (see Peters et al. 2008a, supporting online material, table 10). The main variation between individual researchers is thus not the perception of a positive versus negative influence of scientific norms on their motivation to engage in media relations. Researchers rather tend to perceive scientific norms regarding media interactions as ambivalent and differ in the degree to which they rate them as more versus less relevant.

A likely explanation of the diagnosed ambivalence is that scientific norms and peer responses to media visibility are context-sensitive. It is probably dependent on certain conditions – situation, characteristics of scientist and type of newspaper or TV program, for example – whether scientific communities accept public visibility of their members. This conforms with the findings of Rödder (Chapter 8), who also diagnosed an ambivalent relationship between scientific norms and media interactions. She identified three reasons making media contacts acceptable among peers: (1) rooting in scientific substance, (2) contact resulting from media initiative, and (3) motivation by institutional contexts and goals.

Our mixed findings may be interpreted as resulting from the combined effect of a lingering latent skepticism about media interactions in the scientific community noted by Boltanski and Maldidier (1970), for example, and the increasing salience of justifying reasons in view of the *de facto* need to secure public legitimacy for science, comply with organizational expectations, or address possible “clients” of scientific expertise (patients, citizens, policy-makers) effectively by media communication. The practice of public communication seems to be changing from an exceptional activity requiring justification to a default activity that is accepted as an integral part of the research process (for evidence, see Section 11.3.2).

Research organizations are crucial in providing the economic means and organizational environment required for scientific research as a professional activity depending on the availability of jobs, labs, equipment, research infrastructures and favorable regulation framework. In order to secure the economic resources and political support necessary for their existence, research organizations have to demonstrate their usefulness, excellence, and public support to their (public) funders. The general legitimacy problem of science (Weingart 2001) is thus transformed into a legitimacy challenge for research organizations that – in a media society – also has to be addressed by public communication. Almost all universities and other publicly funded research organizations in the major democratic knowledge societies have public communication departments that deal with "research" as a core organizational activity (besides teaching students in the case of universities, for example) and with "research results" as a major output. High-impact journals addressing a wider audience and also aiming at "impact" in the social environment of science, in particular *Science* and *Nature*, accompany each issue with extensive and elaborate public relations (Chapter 17). Large scientific associations, such as the American Association for the Advancement of Science (AAAS), and more specialized associations and bodies such as the British Science Association (see Chapter 12) and the German *Science in Dialogue* initiative, as well as many foundations worldwide, have programs to promote the public communication of science and technology. A large number of organizations are thus strategically shaping the relationship of science and the public – usually with the goal of intensifying contacts between the two realms and motivating researchers to engage in public communication activities. Taken together, these structures represent an important dimension of institutionalization of media relations on the organizational level.

The above-mentioned organizational contexts confront individual researchers with normative expectations regarding their interactions with the mass media that may differ from those of their scientific communities. Indeed, it is quite likely that individual researchers who depend on resources managed by these organizations – publication chances, jobs, budgets, research grants – somehow integrate these expectations into their role as researchers. Besides the interdisciplinary journals *Science* and *Nature*, whose criteria have to be considered by *scientists as authors*, *scientists as employees* of research organizations are subject to formal organizational policies regarding media contacts as well as to informal expectations on the part of peers, superiors, PR department and management. Interdisciplinary scientific journals also aiming at impact outside the peer communities of scientists, and research organizations that depend on public funding and favorable regulation are thus likely catalysts of a medialization of the scientist's role and effective protagonists of an institutionalization of media contacts by scientists as routine behavior.

The INWEDIS survey shows that media contacts among researchers are quite common and not the domain of relatively few highly "visible scientists" that represent the respective research fields in public. (The instances of highly visible scientists, some of them analyzed by Weingart (2001), may even present special cases in terms of motivation and media logic leading to them.) On average, 59 percent of the surveyed biomedical researchers, with little variance between the five countries included in the study, said that they had had contact(s) with the media in the past 3 years, about 30 percent of the respondents claimed six or more contacts within that time period (Peters et al. 2008a). Another indication of the institutionalization of such contacts is that subjective beliefs and attitudes regarding public communication are only weak predictors of frequency of contact. Strong predictors are, in contrast, organizational leadership position (being principal investigator or institute director, for example), and scientific productivity (number of peer-reviewed publications). This is congruent with a French study that analyzed the yearly activity reports of CNRS scientists and found a positive correlation between scientific productivity and involvement in public communication (Jensen et al. 2008). Contrary to the perception of researchers surveyed by us, Jensen et al. concluded that the dissemination activities of the French researchers had no strong impact on their careers although the existing small effects tended to be positive too – in particular in the life sciences. A likely explanation of the difference in findings is that the crucial career factor is not the researchers' involvement in dissemination activities in general (which Jensen et al. correlated with career development), but that specifically the researchers' ability to generate media response is rewarded by research organizations.

To conclude, on the organizational level as well as on the individual level we find clear evidence of an institutionalization of interactions between science and the mass media. Almost all research organizations have developed departments specializing in public communication, in particular with respect to media relations. There is furthermore a public science communication infrastructure consisting of specialized organizations and programs that integrate scientists in their communication activities. With respect to public communication, scientists are obviously confronted with relevant expectations from different contexts – scientific community, the research organizations employing the scientists, scientific journals, and mass media – which are not always free from contradictions and ambivalences. The resulting role, however, seems to allow or even require scientists, especially leading scientists, to consider media interactions as part of their professional duties. As Schäfer (2008) argues, medialization of science is contingent and differs between research fields. Our conclusion is probably not equally true for all contexts and all research fields, but certainly for the "medialized" biomedical research fields on which this chapter focuses, and probably in other areas as well.

11.2.2 Professionalization

PR departments of research organizations, as I will call them in the following although they have different names (e.g., University Communications, Press Office, Media Relations, Public Relations, or even Corporate Communications) and different scopes of responsibility, have been introduced as a response to media interest in science. Strategic motives were, however, present from the outset. Borchelt (2008: 149) summarizes the public information offices' understanding of their task as "...making sure the public knows a lot about science or the scientists, but only the 'right' things the organisation thinks the public should know." A German survey of public information officers in 1983 found that they tended to subscribe to a partnership model of their own relationship with the media and a mediator role in the science-media relationship. Generally, they expressed strong and genuine attachment to both realms, science and journalism, downplaying the possibility of a conflict of interest resulting from their organizational role. However, if they mentioned such conflicts they gave precedence to organizational interests above the journalistic ethics (Peters 1984).

While the INWEDIS survey of public information officers in 2007 did not show a completely different picture of science PR than that of the 1983 survey, some trends are nevertheless detectable. The public information officers interviewed still emphasized the need to anticipate media criteria in the selection of stories, to provide a good service for journalists seeking information and sources, and they would generally rank the long-term goal of maintaining trust among journalists above some short-term advantage scientists or management might think to gain by occasional secrecy or misinformation. However, the "proactive" strategic component has become more important in their work. This is evident by a change in the terminology they use to describe their work. Terms like "branding" and "marketing," adopted from professional corporate public relations, are now part of the routine vocabulary of public information officers. The strategic goals have become more specific (see next section) as opposed to the classical general goal of securing a "good public image," This might be seen as part of the general trend towards new (more corporate-like) models of organizational management in universities and other research organizations (see Maasen and Weingart 2006). However, public information officers in research organizations are still in search of a clear professional identity in the triangle of science, journalism and public relations, with two transitions having taken place from basic identification with science at the beginning to identification with science journalism, and finally identification with strategic PR. However, even today different orientations and approaches coexist and are often intermingled in ambiguous identities. Many public information officers in research organizations still tend to consider themselves as a kind of science "journalist" working outside the media but feeling themselves to be members of the journalistic profession. This is probably the case because most public information officers have received some kind of training in journalism or may even be former journalists. Many public information officers are members of journalists' associations such as the US *National Association of Science Writers* (NASW) or the German *Technisch-Literarische Gesellschaft* (TELI). Journalists' associations do not always wholeheartedly embrace public information officers as members since journalists are clearly critical of an interpenetration of journalism and public relations (e.g., Stollorz 2005). Unlike TELI, the *Wissenschafts-Pressekonferenz*, the other German association of science journalists, explicitly excludes public information officers from membership. The NASW nowadays accepts them as full members but still precludes them from the inner executive committee.

There are some indications of a growing professional identity of public information officers in research organizations, however, and that identity might be closer to that of the public relations profession than before, although many (but not all) public information officers still have reservations about seeing themselves simply as part of the general PR business. So far, professionalization in terms of institutionalization has lead to the evolution of a few specialized associations such as the German *Bundesverband Hochschulkommunikation* (established in 1969) and the *European Universities Public Relations and Information Officers Association* (EUPRIO, established in 1986), rather than public information officers of research organizations joining existing PR associations like the *Public Relations Society of America* (PRSA) and forming a special section on science PR there.

The INWEDIS surveys of biomedical researchers as well as of public information officers included questions regarding the relation of organizational PR and researchers' media contacts. Generally, in academic research organizations the PR departments try to motivate researchers to interact with the media using various means, even creating a competition among institutes, as the quote from a university public information officer shows:

> [. . .] we analyzed the distribution [of media clippings] for the whole of last year by faculty and institute. [. . .] Well, this analysis is now even being discussed in the president's office. The statistics will also be made available to the faculties. And then they will see for themselves: 'Aha, we are lagging behind the others.' And I think that this will increase the input [of the PR department] quite a lot (translated).

In many instances, the PR department is involved in creating contacts between researchers and the media, e.g., by press releases or by forwarding journalistic inquiries reaching the PR department to researchers. They furthermore try to increase the scientists' understanding and acceptance of science journalism and its selection and quality criteria that differ from those within science. Although researchers are also directly addressed by the media, for example, on the basis of high-ranking publications or talks at conferences, or they address journalists themselves, a large proportion of media contacts by researchers, perhaps the majority in some fields, are initiated by the PR departments or at least with their involvement. 62 percent of the respondents of the INWEDIS survey of biomedical researchers in the five countries covered confirmed that they had interacted with the PR department of their organization within the last 3 years.

Most academic institutions do not aim at strongly controlling scientists' individual interactions with the media, but, organizational policies differ in that respect. On average, about half of the respondents of the INWEDIS survey confirmed that they would not have to ask somebody in their organization if they wanted to talk to a journalist, 35 percent said that they "would have to seek approval" before talking to a journalist. Of course, the need to ask somebody before talking to a journalist is much higher for junior than for senior researchers; Junior researchers are also more often uncertain about the organizations' rules regarding media interactions. Still, about a third of the researchers at the highest management level (dean, director, department head, CEO) said that they would have to seek approval. While there are hardly any differences between countries in that respect, researchers in private or government-controlled research organizations, and – to a lesser extent – those in (university) hospitals, are generally less free to talk to the media without prior approval than researchers in universities and public non-university research institutions. (The latter category comprises different types of research organizations with different media policies.) Most researchers surveyed knew or expected that receiving organizational approval for talking to a journalist "would be easy." This shows that organizational rules of getting approval before media contacts are not so much intended to prevent such contacts as to allow cautious supervision by PR departments to make sure such contacts are not going to violate the organization's interests.

As an indication of the rank and power of the PR departments within research organizations we asked those researchers who indicated that they needed organizational approval for media interactions whom they would have to ask – their superiors, the PR department or some other department. On average, 45 percent of the respondents needing approval for media contacts said that they would have to ask the "person or department responsible for public communication (e.g., public relations department)," 52 percent needed approval from their "superiors (e.g., dean, department head, head of organization)." There are remarkable country differences in responses to this question, indicating different degrees of influence of a centralized organizational public communication policy on media interactions of scientists (Fig. 11.1). In the United States and the United Kingdom, PR departments tend to have a more influential position than in Germany, France and Japan.

An indicator of the "professionalization" of media contacts at the individual level is the participation of scientists in communication training courses. Many research organizations, professional associations, foundations and funders (such as the EU) offer such training courses to improve the researchers' communication skills in public communication, in particular in media communication. On average, 31 percent of the surveyed biomedical researchers indicated that they had participated in a communication training workshop of some kind; 11 percent said that this had been training specifically for communication with the "mass media." Again there are considerable country differences indicating a higher professionalization of media interactions for the Anglo-Saxon researchers (Fig. 11.2). British and US American researchers tend to have received formal training for media contacts more often than their colleagues from Germany, France and Japan. This might indicate a higher perceived relevance of media contacts and thus a stronger medialization, but it

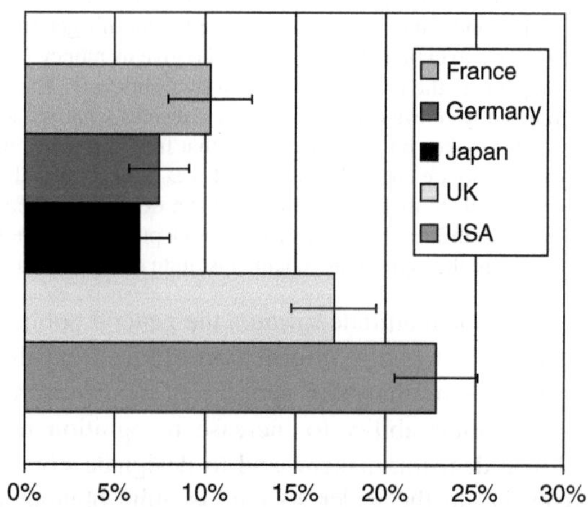

Fig. 11.1 Proportion of biomedical researchers requiring approval from their PR department before talking to the media
Source: INWEDIS survey of scientists

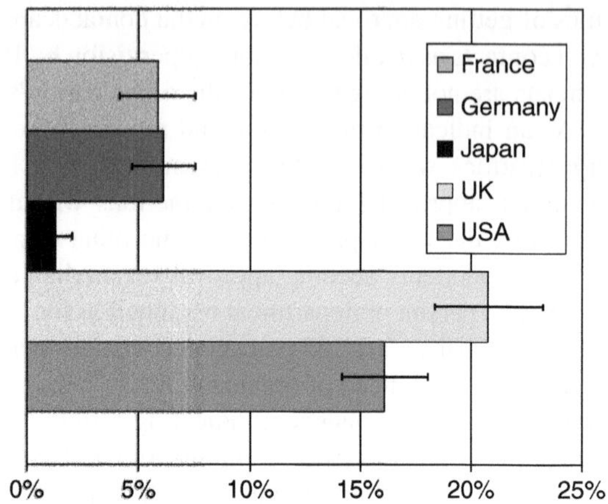

Fig. 11.2 Proportion of biomedical researchers having received media training
Source: INWEDIS survey of scientists

may also reflect a stronger desire to learn strategies of how to deal with journalistic inquiries while regaining control of coverage rather than simply adapting to journalistic expectations.

11.2.3 Strategic Utilization

Based on the almost unanimously expressed conviction that media visibility – as long as it is not negative – serves organizational interests, the public information officers try to increase the volume of media coverage mentioning their organization. However, providing the media audience with information about science serves as a means rather than the actual end of their communication activities, as the quote taken from Peters (1984: 92) illustrates:

> So I have a very special motivation for doing that. The astronomer has to make his sovereign understand what he does; otherwise he doesn't get any money. That was always the case. We have to explain to the people who give us money [what we are doing]. And in a democracy this is the people represented by politicians. The final goal is that mainly politicians, politically relevant people, do not consider what we are doing to be complete nonsense. [. . .] It makes no difference to us that in doing so we have to inform the entire proletariat [*sic*!] who are entitled to vote. If it was only about public information we would not be so motivated. In that case we would have decided to become teachers. No, for [us] PR work is about information, because without public information we won't get any money, and therefore the appropriate channel is quite clear (translated).

The cynical attitude towards the general public expressed in this quote is clearly not typical of public information officers, but the strategic reasoning of addressing politics as the main sponsor of academic research via the mass media thus utilizing their ability to increase recognition by relevant target groups such as political decision-makers, and to designate organizations, activities and results as "relevant" to the wider society, is quite obvious and was also observed by other scholars (see Borchelt 2008: 152; Nelkin 1987). Anecdotal evidence as well as the

INWEDIS survey of decision-makers (Petersen et al. 2010) suggests that in view of medialized politics the implicit media effects model of public information officers underlying this strategy is probably valid.

As Borchelt (2008: 149) notes, "publicity" as the main goal in the initial phase of PR is important but not sufficient to legitimate an organization; and current approaches of PR largely follow the "explanatory PR" model. Legitimation is based on matching the relevant challenges and organizational profile. The logic of strategic organizational communication is thus twofold: First, to demonstrate by public visibility, activity profile and achievement record that an organization is best suited to meet a certain challenge or a portfolio of challenges (i.e., "branding" the organization respectively); and, second, to increase the public salience of those challenges by means of agenda setting. The challenges can be rather general such as excellence of research or excellent study conditions, or specific such as dealing with an increasing rate of dementia, adapting society to climate change, or finding the Higgs boson ("the God particle"). Borchelt (2008: 152) sees the need for yet another phase of science PR dealing with the management of the "trust portfolio," i.e., managing the relationship between an organization and its societal environment using a more symmetrical communication approach. Today, in this respect science still relies on the high general trust it enjoys in Europe and the USA in contrast to political and economic institutions. Specific efforts to manage the trust portfolio thus seem only necessary in areas of public controversy.

Besides the universal goal of legitimating the organization by means of publicity and branding, public information officers in the INWEDIS survey mentioned several specific goals that can be grouped into three categories: marketing, political communication, and public education. Depending on their type, research organizations operate in several service markets and have to survive the competition there. Universities offer places for students, university hospitals offer advanced medical diagnosis and therapies to patients, and many research organizations offer R&D services to industry and other clients. Public science communication thus becomes a means of support for *marketing* these services by establishing an organization's name as a "brand," recognized and valued in these markets. Research organizations' involvement in *political communication* can be based on organizational self-interests of securing political support in terms of, for example, funding, siting of research facilities, or favorable regulation, as in the case of the German law on importing human embryonic stem cells. Sometimes, however, the goals of research organizations' involvement in political communication are not based on their self-interest in a narrow sense but on political goals related to their expertise – the lobbying of the German cancer research center in favor of a stricter Government policy protecting nonsmokers may serve as an example. Finally, the goal of *educating the general media audience* regarding their health-related behavior, for example, was also mentioned occasionally.

Legitimacy-related motives of media contacts are also dominant among individual researchers, although probably not so much focused on the organizations' benefit as on that of their own research, research field or career. Positive motivators that are rated very or somewhat important by most respondents are "A more positive public attitude towards research" (93 percent), "A better educated general public"

(92 percent), “Influence on public debate” (85 percent) and “Increased visibility for sponsors and funding bodies” (77 percent) (see Peters et al. 2008a, supporting online material, table 8). The motive of “a better educated general public,” which could be interpreted in different (even emancipatory) ways, has to be considered in connection with the statement “If the public only knew more about research, it would be more positive about science.” This core belief of the so-called deficit model of public science communication, according to which a knowledge deficit produces negative attitudes towards science and a scientifically better informed public would have more positive attitudes (see Sturgis and Allum 2004), is held by a majority of respondents (71 percent). The most likely interpretation of the strong motive of helping to educate the public therefore is that researchers expect benefits in terms of a more positive public image of science.

The biomedical scientists surveyed in all countries and across several indicators rate their own encounters with the media overwhelmingly positive (Peters et al. 2008a, b). This is rather surprising given the many analyses pointing to communication problems between science and the media mentioned in the introduction. There are some indications that scientists’ assessments of their interactions with the media have become more positive in recent decades. Peters et al. (2008b: 270–272) point to several possible explanations for the (increasingly) positive evaluation. As one of the likely reasons they mention a change in the salience of scientists’ satisfaction criteria – resulting, first, in greater acceptance of media rules and, second, in a move away from accuracy-oriented criteria deriving from the scientific focus on “truth” towards strategic criteria of beneficial impact, i.e., publicity and positive coverage. The positive assessment of their interactions with the media would thus not in the first place indicate scientists’ satisfaction with the “content quality” of media stories about their work, but rather satisfaction with their expected impact in terms of creating positive recognition and mobilizing support. The adaptation of scientists to public communication has resulted in a gradual shift from a communication model viewing public science communication as a popularized “extension” of inner-scientific communication, in which the criteria of inner-scientific communication still apply, even if in a reduced form because of the requirement of comprehensibility for a lay audience, towards an effect-oriented model of strategic communication – similar to that of PR departments – which implies acceptance of the media logic. With respect to medialization, a crucial question is to what degree this model of strategic public science communication also modifies inner-scientific communication and research processes, or, in contrast, reinforces the boundary between inner-scientific and public communication (see Chapter 17).

11.3 Consequences of Medialization

The previous sections provided strong evidence that science responds to and actively seeks to stimulate media attention. Science has developed an institutionalized interface to the mass media consisting of public relations strategies of research organizations, foundations, scientific associations and journals, and also of a set of norms, motivations and interaction routines of the majority of researchers, who are nowadays generally prepared to talk to journalists about their research, and of a number of highly "visible scientists" representing disciplines or research fields in a broader sense. Motivated by a number of organizational goals and individual motivations of researchers, above all by such goals related to the survival needs of scientific organizations and research fields in a media society, the "medialization of science" has intended and unintended consequences for the social environment of science, and possibly also for science itself. In the following, three aspects of these consequences are discussed: effects on public constructs of science, repercussions on the scientific process, and effects on the science-politics relationship.

11.3.1 Public Constructs of Science

The most obvious consequences of the increasing anticipation of media expectations by scientific sources, increasing salience of strategic criteria, and increasing influence of organizational communication concern the media constructs of science. Four likely effects of medialization on public constructs of science are: (1) overemphasizing the immediate practical utility, (2) downplaying the scientific logic and autonomy, (3) focusing on the competitive character, and (4) framing scientific advances as organizational output. Public communication thus becomes part of the competition between research organizations (see Münch 2009).

(1) The major selection principle in the journalistic observation of science is that scientific events have some relevant meaning for the general media audience, i.e., relevance outside science (e.g., Kohring 2005). While this does not preclude basic research from news coverage (as public relevance can be constructed in many ways), an easy way of capturing the attention of non-scientists for scientific events is referring to their practical implications in terms of applications (e.g., medical therapies) or scientific expertise relevant for policy issues (e.g., climate change). Furthermore, showing the use of research results for innovative technologies or as decision-related expertise allows scientific communicators to refer to generally accepted cost-benefit considerations when trying to demonstrate the value of research performed in their organization. Although this is not the only legitimating argument that research organizations can put forth, it is probably the most straightforward one. In comparison, the persuasive effect of possible alternative arguments stressing the cultural value of science, its excellence, its contribution to national prestige or (speculative) indirect and future benefits of basic research is more limited in reach. The public information officers interviewed in the INWEDIS survey therefore confirm a preference for "applied research" in their communication activities, as in the following quote:

> [. . .] not all press releases concern scientific results but there is always a demand for result-oriented press releases, particularly since we naturally select those [topics] that are as close as possible to practical application. All the topics close to basic research

> are not so easy to communicate. But those close to practical application are in great demand (translated) (Kallfass 2009: 142).

The typical way scientific sources in public communication create public relevance thus tends to reinforce a utilitarian construct of science.

(2) According to Borchelt (2008: 152) "managing the trust portfolio" is the major current challenge of science PR. Among other factors, deference to authority and a shared culture are usually considered to increase trust. As deference to authority might be generally decreasing in our societies, the demonstration of a shared culture is the obvious alternative strategy: portraying scientists as ordinary people with similar interests, lifestyles and values to those of non-scientists, showing that research organizations cultivate general social values such as gender equality, and suggesting that researchers and research organizations are motivated by valued social goals such as health, safety, environment, justice, for example, and not just by genuine search for truth. However, at least at the micro-level, the culture of scientific research is very different from everyday culture or from the culture of industrial companies. Research goals are often incomprehensible to non-scientists and their relationship to public values obscure. Many hours of work are invested in researching and preparing a manuscript that might then be rejected in the peer-review process, or – if the researcher is really lucky – is maybe read and cited by a few dozen colleagues. Viewed from the outside, the culture of science is strange in many respects. In order to adapt to normative public expectations regarding shared goals and values, there might be a temptation to present research to the media in such a way that deceives the audience about the esoteric, unfocused and erratic character of science and its demand for autonomy, and rather present it as an activity that conforms to general social routines, values and norms.

(3) In her analysis of media constructs of stem cell research, Jung (Chapter 6) identified a meaning pattern "science as sport," among others, that in several variants seems to be rather widespread in public science communication generally (e.g., Nerlich 2009). Many news stories and press releases rely on framing research results or new instruments as "victories" that make the researchers or the research organizations appear as the winners of a competition between countries, research organizations or researchers who have discovered, acquired or produced, for example, the lowest temperature, the highest pressure, the fastest computer, the deepest hole, or the first anti-proton. This framing of science is in tune with legitimation and persuasion needs since strategic arguments of different kinds can easily be constructed: Individual researchers and research organizations can demonstrate excellence by claiming a "victory," and national representatives of science can point to the unfairness of political regulation by referring to the competitive disadvantages of the German "team" arising from such a "handicap." The press release from the German Research Foundation (DFG) in favor of less strict regulation of imports of human embryonic stem cell lines may serve as example:

> In recent years, international stem cell research has produced important new findings. [...] Because of the legal framework, scientists in Germany can currently only make a limited contribution, however. [...] Cancellation of the deadline regulation would considerably improve the competitiveness of German scientists in the field of stem cell research (DFG, 10 November 2006, translated).

Using the "sports" pattern shifts the focus of public communication from scientific processes and results to scientific "performance" relative to competitors. It also hides the patterns of cooperation across borders of research organizations and countries that are typical of science (see below). In the end, professionalism as the dominant structure of science is downplayed and a business-like market-competition model of science is proliferated.

(4) For organizational PR it is crucial that scientific achievements are attributed to one's own organization. Scientific achievements, which according to the mainstream view of the sociology of science are created by creative individuals in the intellectual context of scientific communities, are "appropriated" by research organizations, framed as their achievements and marketed as their output, as shown in the following quotes from public information officers:

> In terms of Public Relations work we aim to celebrate the successes and the achievements of the University [...].

> [...] but there are mass media that quite intentionally do not mention the originator, [...] this is then completely worthless for me – [for example] a story 'Scientists have discovered [...]'. No, it must read '[Scientists] of the <name of organization> [...]', only then is it of any use to me (translated).

The result is a construct of science in which research organizations become the originators of knowledge, and research becomes a straight production process of transforming financial inputs into knowledge as output. Another likely consequence is that research infrastructures tend to be emphasized to the detriment of intellectual resources since research organizations can plausibly point to their provision as a specific organizational merit.

To summarize, adaptation to public expectations in science communication because of strategic needs and the increasing influence of organizational communication goals tend to promote a public image of science as a directed process of efficient production of socially relevant knowledge as the output of research organizations that compete (like companies) in a global market. There are, of course, limits to the consistency of such a construct because the goals of science communicators may vary, and several compromises have to be made to maintain the credibility of the advocated construct. Furthermore, journalists have access to inner-scientific events (e.g., publications, congresses) enabling them to challenge the self-presentation of science towards the media, and deviations from a business-like construct of the scientific process may be particularly newsworthy. Nevertheless, the esoteric and erratic character of science, its cross-boundary cooperation patterns, the reliance of its proper functioning on a certain degree of autonomy, and the importance of creative individuals as researchers may be understated. One may argue that promoting such a public construct of science will do no harm but rather will contribute to the public legitimacy of science. However, portraying science as if it were always guided by a well-structured plan may create or reinforce exactly such expectations in the social environment of science and might finally motivate governance measures with the aim of "improving" science in this respect. The public construct of science – advocated by the self-presentation of science in public communication – may well have unintended consequences for the expectations of the societal environment of science and the resulting governance processes in that it undermines, for example, the legitimacy of scientific autonomy.

11.3.2 Repercussions on the Research Process

To empirically prove repercussions of anticipated media responses on the research process and its output (i.e., knowledge and publications), to measure their strength and assess their consequence for scientific quality is rather difficult. While case studies might provide conclusive evidence for the individual cases investigated they are not able to determine how common such cases are. Representative surveys of scientists, on the other hand, depend on scientists' perceptions of media impacts on the research processes, and the researchers' answers may furthermore be biased by social desirability effects thus making them reluctant to admit the relevance of anticipated mass media responses for scientific decisions. Further research is thus needed to better understand the conditions under which medialization of the research process is likely to occur, and its impacts on the direction of science and the characteristics of scientific knowledge. In the following, we present some limited evidence that researchers do sometimes anticipate media responses when making scientific decisions – first, from a "case study" based on information from a university public information officer who had co-operated with a researcher to increase the public impact of his research; second, from our surveys of scientists and public information officers.

In the early 90 s, Sean B. Carroll, a genetics researcher at the University of Wisconsin-Madison, who studied the genes determining the development of body patterns, in particular the development of insect wings and their color patterns, decided to switch the model organism he and his team used for their research from the usual drosophila to butterflies. According to Terry Devitt, Director of Research Communications at University Communications, University of Wisconsin-Madison, this change of model organism was partly motivated by expected higher public impact of research on colorful and admired butterflies as compared to research on uncharismatic fruit flies[2]:

> You can imagine it is much easier to make news if you have found the genes that govern pattern and color in butterflies than in fruit flies. I believe Sean recognized that in advance. I also believe that helped ensure the paper would publish in a high-profile journal. The capacity of a paper to generate news is not lost on journal editors (Devitt 2007).

---

[2] I want to thank Terry Devitt for sharing that information with me, and for agreeing to be quoted.

Carroll's research was published in the science section of the *New York Times* (5 July 1994) and in *Science* (22 January 1999) as well as in other scientific journals and mass media. In both cases, the publications made use of attractive graphics showing the butterfly's wing. Devitt's comments quoted above show that he believes that taking the criteria of the mass media into consideration will also increase the chances of publication in a high-ranking scientific journal (see Chapter 17). He also argues that the anticipation of mass media criteria by Carroll in his research did not at all compromise the quality of the science or the functioning of the research process:

> In short, Sean made an interesting and important discovery that was amplified by his choice of animal model. I don't believe the science was changed or publication was delayed in any way by a chance to make news (Devitt 2007).

Rather than adopting the antagonistic model of media orientation and science orientation advocated by Weingart (2001) and Franzen (Chapter 17), for example, Devitt, along with probably most members of the science-related public information profession and media-oriented scientists, justifies using the attention criteria of the mass media in scientific research by a kind of "synergy model": The combination of quality of science and anticipation of criteria of media attention produces the highest "impact" in the public but also in the scientific realm.

The INWEDIS survey of scientists included a question about the anticipation of media response by researchers when making decisions in the research process. We assumed that scientists would be reluctant to admit such influences as they might contradict scientific norms. In order to minimize the social desirability effect and encourage the respondents to report their relevant observations, we did not ask about the respondents' own decisions but about their knowledge of colleagues who "because of anticipated media publicity" modified decisions such as choosing or avoiding particular research questions, methods, funding sources or collaborators, or timing of publication. The answers have thus to be considered with caution. They are not necessarily based on first-hand knowledge about the decision situation the respondents had in mind when answering the questions, but rather reflect the empathic attribution of criteria and motives to decisions by their peers. Furthermore, we used an introductory text in the question with the aim of making media influences appear more "natural." The results show that a majority of the biomedical researchers surveyed feel that in their scientific community decisions during the research process are (partly) influenced by media criteria (Fig. 11.3).

In all five major knowledge societies studied, the surveyed biomedical researchers assumed that scientific decisions of different kinds (see Fig. 11.3) would be influenced by anticipation of media response. On average, 64 percent of the surveyed biomedical researchers claimed to know colleagues who had been influenced by media criteria in at least one of the six types of decisions mentioned in the question. While there are some statistically significant country differences regarding research methods, research collaborators, and presentation of papers at conferences, country differences of the aggregate measure ("One or more of the

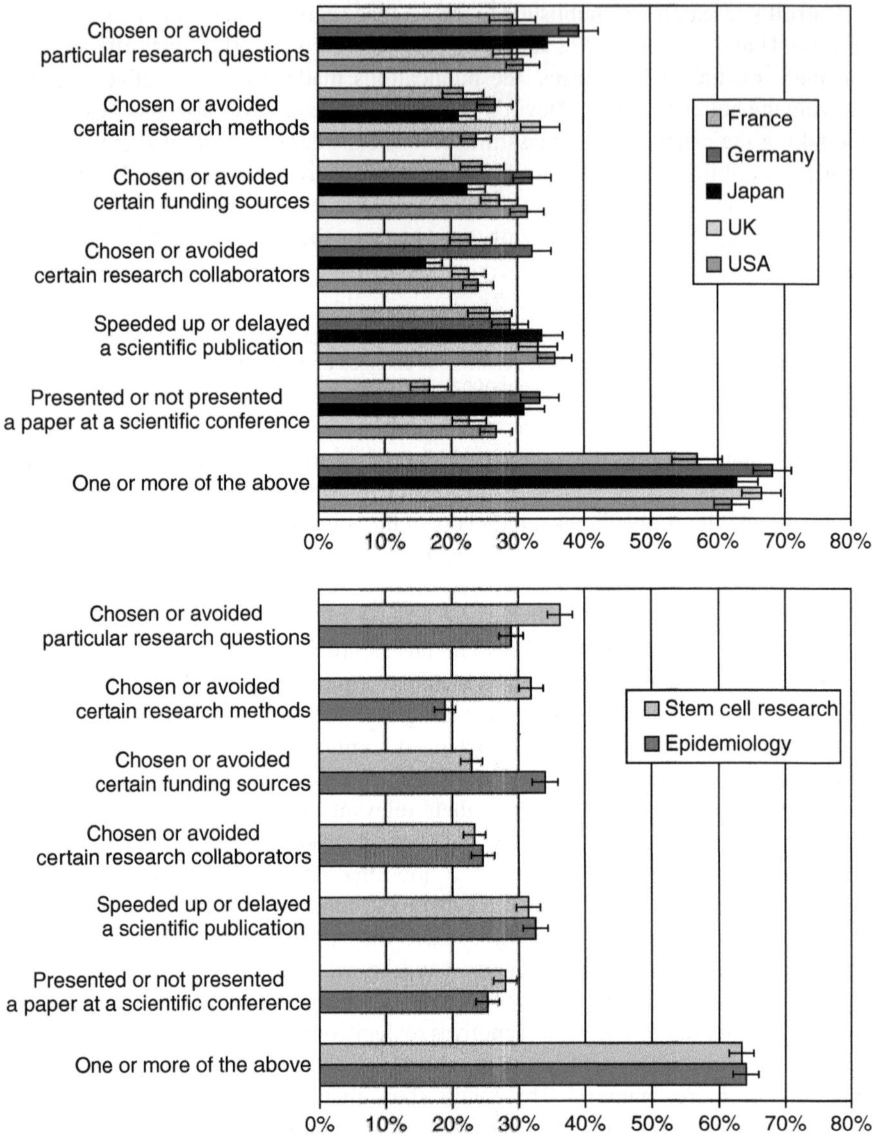

Fig. 11.3 Proportion of respondents assuming that colleagues have modified decisions in the research process in anticipation of media response. The question read: "Scientific decisions are sometimes influenced by factors outside science, such as the availability of funding, the organization's research agenda, or the usability of results. Another factor could be the anticipation of positive or negative media publicity. The statements below pose some possible effects. (Check all alternatives that apply.) Because of anticipated media publicity, scientists I know have . . ."
Source: INWEDIS survey of scientists

above") are not significant ($F = 1.80$, df $= 4$, $p = 0.13$). However, there is an interesting pattern of statistically significant differences between stem cell research and epidemiology. Stem cell researchers claim more media influence on decisions about research questions and methods while epidemiologists claim more media influence on the selection of funding sources (all $p \leq 0.01$, t-test). In view of the respective public images of stem cell research and epidemiology (see Chapter 6), this pattern could suggest a plausible interpretation, supporting the assumption that the scientists' responses to that question are indeed meaningful and not just arbitrary. It is quite likely that the particular sensitivity of stem cell researchers to media criteria related to research questions and methods is a consequence of the public controversy about the use of human embryonic stem cells and the choice between research strategies focusing on adult or embryonic stem cells. The specific sensitivity of epidemiologists regarding funding sources, on the other hand, probably results from the endangerment of trust caused by industry funding of epidemiological studies.

According to the INWEDIS survey of public information officers, almost all research organizations closely monitor the media coverage of their own organization and of relevant other subjects. This effort shows that media coverage is important to them and is probably a factor considered in decision-making. One of the respondents gave an example of such an influence. The respondent said that critical media coverage of industry funding regarding research on tobacco risks led to a new organizational policy for research collaborations:

> And that has resulted in a new code of ethics that prohibits money being accepted from the tobacco industry. That wouldn't have happened without the media. They would have said: '[I] don't care where the money comes from'. So, this actually has had an impact (translated).

This quote shows, as argued above, that organizations are sensitive to media coverage and thus serve as effective mediators of normative public expectations to researchers.

11.3.3 Science-Politics Relationship

As we have argued elsewhere, making use of a general thesis of Imhof (2006) regarding stakeholder strategies in the "media society," the medialization of science can plausibly be understood as a consequence of the medialization of politics (Peters et al. 2008c). The argument is therefore that political developments lead to a modification of the science-media relationship which in turn modifies the political effects of mediated science. Insofar as public visibility of science is affected by the adaptation of science to the media and its orientation towards legitimation goals, political effects of "mediated" science may be plausibly attributed to the medialization of science.

According to the INWEDIS survey of 40 German political and/or administrative decision-makers in the biomedical field (see Petersen et al. 2010), politics in science-related fields pays close attention to journalistic accounts of science and scientific knowledge. The analysis of the interviews revealed several mechanisms by which journalistic accounts of science become politically effective: (1) media visibility is widely interpreted as a "relevance cue" pointing to the necessity of responding or the possibility of relating one's own activities to the events covered (the basis of the agenda-setting effect), (2) journalism pragmatically re-contextualizes science exposing its possible practical uses and regulation demands thus making it compatible with the political logic, (3) media coverage provides a repertoire of facts, arguments, metaphors and symbols originating from science for convenient use in political rhetoric, and (4) decision-makers use stories of trusted media and journalists in forming their personal opinion about science-related policy issues.

Two cases of political effects of medialization of science can be distinguished: first, the (de)legitimation of scientific practices, goals, applications and use of public resources, and, second, the application of scientific expertise in policy fields such as health, risk, environment and global climate change, for example. In the first case, science itself is the governed field and scientific actors are thus stakeholders; in the second case science as expertise contributes to the governance in other policy fields.

Within politics, non-negative media visibility of actors and initiatives is interpreted as an indicator of political success, since media attention suggests relevance for their voters. Accordingly, media visibility of research findings, research fields, research organizations, research institutes or research projects serves as a cue that these are in some way or another meaningful for the media audience and political actors might improve their public image if they respond to public science and relate their own political activities to it. That is the assumption underlying the publicity strategies of scientific communicators. They calculate that their political addressees might be inclined to support successful – i.e., marked as "relevant" by media visibility – research and research organizations in order to increase their own political standing. Of, course, by the same mechanism negatively framed science in the media might also stimulate science governance activities with the goal of increasing congruence between scientific practices and public expectations.

While there are several ways of constructing public relevance, linking scientific knowledge and research results to existing policy issues, or creating new challenges to politics by identifying new risks, for example, is a very effective way of attracting journalistic attention and presenting oneself as addressing wider societal goals. Pointing to the need of scientific communicators to connect science to extra-scientific issues in order to meet journalistic selection criteria, and to its reception in the context of medialized politics, Petersen et al. (2010: 869) argue "that scientific expertise in the mass media is observed by political decision-makers and effectively enters the policy-making process." They consider mass-mediated scientific expertise as an important mechanism for increasing its political impact – as a complement to the many direct ways of science-policy interactions (e.g., expert commissions, reports or hearings). One may even go a step further and hypothesize that this mechanism of constructing public relevance by relating it to policy-making does not only lead to public communication of expertise, but even stimulates its creation. Journalism tends to pragmatically re-contextualize science and integrate it with case-specific knowledge (see Spinner 1988), and thus motivates scientific sources to anticipate this demand.

11.4 Discussion

The analysis of the orientation of individual and organizational scientific communicators towards the mass media has resulted in evidence of medialization of science in the sense that scientific communicators anticipate media criteria and media routines. On the individual and organizational level, we found indications of institutionalization, professionalization and strategic utilization of media communication motivated above all by the goal of securing public legitimacy and thus societal support for science. We also found some evidence that the orientation towards media criteria is not limited to public communication contexts but competes or intermingles with scientific norms and routines, thus constraining scientific autonomy. The question that still remains open is whether concrete examples of repercussions of media criteria on research processes and results can be found, are these isolated exceptions perceived by peers as a violation of scientific norms and mobilizing sanction mechanisms of the scientific community, or are they "tips of the iceberg," i.e., visible indicators of a mostly hidden but rather general trend of changing (or corrupting) scientific norms? It seems useful to distinguish different possible "shells" of medialization from the surface of public self-presentation to the core of knowledge creation in future studies of medialization. On the outer shell, public communication behavior of scientific communicators would be guided by media criteria, in the medium shell scientific publication, research processes and allocation of resources would be affected, and in the inner core scientific knowledge itself, i.e., the scientific construction of facts and theories, would be modified by the aim of increasing their compatibility with media criteria of newsworthiness. The empirical data presented in this chapter show medialization effects in the outer and medium shell; but it cannot prove effects (or their absence) in the epistemic core of science.

The implications of medialization for scientific autonomy may be ambivalent, however. In analogy to the concept of "symbolic politics" (Sarcinelli 1987), one might assume that there is a growing differentiation between the presentation of science in the media and the actual processes in science, and a limitation of medialization effects to the presentation. According to Merten (2000), the very essence of public relations is the construction of a "desirable reality." Besides the possible effect of a diffusion of media criteria from the outer shell towards the core of knowledge production, self-presentation of science to the media by efficient public relations (in a broader sense) showing its conformity with media criteria may actually result in a defense of scientific autonomy, in that it hides its esoteric and erratic nature that might otherwise give rise to provocations in the social environment and stimulate governance activities. The "branding" concept in the self-description of organizational science PR may be understood in a similar way. The ascription of symbolic properties matching public expectations to the media image of organizations makes their appreciation relatively independent of their "real" inner properties. In the same way as consumers are expected to respond to the quality promise of a "brand" in making purchase decisions, rather than examining the quality of the specific products they intend to buy, citizens and decision-makers are expected to respond to the corporate "brand" of a research organization rather than to critically examine and interfere with actual processes and outcomes in detail. Establishing a "brand" for research organizations that meets public expectations may thus mean relieving research organizations of specific external demands. Our data does not allow us to answer the question of which of the two possible effects – shielding against external influences or transferring media criteria to inner-scientific decision-making – is dominant; very likely the answer would vary with the kind of science-society relationship.

While Weingart (2001), focusing on science, sees mostly the negative impacts of medialization and points to risks for scientific autonomy and quality, from a science governance perspective an alternative – more positive – view on the medialization of science is also possible. Scientific autonomy is limited by several societal influences such as collaborations with industry and the medical system, and science and technology policy, for example. Each form of science governance in one way or another has to interfere with decisions made by scientists and research organizations and thus limits their autonomy regarding the setting of research priorities, use of public resources, acceptance of research methods and research findings. One might view the medialization of science as an informal governance mechanism, based on the potential of the mass media to legitimate or de-legitimate research. Whether medialization is more intrusive (and corruptive) than other forms of science governance that are based on criteria of political or civil research funders, economic utility or the principles of "New Public Management" still remains to be shown. Since the media refer to the expectations and values of their audience in their observation of science, the media-based governance mechanism might even be considered to be participatory in nature, involving the "public" in an indirect way in the governance of science.

References

Boltanski, L. and P. Maldidier (1970). Carrière scientifique, morale scientifique et vulgarisation. *Informations sur les sciences sociales*, 9(3), 99–118.

Borchelt, R. E. (2008). Public relations in science. Managing the trust portfolio. In M. Bucchi and B. Trench (eds.), *Handbook of public communication of science and technology*. London: Routledge, pp. 147–157.

Devitt, T. (2007). Personal communication (email, received 10 August 2007).

Dunwoody, S. and M. Ryan (1985). Scientific barriers to the popularization of science in the mass media. *Journal of Communication*, 35(1), 26–42.

Goodfield, J. (1981). *Reflections on science and the media*. Washington, DC: American Association for the Advancement of Science.

Imhof, K. (2006). Mediengesellschaft und Medialisierung. *Medien und Kommunikationswissenschaft*, 54(2), 191–215.

Jensen, P., J.-B. Rouquier, P. Kreimer, and Y. Croissant (2008). Scientists who engage with society perform better academically. *Science and Public Policy*, 35(7), 527–541.

Kallfass, M. (2009). Public Relations von Wissenschaftseinrichtungen – explorative Studie in Deutschland, Frankreich und Großbritannien. In H. P. Peters (ed.), *Medienorientierung biomedizinischer Forscher im internationalen Vergleich*. Jülich: Forschungszentrum Jülich, pp. 101–175.

Kohring, M. (2005). *Wissenschaftsjournalismus. Forschungsüberblick und Theorieentwurf*. Konstanz: UVK.

Maasen, S. and P. Weingart (2006). Unternehmerische Universität und neue Wissenschaftskultur. *die hochschule*, 15(1), 19–45.

Merten, K. (2000). *Das Handwörterbuch der PR*. Bd. 1. Frankfurt/M.: F.A.Z. Institut.

Münch, R. (2009). Die Inszenierung wissenschaftlicher Exzellenz. Wie der politisch gesteuerte Wettbewerb um Forschungsressourcen die Wissenschaft den Darstellungszwängen der öffentlichen Kommunikation unterwirft. In H. Willems (ed.), *Theatralisierung der Gesellschaft. Band 1: Soziologische Theorie und Zeitdiagnose*. Wiesbaden: VS Verlag für Sozialwissenschaften, pp. 485–497.

Nelkin, D. (1987). *Selling science: How the press covers science and technology*. New York: Freeman.

Nerlich, B. (2009). Breakthroughs and disasters: The (ethical) use of future-oriented metaphors in science communication. In B. Nerlich, R. Elliott, and B. Larson (eds.), *Communicating biological sciences*. Farnham: Ashgate, pp. 201–218.

Peters, H. P. (1984). *Entstehung, Verarbeitung und Verbreitung von Wissenschaftsnachrichten am Beispiel von 20 Forschungseinrichtungen*. Jülich: Kernforschungsanlage Jülich.

Peters, H. P. (ed.) (2009). *Medienorientierung biomedizinischer Forscher im internationalen Vergleich. Die Schnittstelle von Wissenschaft & Journalismus und ihre politische Relevanz*. Jülich: Forschungszentrum Jülich, http://nbn-resolving.de/urn:nbn:de:0001-00542.

Peters, H. P., D. Brossard, S. de Cheveigné, S. Dunwoody, M. Kallfass, S. Miller, and S. Tsuchida (2008a). Science communication: Interactions with the mass media. *Science*, 321, 204–205.

Peters, H. P., D. Brossard, S. de Cheveigné, S. Dunwoody, M. Kallfass, S. Miller, and S. Tsuchida (2008b) Science-media interface: It's time to reconsider. *Science Communication*, 30(2), 266–276.

Peters, H. P., H. Heinrichs, A. Jung, M. Kallfass, and I. Petersen (2008c). Medialization of science as a prerequisite of its legitimization and political relevance. In D. Cheng et al. (eds.), *Communicating science in social contexts*. Dordrecht: Springer, pp. 71–92.

Petersen, I., H. Heinrichs, and H. P. Peters (2010). Mass-mediated expertise as informal policy advice. *Science, Technology & Human Values*, 35(6), 865–887.

Sarcinelli, U. (1987). *Symbolische Politik. Zur Bedeutung symbolischen Handelns in der Wahlkampfkommunikation der Bundesrepublik Deutschland*. Opladen: Westdeutscher Verlag.

Schäfer, M. S. (2008). Medialisierung der Wissenschaft? Empirische Untersuchung eines wissenschaftssoziologischen Konzepts. *Zeitschrift für Soziologie*, 37(3), 206–225.

Singer, E. (1990). A question of accuracy. How journalists and scientists report research on hazards. *Journal of Communication*, 40(4), 102–116.

Spinner, H. F. (1988). Wissensorientierter Journalismus. Der Journalist als Agent der Gelegenheitsvernunft. In L. Erbring (ed.), *Medien ohne Moral*. Berlin: Argon, pp. 238–266.

Sturgis, P. and N. Allum (2004). Science in society: Re-evaluating the deficit model of public attitudes. *Public Understanding of Science*, 13(1), 55–74.

Stollorz, V. (2005). Indecent proposals – How should we (re)act? *WPK Quarterly. Magazine of the German Association of Science Journalists*, November, 8–10.

Weingart, P. (2001). *Die Stunde der Wahrheit? Zum Verhältnis der Wissenschaft zu Politik, Wirtschaft und Medien in der Wissensgesellschaft*. Weilerswist: Velbrück.

Willems, J. T. (1976). *Wetenschapsjournalistiek: Klachten en Communicatiebarrieres*. Nijmegen: Gerderlander Pers.

Chapter 12
Campaigns and Coalitions: Governance by Media

Joan Haran

12.1 Introduction

In this chapter, I discuss the establishment of the Science Media Centre in 2002, "housed within the Royal Institution but independent from it" as an instantiation of the concerns that UK scientists have with the mass media as the primary means for transmitting the right message about science to the public. I suggest that there is an implicit tension – even a contradiction – built into the organisation's mission, although I would certainly not contend that this is a wilful or malicious contradiction. I also want to suggest that some of the most "successful" interventions of the Science Media Centre have resulted from cooperation with scientists predisposed to particular modes of media engagement because of contingent histories. I refer in particular to scientists involved with cloning and stem cell science who have a significant track record of engaging with public debate about embryo research.

In this chapter, I comment on a number of accounts of the Science Media Centre's foundation, mission and function, and suggest that the aforementioned contradictions impact on the way in which it functions, and may – over time – exacerbate the issues it was founded to tackle. I do so from the perspective of a researcher with a critical orientation to issues around the public understanding of science. That is to say that I am interested in the democratic governance of science, and how it might be affected by the science/media relationship, rather than in the rectification of any putative deficit of scientific knowledge. I am interested in considering science "from the citizen's side rather than from that of the scientific establishment" (Irwin 1995: 5), and as such, I am interested at least as much in how citizens or the public are framed, as in how science is framed in media representations.

My concern is that, in its contribution to the setting of media agendas and to the framing of news stories, the Science Media Centre (SMC) potentially obfuscates the distinctions between scientific knowledge, scientific practice and science

J. Haran (✉)
Cardiff University
e-mail: HaranJ@Cardiff.ac.uk

S. Rödder et al. (eds.), *The Sciences' Media Connection – Public Communication and its Repercussions*, Sociology of the Sciences Yearbook 28,
DOI 10.1007/978-94-007-2085-5_12,

governance to which Wynne called attention in one founding moment for Public Understanding of Science in the UK (Wynne 1991, see also Wynne 1992). By conflating anxieties with regard to perceived knowledge and/or trust deficits, models of rational or irrational publics for science are constructed in the process of defining the communication task. This has consequences for both the messages the SMC produces and the publics or audiences to whom they are addressed. Of course, I would not want to suggest that scientific knowledge, practice and governance are entirely discrete domains, but I would argue that some attempt must be made to hold them apart in order for "the public" – and, indeed, policymakers – to be constituted as meaningful actors in relation to science.

Fiona Fox (Chapter 13) locates the origins of the SMC in the House of Lords (2000) Select Committee Inquiry into Science and Technology and the broader context of the BSE crisis. Science studies scholars might point to a more extended context and frame the issues slightly differently (Irwin 2006; Miller 2001).[1] In the chapter that follows I will expand on this suggestion and explore how that shift in context might suggest that there is a fundamental mismatch between the definition of the problem and the proposed solution to the "decline of trust in science". I do so speculatively, following Donna Haraway's strategy of retelling origin stories in order to disrupt them (Haraway 1984).

The SMC is undoubtedly very effective in its press relations, but I want to consider whether this very effectiveness allays or exacerbates the concerns it was set up to address. In asking these questions, I will note that the success story about which Fiona Fox writes in this collection – the media coverage of debates about legislating to permit hybrid embryo research under the 2008 Human Fertilisation and Embryology Act – was contingent on the prior existence of a strong and vocal community of scientists and clinicians keen to secure their continued license to practice, and with a successful track record of so doing. In preparing this chapter, I have focused on the public and mediated accounts of the SMC's mission, as well as evidence of its activities that can be inferred from comparisons of the briefings and press releases published on its website with science news stories in UK broadsheets and tabloids. In addition to Fox's essay, previously published in *Communicating Biological Sciences*, I also draw on material published on the SMC's website and the Third Report of the House of Lords Select Committee on Science and Technology, as well as my own research on the mediation of, and public consultation around, embryo science conducted since 2004.[2]

---

[1] Indeed Walter Bodmer, although taking issue with some of the critiques cited in this chapter, has recently reflected on a quarter century of Public Understanding of Science work in the UK (Bodmer 2010).

[2] This chapter emerges from research conducted between 2004 and 2010, at the ESRC Centre for Economic and Social Aspects of Genomics (Cesagen) on media representations of genomics, particularly at the intersection of genomics and assisted reproductive technologies. The research began with the assembly of an archive of UK press news coverage of genetic science in 2004. This time period including a number of news events relating to embryo research to which the SMC co-ordinated media responses. This initial intensive sampling has been supplemented by purposive

12.2 Engaging the Origin Story of the Science Media Centre

Fox accounts for the foundation of the Science Media Centre as a response to the Third Report of the House of Lords Select Committee on Science and Technology (2000).[3] However, the evidence that was collected for the report and its terms of reference may be more expansively understood as further steps in an ongoing Public Understanding of Science movement, rather than the start of something entirely novel. As Steve Miller (2001: 115) notes:

> It is always dangerous to date the start of a historical process. But for the recent movement for public understanding of science within the United Kingdom, the publication by the Royal Society, in 1985, of a report entitled 'The Public Understanding of Science' – known as the Bodmer Report after the chair of the working group, Sir Walter Bodmer – is a reasonable place to begin a short survey of recent public understanding of science activity.

The publication of the Bodmer Report and the subsequent establishment of the Committee for the Public Understanding of Science (CoPUS)[4] does indeed seem to be a salient marker in relation to UK attitudes to the science/society relationship. Social scientists in the UK might further point to the Economic and Social Research Council's funding of multiple social science research projects in this area beginning in 1988, or the establishment of the journal *Public Understanding of Science* in 1992.[5] Unlike the 2000 Report, the Bodmer Report was initiated from within the UK's scientific learned societies and one interpretation of the later Inquiry might be that its production from within the Houses of Parliament was an index of the success of the initiatives sparked by Bodmer, such that the importance of the Public Understanding of Science is now on the agenda of policymakers, one of the five "overlapping functional categories" of the public defined in the Bodmer Report (The Royal Society 1985: 7). In a paper delivered at a conference about the Royal Society and science in the twentieth century, Sir Walter Bodmer's (2010) interpretation was somewhat less sanguine, viewing the Third Report as at least an implicit critique of his earlier work. He also indicated that he believed that social science scholars in the PUS field had misunderstood or misrepresented the thrust of his working group's interventions which he summarised as being "directed at the need for scientists to

---

sampling of subsequent key news events in the area, including those associated with the Hwang scandal and those related to the reform of the Human Fertilisation and Embryology Act (Haran 2007; Haran and Kitzinger 2009).

[3] The Third Report is sometimes known briefly as "Science and Society", but for clarity in this chapter I will refer to is as The Third Report.

[4] The Committee for the Public Understanding of Science, formed on the recommendation of the Bodmer Report, had representation from each of the three bodies of the Royal Society, the BA (now known as the British Science Association), and the Royal Institution, thanks in part to the then President of the Royal Society, George Porter who was also – uniquely – President of the BA and Director of the RI. (See Bodmer 2010 for one account of the formation of COPUS).

[5] Bodmer (2010) points out that the funding of PUS research through the UK Research Councils was actually one of the recommendations of his 1985 Report.

learn how to communicate with the general public in all its guises, and to consider it a duty to do so" (Bodmer 2010: 1).

A number of scholars in the social studies of science and in the field of Public Understanding of Science (see McNeil (2012) for an account of how these fields do and do not intersect) have indeed suggested that the House of Lords Third Report (2000) can be interpreted in relation to the Bodmer Report and the successes and failures of CoPUS[6] (for example, Miller 2001). Others have noted that the language of Science and Technology Studies, albeit in an attenuated form, has been taken up as the language of science policy (for example, Irwin 2006). They note that concern for the public's cognitive deficit in relation to science has, to a variable extent, been substituted by a concern for the public's trust deficit in relation to science. Writing shortly after the publication of the Third Report, Miller (2001: 117) is keen to witness the dawn of a new age for Public Understanding of Science that he sees embodied in the report:

> In contrast to Bodmer, bemoaning the level of public ignorance and the fickle nature of the media is almost totally absent. Instead, the report is peppered with calls for dialogue, discussion and debate. A new era, which perhaps really opened when Science Minister Lord Sainsbury pronounced the 'demise of the deficit model' at the 1999 meeting of the BAAS,[7] is being ushered in. So what will this 3D – dialogue, discussion, and debate – world of newPUS (*sic*) look like?

Miller was considering the 2000 report in its entirety, and not simply the section on science and the media, but it is interesting to consider the origin stories of the Science Media Centre in the context of this apparent new age for Public Understanding of Science, or what Irwin (2006) calls the "new" scientific governance.

12.3 Science Media Centre Origin Stories

There are three key versions of the origin story of the Science Media Centre to which I will refer in this chapter. One is embodied in the Science Media Centre Consultation Report from March 2002.[8] The second forms part of an earlier version of Fiona Fox's chapter in this volume (Fox 2009b). The third is the aforementioned House of Lords Select Committee on Science and Technology's Third Report (2000) which is heavily referenced in the SMC Consultation Report. I will refer mainly to the first two versions, noting the ways in which they refer to the third as justification or rationale.

---

[6] The acronym is sometimes rendered as CoPUS (see, e.g., Miller 2001) and sometimes COPUS (see, e.g., Bodmer 2010).

[7] Miller uses the acronym BAAS for the British Association for the Advancement of Science, although it was generally known – somewhat confusingly – as the BA. The organisation was relaunched in January 2009 as the British Science Association.

[8] Report downloaded from http://www.sciencemediacentre.org/consultation.htm 19/09/05 Now available at http://www.sciencemediacentre.org/pages/publications/last accessed on 30 November 2010.

The Science Media Centre's Consultation Report is still available on their website, although it has a less prominent location than earlier in the organisation's life. The Introduction to the report claims that it

> seeks to show how the service that the SMC will offer has emerged from a general consensus among key stakeholders as to where the real problem lies and how the SMC can best direct resources to make a positive contribution to media coverage of science (SMC 2002: 2).

The stakeholders identified in the report were limited to scientists, journalists, press relations professionals and "politicians and policy makers with an interest in science". This suggests that rather than providing information on science issues that could be weighed up in democratic debate, the media relations task of the SMC is understood as advocating on behalf of stakeholders. Alternatively, a more generous reading would be that the interests of some key stakeholders are weighed in relation to each other, but the stakeholders are limited to scientists, policymakers and science communications professionals. This is hardly an audience, consumer, or citizen focused dialogic process, but rather a knowledge-producer, and indeed a science-centred focus that seems to view public understanding as just one phase in governance by, rather than governance of, science. Indeed, the consultation report concludes by describing the SMC's vision as: "Good public policy decisions on science based on a more balanced, rational, accurate debate within the news media about science issues" (SMC 2002: 23). It is really not clear from such a vision where the public might come in, except in a limited sense as eavesdroppers on a conversation between scientists and policymakers mediated by press officers and journalists. So, those chapters of the *Third Report* which embodied a more inclusive voice – such as Chapter 5 "Engaging the Public" which refers to "a new humility on the part of science in the face of public attitudes, and a new assertiveness on the part of the public" – are set aside for the scientistic voice which attributes rationality and accuracy to the experts. In his own analysis of the report, Irwin references its oscillation between Hagendijk's two voices of inclusiveness and scientism and notes the emphasis in the report on the "presumption of openness as a means to secure science's 'license to practice'" (Irwin 2006: 306). The degree to which the leadership provided by Baroness Greenfield, then Director of the Royal Institution, influenced the SMC in favour of the scientistic rather than the inclusive voice is open to question. However, the incorporation of the following statement in the consultation report seems to be indicative of the imagined relationship between science and society that is at the root of the terms of reference arrived at for the SMC:

> In the twenty-first century science is going to be at the centre of all the things we most cherish: food, health, reproduction, education, business, communications and so on. In order to meet the challenges and minimise the problems, the public need the scientific literacy to engage with all the major scientific developments in these areas (SMC 2002: 2).

So the much critiqued deficit model of the Public Understanding of Science is still at the centre of this imaginary, despite the contributions to the consultation process of "a significant minority [who] saw the growth of public questioning of scientific expertise as a largely positive development and a step towards the widely held goal of 'democratising science'" (SMC 2002: 3). This "significant minority" included social sciences and humanities scholars.

The *Third Report* had enjoined scientists to learn to work with the media, and this is interpreted to a large degree in the consultation document as understanding the hard news agenda. The message that scientists took from journalists' input into the process was that "what appears on the news pages of The Times, Telegraph, Guardian and Daily Mail is what appears in MP's mailbags" (SMC 2002: 5). There is little suggestion in this overview of the consultation process that journalists may have vested interests in talking up the power and reach of their profession and of media outlets. Science correspondents are represented in this report as extremely bullish about media coverage of science. They think that there are plenty of good stories available, they appear to share with scientists a mission to inform and educate those with a deficit in scientific knowledge, and they urge the SMC to focus more on the news correspondents. Tim Radford, then Science Editor of *The Guardian*, is quoted in both the *Third Report* and the *Science Media Centre Consultation Report* giving support to such a perspective. However, post-retirement he gives a slightly different account of the role of science journalists:

> Newspapers follow the same imperative [as Scheherazade]: the day the readers stop reading, they stop buying, and the newspaper dies. So even the science stories in newspapers are just that, stories. They are drawn from the world of science. They are told for serious purpose. But they are told so as to give pleasure. It is not our business to advance the public education in science, except by the way, and as a kind of happy accident. It is our business to be read [...] (Radford 2009: 151–152).

Nonetheless, in the SMC consultation report, science correspondents represent themselves as, and are represented by the SMC, as being extremely well informed and well networked into science and it is the role of more generalist journalists that is emphasised as requiring attention. According to the report:

> The SMC is being lobbied to position itself in the arena of science as news, science as controversy and in areas of public concern about science. This is perceived as the area in which there are the most problems, the most potential, and because to date (*sic*) no other initiative has this as its primary focus (SMC 2002: 11).

The Consultation Report outlines how the SMC intends to respond to this lobbying with what they call pro-active reaction to the news agenda:

> The centre will react to the media agenda by pro-actively promoting comment, interviews and articles from scientists and others when science hits the headlines.
>
> With the help of the Science Advisory Panel and the horizon scanning sessions the SMC will endeavour to anticipate and prepare for the next major news story or controversy in science and be ready for a swift response (lining up interviewees, articles, comments on several subject areas in advance, etc.) (SMC 2002: 12).

If the SMC is endeavouring to position itself in areas of public concern about science, public concerns are notably absent from this account of the SMC's strategy. One might wonder if public concerns are implicitly referenced as something which must be anticipated or pre-empted in the horizon scanning sessions and, if so, whether "pro-active reaction" bespeaks a commitment to dialogue, discussion and debate, or simply a more effective and tightly-focused mechanism for communicating scientists' concerns about science, a much less dialogic approach to public relations than the *Third Report* seemed to be promising in Miller's optimistic account (Miller 2001).

Perhaps prompted by awareness of such reservations, in a more recent account of the SMC's operation published in 2009, Fiona Fox, the Director of the Science Media Centre since its founding, claimed that it "is a press office that is completely independent, eschews PR and has no brand or institution to promote" (Fox 2009b: 109). Each of these three claims is highly contestable. The Science Media Centre could not exist without funding from a range of stakeholders, nor could it function without the co-operation of the scientists who attend its press briefings or provide comments for the press releases or round-ups of commentary written and/or collated at the Science Media Centre. The claim to independence rests on a funding formula which caps the amount that individual funders can "donate". To suggest that independence is simply a financial issue seems odd – naïve or disingenuous – when the centre is as invested in the pursuit of relationship-building as it proudly acknowledges itself to be. Disavowing public relations (PR) is an even odder claim. The centre uses the time-honoured routines of PR – press briefings, press releases, round-ups of quotes – so it is hard to know which aspects of the public relations business are being disclaimed. Finally, to suggest that there is no brand to promote only makes sense if one is extremely literal-minded and equates brands with consumer goods. The Science Media Centre treats "Science" as a brand – albeit one with multiple product lines – and aims to help it compete with other brands – such as religion – in the marketplace of media attention. To be fair, the Science Media Centre is an unusual organisation. It does not follow either the model of a retained public relations consultancy, which exchanges a certain amount of time and labour for a pro rata fee, nor the model of an in-house press office with internal clients from just one organisation. It is rather a somewhat peculiar hybrid of the two, with a multiplicity of potential "clients" and a financial structure that compensates for time and labour but with no direct pathway from client to outputs to compensation. So perhaps it is this disruption in the path between inputs and outputs on which Fox bases her claims of independence. However, without some financial investment from the SMC's client base, and some sense from that client base that the centre represents value for money, it is difficult to see how the organisation could be sustained. Perhaps what she is trying to suggest with the formulation "eschews PR" is that the SMC is not in the business of "puffery" or "spin", but many public relations professionals would understand their role similarly to be about the dissemination of information – leaving persuasion to their colleagues in advertising, marketing and sales promotion – so perhaps in that sense the SMC is not quite as distinctive as she suggests. However, the line between information and persuasion can be a hard one to draw when in the business of securing share of voice or share of audience for a client, so I would suggest that the SMC is as open to these risks as other organisations performing public relations.

Fox writes: "The Centre's official philosophy [is] that 'The media will "do" science better when scientists "do" media better'" (Fox 2009b: 109). She notes that this vernacular philosophy mirrors the sentiments of "the BBC's veteran science reporter Pallab Ghosh who has long called on scientists to 'get off the sidelines, learn the rules of the game and get on the pitch'" (ibid.: 109). This injunction to scientists to learn the rules of the media game and to play by them in the bid to "help renew public trust in science" (SMC 2002: np) takes for granted the following propositions: that there is a deficit in scientists' understanding of the media game; that there are rules to this game that can be learnt; that learning and applying the rules will lead to the media "doing" science better as a corollorary; that this remedial action is the appropriate way to restore public trust in science.

Beginning with the last of these propositions, the spectre of the BSE crisis looms large, but it is important to distinguish between the lack of trust that members of the public might have in government scientists or scientists representing, for example, the interests of food producers, from the trust that they place in publicly funded scientists working on research with prospective medical applications. *The Third Report,* "the hugely influential House of Lords Select Committee Report on Science and Technology (2000), which gave rise to the setting up of the Science Media Centre" (Fox 2009b: 109), claims unequivocally that

> the culture of United Kingdom science needs a sea-change, in favour of open and positive communication with the media. This will require training and resources; above all it will require leadership [. . .] It will inevitably involve occasional embarrassment or frustration. But, if it succeeds, it will pay for itself many times over in renewed public trust (Chapter 7).

The authority with which this claim is stated is rather at odds with its lack of specificity, particularly as the report itself does note the distinction in trust attribution that I mention above. What kind of public trust is being sought? Is it trust that scientists will engage openly and positively with the media? Trust in the contents of scientific knowledge? Trust in the practice of science? Trust in the institutions of science? Is this particular sea-change in "the culture of United Kingdom science" standing in for other kinds of change argued for by science scholars with regard to the institutions of science and science policy and, if so, is it a useful supplement to, or a diversion from rethinking the complexity of the science/society relationship? With regard to the proposition that there is a deficit in scientists' understanding of the media game, the report cites Tim Radford's claim that "the cadre of scientists with 'media savvy' is expanding" (House of Lords 2000: 7.39.) but goes on to note: "So too, however, are the ranks of scientists with bad experiences of dealing with the media, who may be fearful of engaging with them again" (loc. cit.) The Science Media Centre has certainly enrolled a large number of scientists to provide content for its news briefings – over 2,000 according to Fox (2009b), but has this really changed the tendency reported by Sir Robert May and Dr. Farmelo of the Science Museum, for the media to be "too reliant on a small number of highly articulate spokesmen for science"? (loc. cit.) I will suggest in Section 12.4 that there are institutional histories and relationships that still make this over-reliance more likely than not. Further, the ongoing economic pressures on the print news media in particular (Williams and Clifford 2009) means that organisations like the SMC, which can act as intermediaries to lift the burden from overstretched journalists, may simply shift this dependency from a few well-known and articulate scientists to the conduit provided by the SMC, which may suit those individuals and institutions that can buy into what the SMC requires of them, but which still leaves out in the cold those who do not. Fox claims that a

> clear consensus emerged about the weak spots in science media relations where the Centre could make the biggest difference [. . .] the apparent inability of many in the scientific community to react quickly, confidently and effectively to breaking news about controversial science stories (Fox 2009b: 112).

In developing a database of experts who could respond to this challenge,

> scientists and engineers are selected not just for their proven expertise, but also for their willingness and ability to engage with the media when their issues hit the headlines. Given the SMC's focus, those joining our database know they are signing up to our goal of improving the way issues are covered by getting stuck into the media debates rather than shouting from the sidelines. [. . .] they understand that the SMC will call on them to drop everything – even late at night or at weekends – to do media work when their area hits the headlines (Fox 2009b: 115–116).

However, laudable as the ambition might be to ensure that spokespersons with the appropriate expertise and commitment are available to respond to controversy, as Fox notes: "The SMC is certainly not restricted to reacting to the headlines, and since opening we have facilitated scientists to generate their own headlines on many occasions" (Fox 2009b: 117). It is this aspect of the SMC's work that I would suggest is particularly risky if the SMC's core mission is to restore public trust in science. Facilitating clients to generate their own headlines seems to me to be the essence of the kind of public relations from which Fox seeks to distance the SMC. Running the risk of stirring up controversy or creating hype whilst generating such headlines is fraught with the potential for backlash. In the next two sections, I discuss some of SMC's interventions in relation to the field of embryo research in relation to this potential.

12.4 Pro-active Reaction to the Cloning News Agenda

One early noteworthy example of the SMC's "pro-active reaction" to the news agenda was the concerted response by UK science sources to the claims made by Dr. Panos Zavos – about reproductive cloning – in a press conference in London in January 2004. The press coverage of this news event bore all the hallmarks of the pro-active reactivity that the Science Media Centre outlines in its consultation report. It also bore out the distinctiveness of the SMC that is argued for in the report: "By not being a part of any one institution, or promoting its own brand name, the SMC can potentially look to the whole world of science for spokespeople and comment" (SMC 2002: 14). Fiona Fox (2009b) argues that publicity for maverick cloners "was feeding the totally inaccurate view that mainstream science was in a race to clone the first human" and that the SMC wanted to tackle this by encouraging stem cell scientists working on therapeutic cloning to challenge these stories in the media rather than boycotting them. The SMC was successful in mobilising "the leading cloning and fertility experts with whom we had been debating the issue" and thus providing a clear news frame distinguishing reproductive cloning from therapeutic cloning and gaining coverage across the UK national press and television news broadcasts. The strategy also allowed journalists to "have their cake and eat it", as they could report Zavos's sensational claims – with all the entertainment value they might provide (not to mention avoiding the risk of being scooped in the very slim chance the claims had any foundation) – whilst distancing both themselves and the scientific establishment from them.

However, in the case of embryological science and therapeutic cloning, the continued prominence of key spokesmen (rarely spokeswomen) in media coverage is largely a result of the very close working relationships established by the Science Media Centre with this small cadre of media savvy scientists and the SMC's provision of the infrastructure to stage briefings. Arguably, however, there was no need for the SMC to change the culture within this particular niche of the science community so that they would view headlines as a golden opportunity. In the field of embryonic stem cell research and therapeutic cloning, there was a convergence of the experience gained in lobbying and media communications in support of embryo research provision in the 1990 Human Fertilisation and Embryology Act (Mulkay 1997) with the media orientation of the Human Genome Project that Hilgartner (Chapter 10) discusses. Add into the mix cloning controversy and what Hilgartner calls "media orientation as a form of theatrical self-consciousness" seems somewhat over-determined (Haran et al. 2008). Additionally, as Elliott (Chapter 5) notes, leading science journalists seemed to have lost their critical distance from this particular area of science endeavour. He quotes from the same Tim Radford chapter I have already referenced to emphasise the extent to which science journalists were enrolled by the story of the prospective benefits of stem cell research and employed "hype, spin, naivete and ruthless headline grabbing".

12.5 "A Model of How Scientists Should Engage with the Media"?[9]

So when Fiona Fox claims that "In May 2008, our elected representatives reflected public opinion in favour of allowing the [human-animal embryo] research to continue", and highlights that "many of the approaches that the SMC has pioneered were applied to the media work around this contentious issue" (Fox 2009b: 126–127), her claims warrant further investigation. If it seems unfair to focus on one particular news story – albeit one with multiple episodes over 3 or 4 years – to the exclusion of the other stories funneled through the SMC I would note that the SMC itself wishes to represent it as exemplary. Fox points out that

[9] Fox (2009b: 126).

> it is a little known fact that leading stem cell scientists were the first to kick off the media debate about human-animal embryo research when they briefed science reporters at the SMC in June 2005. Stephen Minger, Robin Lovell-Badge and the late Anne McLaren did what all scientists should do: they took one of the most controversial areas in their field and gave journalists an in-depth briefing on the scientific potential as well as the ethical issues around this research; even voicing their concerns about a gaping regulatory loophole (Fox 2009b: 126)[10].

This laudatory summary does not give any accounts of the motives or interests of these scientists which include securing their license to practice this type of research in the light of a government review of the 1990 Human Fertilisation and Embryology Act announced in 2004, including the holding of a public consultation exercise in 2005.

The ongoing news coverage of "animal-human admixed embryos", as they came to be called, certainly seemed to suggest that science journalists were fully persuaded by these briefings, in the same way that Radford suggests they were earlier in regard to human embryonic stem cell research, as they framed the government's public consultation as "biased" and represented any potential opposition to legislating to permit the research as irrational. The briefing in 2005 is not the only one that Fiona Fox mentions on this topic. In her blog, *Science and the Media*, she writes that a further briefing, on 4 January 2007, was planned

> after eagled eyed Evan Harris MP and leading stem cell expert Stephen Minger spotted a sentence in the Government's White Paper on fertility laws published in late December which proposed a ban on the use of human-animal hybrid embryos for research (http://fionafox.blogspot.com/2007/01/stem-cell-scientists-seize-media-agenda.html).

Andy Williams et al. (2009) have conducted detailed content analysis which demonstrates that pro-hybrid scientists were the source group quoted most often in the period between January 2006 and November 2008, particularly in the broadsheets, and that overall newspaper coverage contained more pro-hybrid than anti-hybrid sources. Further, they note that pro-hybrid sources dominated coverage during 2006 and 2007 more than they did in 2008 following the entry of the Catholic Church into the "source struggle". So the SMC was very effective in setting the news agenda and framing the issue. I am particularly interested, however, in stories, such as the ones discussed in this paragraph which are editorial opinion pieces by science journalists. The first piece appeared in *The Times* on January 5, 2007, bylined to Mark Henderson, Science Editor. Its headline is unequivocal and – *pace* received wisdom about sub-editors misrepresenting story content – neatly sums up Henderson's claims in the piece: "Ministers have been spooked by 'frankenbunny' headlines". In the *Independent*, Science Editor Steve Connor wrote:

---

[10] In the January 2007 blog, Fox seems to indicate that the first briefing occurred in January 2006, but her presentation at the Genomics Forum event in March 2009 states that the background briefing on "Chimeras" took place in August 2005, while the background briefing in January 2006 was represented as the SMC's response "to the media's requests for a background briefing about the impact of the [Hwang] crisis on cloning research" (http://www.genomicsnetwork.ac.uk/forum/publications/egneventreportsvideospresentations/title,8496,en.html).

> A public consultation into the issue of creating embryos by combining animal and human material produced a massive response against such research. This is said to have spooked the Department of Health in particular into calling for an outright ban. Critics have argued, however, that the consultation had been hijacked by pressure groups opposed to all research on human embryos (Connor 2007).

Similarly, in the *Guardian*, under the headline "Luddites and moralists", Alok Jha, Science Correspondent, wrote:

> When public health minister Caroline Flint unveiled the white paper on fertility research last month, the clause on animal-human hybrid embryos flew in the face of all of the scientific advice, proposing that it should not be allowed. Flint cited a preceding consultation as justification for the government's reversal of support. But the extent to which these sorts of consultations can be hijacked by pressure groups is well known (Jha 2007).

This framing of members of the public who organize collectively as "special interests" and therefore outwith the mainstream of public opinion accords with Irwin's analysis of the implicit premise in critical accounts of the *GM Nation?* debate that "organized groups represent a problem for this form of public debate and, conversely, that 'true' public opinion must be held by those without 'fixed views'" (Irwin 2006: 312).

It cannot be determined definitively from these newspaper stories whether this framing of members of the public emerges from the Science Media Centre, the scientists at the briefing, or the journalists quoted, but the homogeneity of the approach is striking. Further, the assumption that scientific advice offered by those who wish to conduct the science can conversely be taken on face value without any interrogation of their special interests bespeaks a lack of critical engagement with the wider context of that advice by science journalists. Is this lack of critique a feature that the Science Media Centre should exploit on behalf of its "stakeholders", by promoting easy access to media savvy experts, or should the SMC and its "clients" be more wary of undermining public trust in scientific journalism?

To suggest that public opinion was absolutely measurable, and unequivocally in favour of legislating to permit human-animal hybrid embryo research in this case is arguable. In 2007, the Human Fertilisation and Embryology Authority conducted its own public consultation and concluded:

> Having looked at all the evidence the Authority has decided that there is no fundamental reason to prevent cytoplasmic hybrid research. However, public opinion is very finely divided with people generally opposed to this research unless it is tightly regulated and it is likely to lead to scientific or medical advancements (HFEA 2007).

Further scrutiny of the press release announcing this conclusion makes it clear that this is only the public opinion of "people who do not fundamentally oppose embryo research". In its public consultation exercises, the HFEA is always clear that it works from the premise that what is already permitted in law has been publicly debated and is no longer open to question, although it does consult repeatedly on extending what is permissible. So to represent the outcome of the HFEA's consultation as more legitimate than the consultation undertaken by the Department of

Health obscures the difference in roles that legislators and regulators perform, and the different constituencies they attract.

In terms of access to scientists with "proven expertise" and "willingness and ability to engage", the SMC was able to depend on the participation of a highly motivated and highly articulate community in what Fox called "the battle for human-animal" hybrid embryos (Fox 2009a: Slide 2). In her blog, she also refers to "the enthusiastic backing of their respective press officers" (Fox 2007). Ian Wilmut from Edinburgh, Lyle Armstrong from Newcastle, Anne McLaren from Cambridge and Chris Shaw and Stephen Minger from Kings College, would be able to depend on such enthusiastic backing from their employing institutions because of their respective roles as research leaders and fundraisers for such research, so their participation in such briefings is somewhat overdetermined. However, this participation is not necessarily cost-free to the individuals and it potentially also misrepresents the larger institutional and funding context in which their proposed research must be conducted. For example, Fox claims that participating scientists understand they must be instantly available at any time if their expertise is relevant to a breaking news story (Fox 2009b: 115–116), but if a press campaign is as sustained as the one in support of hybrid embryo research, such a commitment must surely place significant – even punitive – additional workload burdens on those scientists enrolled as key spokespersons. Further, as became apparent once the new Human Fertilisation and Embryology Act passed into law, just because research is permissible, does not mean that it will be funded. On 13 January 2009, under the headline "Funding halted for stem cell research", an exclusive story bylined to Steve Connor, Science Editor of the *Independent*, claimed in its subheading that "scientists say cash for research and existing projects has been cut off for 'moral reasons'" (Connor 2009a, b). The story included quotes from both Stephen Minger and Lyle Armstrong, neither of whom had been able to secure funding for hybrid embryo research and concluded with the paragraph:

> The revelation that the research has been frozen by lack of funding has astonished some observers. Fiona Fox, head of the Science Media Centre in London, who co-ordinated a successful campaign to support the Bill, said: 'I find it remarkable given the unprecedented level of support for this research across the scientific community' (Connor 2009a, b).

Apart from the atypical reference to the Science Media Centre brand in this story, there is the unreasonable assumption that support voiced in the media for the license to practice will translate into anonymous peer-reviewed funding in a highly competitive funding climate. The sources quoted and the framing of the story beg the question whether the SMC has been too successful in brokering close relationships between scientists and science journalists, to the potential detriment both of appropriate scepticism on the part of journalists and to the reputation of individual scientists who, despite being willing and able to engage, may not always be sufficiently cognisant of the shift from "background briefing" to a fully attributable quote in an "exclusive" news story. According to a news story on Nature.com the following day, "When *Nature* spoke to Minger he said the *Independent* misinterpreted his comments, adding he did not have any evidence that moral objections led to his proposal being rejected" (Gilbert 2009).

Steve Connor returned to the story on 5 October 2009 on the front page of the *Independent* with the headline "Vital embryo research driven out of Britain", apparently when he learned that Justin St. John, one of the three holders of HFEA licences to create animal-human hybrid embryos, was leaving the UK to work in Australia. The SMC's website reveals that, in response, it produced a press release with comments from senior executives within the MRC and BBSRC, pointing out that embryonic stem cell research is extensively funded by both their councils and reiterating the points made earlier in the year about peer review of funding applications and the non-equivalence of licensing and funding. Both Justin St. John and Keith Campbell provided comments about the different research projects they were pursuing which they saw as both more scientifically promising and better use of resources.[11] Does this front page story and the availability of such sources to provide critical commentary on it demonstrate that the SMC is successful in its mission? Or might it lead to re-evaluating the strategy to focus attention so heavily on a single area of enquiry to the exclusion of the rest of a diverse research field? Was the public – whoever they might be beyond the limited target audience of legislators drafting and voting on the Human Fertilisation and Embryology Bill – served by extensive and repetitive coverage of just one of myriad pathways to researching disease and its cures? Or might this approach skew understanding of the field to an extent that ultimately undermines trust in scientists to pursue research that addresses public concerns rather than their own? Did this campaign, which fused representations of scientific knowledge with claims upon science governance and an overly simplified vision of scientific practice, generate public understanding of and trust in science or did it reveal some of the pitfalls in equating positive media coverage with public understanding of and trust in science?

12.6 Conclusion: Unashamedly Pro-Science

The Science Media Centre is positioned in accordance with Baroness Susan Greenfield's original vision of the centre as "unashamedly pro-science" (SMC 2002: 14). Cited in support of this decision is the likely disproportionate media exposure that accrues to "the voices that are often critical of science" (loc. cit.) suggesting that scientists' self-assessment of their relative power and effectiveness in society is radically at odds with how they are perceived by others, particularly if they are promising cures.

The adoption of public relations techniques and message simplification by the Science Media Centre and their roster of media-savvy scientists, to attempt to meet the news agenda head-on, conceptualised as "open and positive communication"

[11] See http://www.sciencemediacentre.org/pages/press_releases/09-105_hybrid_research_funding.htm.

(SMC 2002: 23), could be alternatively framed as an attempt to impose closure on the messages transmitted by the media and to retain the authority to speak for – and about – their profession and its practices. Although this is a plausible and time-tested strategy for media relations in other professional and commercial sectors, in the bid to avoid misunderstanding or misrepresentation, it is arguably at odds with the attempt to restore public trust in science by engaging members of the public in informed debate. Whether informing and engaging the public about science could ever be effected through media relations is an open question, but the House of Lords report which prompted the founding of the Science Media Centre certainly viewed this as a possibility. In terms of democratic governance of science, however, there was always already a tension inherent in the challenge issued through this report, conceptualised in the SMC's Executive Summary as "improving science communication as a means to 'secure science's license to practice not to restrict it'" (SMC 2002: np) as it erases the public as stakeholders through its narrow focus on policymakers and communications practitioners.

Acknowledgments The support of the Economic and Social Research Council (ESRC) is gratefully acknowledged. This work is part of the Research Programme of the ESRC Genomics Network at Cesagen (ESRC Centre for Economic and Social Aspects of Genomics).

References

Bodmer, W. (2010). Public understanding of science: The BA, the Royal Society and COPUS. *Notes & Records of The Royal Society*. Published online July 14, 2010, doi: 10.1098/rsnr.2010.0035 (last accessed on 30 November 2010).

Connor, S. (2007). Should we allow the creation of embryos which are animal-human hybrids? The big question. *Independent*, January 10, 2007.

Connor, S. (2009a). Funding halted for stem cell research. *Independent*, 13 January 2009.

Connor, S. (2009b). Vital embryo research driven out of Britain. *Independent*, 5 October 2009.

Fox, F. (2007). Stem cell scientists seize the media agenda on human animal embryos. *On Science and the Media*. Blog entry. http://fionafox.blogspot.com/2007/01/stem-cell-scientists-seize-media-agenda.html (last accessed on 30 November 2010).

Fox, F. (2009a). Slide presentation at Human Fertilisation and Embryology Act: A retrospective, 12 March, 2009. http://www.genomicsnetwork.ac.uk/forum/publications/egneventreportsvideospresentations/title,8496,en.html (last accessed on 30 November 2010).

Fox, F. (2009b). Science communications and ethics – Trying to get it right: The science media centre – A case study. In B. Nerlich et al. (eds.), *Communicating biological sciences: Ethical and metaphorical dimensions*. Farnham: Ashgate, pp. 109–128.

Gilbert, N. (2009). *Moral objections to hybrid embryo research claims rejected*. Blog entry on The Great Beyond at Nature.com. January 14 2009. http://blogs.nature.com/news/thegreatbeyond/2009/01/moral_objections_to_hybrid_emb.htm (last accessed on 30 November 2010).

Haran, J. (2007). Managing the boundaries between maverick cloners and mainstream scientists: The life cycle of a news event in a contested field. *New Genetics and Society*, 26(2), 203–219.

Haran, J. and J. Kitzinger (2009). Modest witnessing and managing the boundaries between science and the media: A case study of breakthrough and scandal. *Public Understanding of Science*, 18(6), 634–652.

Haran, J., et al. (2008). *Human cloning in the media: From science fiction to science practice*. London: Routledge.

Haraway, D. J. (1984). Primatology is politics by other means. *PSA: Proceedings of the Philosophy of Science Association*, 2, 489–524.
Henderson, M. (2007). Ministers have been spooked by "frankenbunny" headlines. *The Times*, January 5, 2007.
HFEA (2007). HFEA statement on its decision regarding hybrid embryos, 5 September 2007, available at http://www.hfea.gov.uk/455.html (last accessed on 30 November 2010).
House of Lords Select Committee on Science and Technology (2000). Third Report of Session 1999–2000 Science in Society.
Irwin, A. (1995). *Citizen science*. London: Routledge.
Irwin, A. (2006). The politics of talk: Coming to terms with the 'new' scientific governance. *Social Studies of Science*, 36(2), 299–320.
Jha, A. (2007). Luddites and moralists: A ban on the use of hybrid embryos will be the consequence of ill conceived pressure. *The Guardian*, January 11, 2007.
McNeil, M. (2012). Between a rock and a hard place: The deficit model, the diffusion model and publics in STS.
Miller, S. (2001). Public understanding of science at the crossroads. *Public Understanding of Science*, 10, 115–120.
Mulkay, M. (1997). *The embryo research debate: Science and the politics of reproduction*. Cambridge: Cambridge University Press.
Radford, T. (2009). A workbench view of science and communication. In B. Nerlich et al. (eds.), *Communicating biological sciences: Ethical and metaphorical dimensions*. Farnham: Ashgate, pp. 145–152.
Science Media Centre (2002). Consultation Report. March 2002. Available at: http://www.sciencemediacentre.org/uploadDir/536adminconsultation_report.pdf (last accessed on 30 November 2010).
Science Media Centre (SMC) (2009). Expert reaction to Independent story on funding of hybrid research, 5 October 2009. http://www.sciencemediacentre.org/pages/press_releases/09-10-05_hybrid_research_funding.htm (last accessed on 30 November 2010).
The Royal Society (1985). *The public understanding of science*. Report of a Royal Society ad hoc Group Endorsed by the Council of the Royal Society.
Williams, A. and S. Clifford (2009). *Mapping the field: Specialist science news journalism in the UK national media*. Research Report. The Risk, Science and the Media Research Group, Cardiff University School of Journalism, Media and Cultural Studies.
Williams, A., et al. (2009). *UK National Newspaper coverage of hybrid embryos: Source strategies and struggles*. Research Report. The Risk, Science and the Media Research Group, Cardiff University School of Journalism, Media and Cultural Studies.
Wynne, B. (1991). Knowledges in context. *Science Technology & Human Values*, 16(1), 111–121.
Wynne, B. (1992). Public understanding of science research: New horizons or hall of mirrors? *Public Understanding of Science*, 1, 37–43.

Chapter 13
Practitioner's Perspective: The Role and Function of the Science Media Centre

Fiona Fox

13.1 Introduction

The Science Media Centre (SMC) is a unique experiment in a new kind of media relations activity. It is a press office that is completely independent, eschews PR and has no brand or institution to promote. Instead, it works on behalf of the whole of science to improve the quality of the reporting – and thus the information the public receive – when science hits the headlines. The Centre's official philosophy that "The media will 'do' science better when scientists 'do' media better" mirrors the conclusions of the House of Lords Select Committee Report on Science and Technology (2000), which gave rise to the setting up of the Science Media Centre:

> The current high level of media interest in science-related issues is to be welcomed. While it sometimes makes for public dialogue in terms which are unsatisfactory to some of the players this is still much better than no dialogue at all. Scientists must indeed take the rough with the smooth, and learn to work with the media as they are.

In this chapter I lay out the philosophy, values and approach to science media relations that the SMC has developed since it opened for business in April 2002. I and many others believe the Centre has had a huge influence on the coverage of science in the media and perhaps more significantly, has contributed greatly to creating a new culture within science, which regards engaging with the big topical controversies in the media as part and parcel of the scientist's role.

What I do not have the space to do here is present the thousands of examples of science stories that we have influenced or indeed generated, and I would therefore urge readers to consider this chapter alongside a visit to our website, which shows how we have put our philosophy into practice on a daily basis for the past 9 years (http://www.sciencemediacentre.org).

The chapter is a revised reprint of a chapter published in Nerlich, B., R. Elliott, and B. Larson (eds.), *Communicating biological sciences – Ethical and metaphorical dimensions.* Ashgate 2009.

F. Fox (✉)
Science Media Centre
e-mail: fiona@sciencemediacentre.org

S. Rödder et al. (eds.), *The Sciences' Media Connection – Public Communication and its Repercussions*, Sociology of the Sciences Yearbook 28,
DOI 10.1007/978-94-007-2085-5_13, © Springer Science+Business Media B.V. 2012

13.2 History

The SMC has its roots in the House of Lords Select Committee Inquiry into Science and Technology (Science and Society), reported in February 2000. The backdrop to this inquiry was the BSE ("mad cow disease") affair during which leading commentators had started to openly challenge the authority of science, placing the scientific community alongside other institutions like the Church and the family that were suffering from their own crises of trust. It became clear from the evidence presented to the Committee, that, while scientists were getting better at speaking to trusted journalists working for specialist science media like *Horizon* or *New Scientist*, they were still struggling to engage effectively with the demands of a hungry, often brutal, 24-h news machine splashing on a science story that had become the subject of national controversy. The need for scientists to get better at the latter was summed up in the conclusions of the Report as the "great challenge", and recommendations called for new resources and leadership to focus on this specific and crucial area. The first person to step up to provide those resources and leadership was Baroness Susan Greenfield, Director of the Royal Institution, who believed that the RI's historic role of disseminating science to the public, combined with its independence from Government, made it suitable to lead any new drive to improve the media's coverage of science. Baroness Greenfield set up an Advisory Committee (Science Media Centre Consultation Report 2002) comprising leading scientists, journalists and editors who over the course of a year in 2000–2001 delivered the concept of a SMC, which would "help renew public trust in science" and bridge the gap between the cultures of science and the media.

A number of bold decisions made by the Advisory Committee defined the core principles of the Centre, namely:

> that the Centre would be independent from any one scientific institution and therefore officially operate on behalf of the whole scientific community. While the Centre would be housed within the Royal Institution it would operate independently from the RI with its own governance structures

and

> that the funding for the Centre would reflect the principles of independence seen as critical to earning and keeping the trust of journalists and scientists alike. As such the decision was made to put a cap on all donations ensuring that no one sponsor can give more than 5% of the overall running costs of the Centre.

In addition it was agreed that the Centre would make it clear that the Centre is independent form all its sponsors and that all donations are unconditional.

As a registered charity with close links to, but operational independence, from the RI, it was agreed that the SMC would be governed by a Board of Advisors – made up of influential scientists, editors and media relations experts – in addition to calling on a list of leading scientists who agreed to sit on the Centre's Scientific Advisory Panel. Funding was and continues to be sought from the SMC's key stakeholders: scientific institutions, science-based companies, media owners and the Government.

The legacy of poor relations between science and the media made it difficult for the Committee to find a founding Director with both the scientific credentials and extensive experience of media relations they asked for. Baroness Greenfield frequently re-tells the story of my interview, which she had originally objected to on the basis that I had no scientific qualifications. She then tells how I won the unanimous backing of a panel consisting of ten eminent scientists, including several Lords and Ladies, who agreed with her that a passion for and commitment to the issues, alongside my background in media relations were what would make the Centre work. We then ensured that the two members of staff I recruited had the scientific credentials deemed necessary for the job.

It was clear when I took up the position of Director in December 2001 that a period of intense consultation was needed to flesh out the specifics of what was originally an incredibly broad concept. Baroness Greenfield's widely reported vision that the Centre would be "unashamedly pro-science" and "help renew public trust in science" went some way towards positioning the Centre, but said little about what it would actually do and there were plenty of scientists, journalists and press officers lining up to offer their advice about what was required. In the first weeks of consultation with key stakeholders, it sometimes felt as though every individual we approached had different and often conflicting ideas about what the Centre should do, with leading scientists encouraging us to focus on issues from improving the science in gardening programs to acting as a base for visiting science reporters from overseas. But after 3 months of consultation a clear consensus emerged about the weak spots in science media relations where the Centre could make the biggest difference. This consensus revolved around the very same weakness identified in the House of Lords report referred to as the "great challenge" – the apparent inability of many in the scientific community to react quickly, confidently and effectively to breaking news about controversial science stories. I leave the philosophical discussions about the "Two Cultures" for others to debate, but suffice to say that even a cursory glance at the way science operates reveals the gulf between the needs of science on the one hand and the news media on the other. Having worked for campaign groups and charities in the past, I can honestly say that it is hard to imagine two disciplines with as little in common as science and the media. This culture clash – at its most intense in relation to breaking news – raised its head throughout almost all discussions with key stakeholders. Almost every journalist admitted particular difficulties finding scientists to speak to when a controversial science story broke, press officers repeatedly emphasised how hard they found it to convince scientists to drop everything to engage with a breaking story on their area of expertise, and scientists admitted that while they were happy to speak to trusted specialist reporters, they were reluctant to drop everything to appear live on air and engage in debate in the heat of a media storm.

This focus on breaking news and controversy within the UK science media environment is key to understanding the SMC. While many initiatives fail because it is not obvious what problem they need to solve or what their role is, the role and position of the SMC are clear, unique and very specific. The Centre is part of a broader effort to overcome the cultural clash between science and news, and its role is to work with the scientific community to adapt the very best science to the demands and needs of the news media when controversial stories arise. Nine years later, it has allowed us to build up an incredible bank of expertise and skill in dealing with controversial science issues in the media, which is now used to great advantage by the scientific community.

The scientists, journalists and press officers approached during the consultation were in no doubt as to the kinds of stories that needed the SMC's attention; they were those such as GM and MMR. Just months after the House of Lords Select Committee began hearing evidence about the way the media covered science, GM crops exploded onto the headlines with tabloid front pages screaming out "Frankenfood". Furthermore, throughout 2001 Tony Blair's refusal to confirm whether his young son Leo had been vaccinated with MMR fuelled the media row over a possible link between the jab and autism. There is no room to consider both stories in detail here, and much has already been written about the way scientists and the media dealt with both stories, providing useful insights (see POST Report 1993; Boyce 2007). However, it is fair to say that almost all those involved in these issues felt that one of the key roles of the new SMC would be to work with scientists to enable them to engage more effectively with headline stories, which in the hands of effective campaigners and a sensationalist media had become some of the most contentious issues of our times.

It was the GM debate in particular which informed the values and philosophy of the Centre very profoundly. At visits to UK plant and agricultural science institutes in early 2001, we met a succession of the scientists who knew more than politicians, campaigners and journalists put together about this new plant breeding technique – they were the real experts. Yet, with a handful of exceptions, they had remained silent in the national media debate on GM. There were many reasons why, but looming large was the fact that these mild-mannered scientists were bewildered by the nature and size of the debate. Few journalists, especially from tabloid populist media, had ever shown any interest in the finer points of their research before, and the prospect of dropping their research for several hours to take part in a potentially hostile interview with an anti-GM campaigner on breakfast television news in front of millions was a completely alien one. The vacuum left by the failure of the best experts to engage was quickly filled by a mixture of politicians, campaigners, protest groups, representatives of the GM Industry, columnists and high-profile scientists, who were not necessarily the true experts in this area but felt compelled to wade in to defend science. The views of these groups are of course all legitimate and formed a critical part of national debate, but what was missing were the experts – the scientists who had actually developed the techniques and who were involved in numerous research projects to identify risks and benefits, strengths and limitations. Those who have heard me speak on this subject will be well-accustomed with my verdict on the GM debate but I make no apology for repeating it as it lies at the heart of the role of the SMC. I genuinely do not care whether the British public say no to GM as long as they do so after access to a balanced, accurate and informed debate in which the best plant scientists have had their say. Herein lays the basis for the SMC: It is our job to ensure that never again will the UK have a national media debate about an area of science without the best scientists taking up their rightful part in that debate. If at the end of discussions the UK still says no to GM then that is simply a healthy democracy.

The GM debate also influenced the values and philosophy of the SMC in that it was clear to the newly formed team at the SMC that the debate could have played out very differently if the best plant scientists in the country had viewed it as an opportunity rather than a threat. From this point onwards, the SMC set out its stall to change the culture within the scientific community so that future headlines would be viewed as just that – a golden opportunity.

13.3 How the Science Media Centre Operates

The SMC has worked closely with journalists, press officers and scientists to develop a number of services which are effectively delivering our goals. These include our database of media-friendly scientists, Rapid Reaction service, Round-Up press releases and our regular press briefings. The following is a summary of these activities, the latest examples of which can be found on our website.

13.3.1 The SMC Database of Experts

The SMC's database is not quite like any other database of experts. It is not a searchable database for the media to use, but a resource for the SMC to find the right expert on the right subject within the right timeframe. Scientists and engineers are selected not just for their proven expertise, but also for their willingness and ability to engage with the media when their issues hit the headlines. Given the SMC's focus, those joining our database know they are signing up to our goal of improving the way issues are covered by getting stuck into the media debates rather than shouting from the sidelines. Like a dating agency, we pair the right experts with the right journalists when an issue hits the headlines, and at the last count there were over 2000 scientists and engineers signed up to our database. The quality of our experts is extremely important and many people ask how the SMC selects the experts on our database and how we check for quality and guard against bias. This is one of the biggest challenges for the SMC and, given how many experts we deal with, we may make mistakes along the way. However the criteria we use to recruit experts has so far proved extremely effective at ensuring the SMC is putting forward credible experts and good quality science and engineering. These criteria include ensuring that scientists are from respected scientific institutes, have a distinguished track record of research and publish in peer reviewed journals. Scientists are generally recommended to us by press officers from universities or scientific institutions as being both media friendly and recognised experts on issues that are likely to hit the headlines. We then set about approaching these scientists to ensure that they fully understand what it means to be on the SMC database. This involves ensuring that the scientists sympathise with our broad values on the need to engage, and that they understand that the SMC will call on them to drop everything – even late at night or at weekends – to do media work when their area hits the headlines.

13.3.2 Rapid Reactions

The demands of the 24-h media machine mean that news journalists often do not have the luxury of time to track down the best scientists when a science story lands on their desks, which means that availability can sometimes win out over expertise. This is where the SMC steps in. Once alerted to a breaking story, we track down the right experts and, with their permission, offer them to all the national newspapers and news programmes. Our Rapid Reaction service is a real example of a win-win scenario for science and the media: We help frantic news journalists, delighted to be given great scientists to interview, and we enable the best scientists to get their voices heard at the very time that their issues are in the headlines and therefore on the public's mind. Whether it's the latest food scare, an outbreak of foot-and-mouth disease or a controversial medical breakthrough; ensuring scientists engage in the story directly influences what the public see and hear, and in many cases can defuse an unnecessary media scare story. In recent years a "fact sheet" has been added to this Rapid Reaction service after pressure from journalists being asked to deliver fact boxes very quickly for their ever growing online operations. Typically, the Rapid Reaction service from the SMC will include three things: contact details of leading experts who we have established are both available and willing to do interviews immediately; written quotes from these experts and others who are not in a position to do much media work but can give us their informed reaction; and a fact sheet with a list of accurate, well-established scientific facts about the issue.

13.3.3 Round-Up Press Releases

The Round-Up press release is a similar but distinct part of the SMC service. It ensures that the media have access to scientists and engineers in advance of a story breaking, by offering journalists a variety of comments from experts reacting to embargoed stories that we predict will make the headlines. These generally originate from the major science journals that the SMC has negotiated special advance access to, including Nature, The British Medical Journal and The Lancet. We also have permission to share embargoed papers with trusted contacts to elicit reactive comments, which can then be sent to reporters to feed into their coverage. These comments are often used by journalists to balance coverage and help them to put a "breakthrough" into a more measured context for their sometimes overexcited news editors. We also issue Round-Ups on Government white papers, Select Committee reports, and other reports that we are given embargoed access to by our friends throughout the scientific community and beyond. When you see an independent scientist quoted reacting to a breakthrough reported in The Lancet, there is a high chance that the SMC provided that comment. Round-Ups differ from our Rapid Reaction service as scientists have more time to read scientific papers or reports in detail rather than having to react on the spot; these unique press releases provide busy journalists with great sound bites they can use in their articles, while simultaneously informing the broadcast media which scientists are available for interview and how they are reacting.

13.3.4 SMC Press Briefings

The SMC is certainly not restricted to reacting to the headlines, and since opening we have facilitated scientists to generate their own headlines on many occasions. These come from the Centre's regular press briefings – often held when scientists and science press officers feel that good, accurate science is being lost in the public and media debates around certain subjects. These briefings take a variety of forms. Many are "Background Briefings" introducing the national media's science, health or environment correspondents to the best experts and science on controversial issues like nuclear waste, nanotechnology, stem cell research, animal research, etc. Or they can be "News Briefings" where we work with scientists to give the national media a new story on developments within science – whether it's an IPCC (Intergovernmental Panel on Climate Change) report, a paper from the journal Nature on stem cells or science funding cuts in the Budget report. The SMC also encourages leading scientists to "speak out" to the media about policy developments they believe may pose a threat to research – not something science has been renowned for in the past. Over the years the SMC has drawn national media attention to scientists' concerns about many issues, including the early drafts of the Human Tissue Bill, the attempts to ban research on human-animal embryos, the moves to repatriate collections of ancient remains being used in research, the deliberate destruction of GM Crop field trials, the EU Clinical Trials Bill and the EU Animal Research Directive. The SMC has established such a good reputation for running professional press briefings, popular with the national media, that many organisations choose to work with the Centre to launch their stories – especially if they have the potential to be controversial and make headline news. Examples include the launch of the Farm Scale Evaluation of GM crops, the launch of the Royal Commission on Environmental Pollution's report on nanotechnology, the Committee on Radioactive Waste Management's final report, as well as many of the best science stories published, including *The Lancet*'s publication of the first successful human trachea made from the patient's own stem cells.

13.3.5 "Crap Busting"

The SMC would love to be able to change the media so fundamentally that they never covered terrible science stories, but while that remains an unrealistic prospect at this time we offer the media a specialist and affectionately termed "crap busting" service. Rather than hectoring journalists for covering stories that their editors tell them to cover, we instead find them leading experts willing to offer quotable comments rubbishing the story that can be included in their copy. Reassuringly, we have countless examples of how these comments often do help specialist journalists to persuade their news editors not to run the story. We also have many more examples of how the comments we have supplied have been used prominently in the articles in a way that provided much needed balance and added a note of scepticism for readers. One example amongst others is the reactions we gathered on a press release about how watercress could reduce the risk of cancer, based on a study funded by the Watercress Alliance. The SMC went to one of the world's leading cancer experts whose measured and informed comments included the fact that to get such an effect we would need to eat so many tons of watercress that we may well turn green. The story did run in every national newspaper, but with our expert's comments encouraging readers to view it with a very large pinch of scepticism.

13.3.6 Supporting Scientists

The SMC would be nothing without the thousands of scientists and engineers who have agreed to join our ever-growing database of experts. Many of them are new to media work and therefore much of what we do at the Centre involves us providing them with all of the support they need. We produce information leaflets that give advice about when and how to engage with the media, including Top Tips for Media Work, Communicating Uncertainty in a Soundbite and When Animal Research Hits the Headlines. A new leaflet is also being written specifically with engineers in mind, to help them work with journalists on the issues closest to their areas of expertise. We also run our hugely popular Introduction to the News Media events several times a year, with at least one event each year specifically put together with engineers in mind. Designed to give those who are considering media work an idea of how the UK national news media operate, these events give attendees the opportunity to meet a panel of journalists and discover what the working life of their media counterparts is really like. These sessions always result in lively debate about why experts should engage with the media, and attendees leave feeling much more comfortable about doing so. In addition to these activities, we are always on-hand on a day-to-day basis to answer our experts' queries and concerns, and facilitate their media work during what can often be extremely challenging times.

13.4 Spinning for Science

Not having a brand name, institutional message or corporate identity to promote in the media makes the SMC a radical and unique new experiment in media relations. Writing about PR in his seminal book *Flat Earth News*, Nick Davies (2008: 89) suggests, "Fabrication is at the heart of PR, the fabrication of news which is designed to open the media door. PR is clearly inherently unreliable as a source of truth simply because it is designed to serve an interest." The fact is that the SMC is a media relations outfit that is interested in what Nick Davies is interested in: improving the quality of science journalism and truth telling. While the Centre works hard to advise scientists on how to get their science and messages across in a way that will guarantee the best quality coverage, it does not attempt to influence those messages. In that sense the Centre's media work is un-spun, a feature often commented on by journalists who find it refreshing to be allowed to question leading experts at the Centre with no restriction. To maintain that reputation, the Centre has, on occasion, refused to work with institutions – including Government departments – who have attempted to run briefings off the record or with other restrictions that are in our view not justified. Philosophically we see the Centre as a facilitator, a place where key journalists can meet and question key scientists on their work. This feature is reflected in our decisions not to allow briefings to be chaired by external organisations. Instead a panel of scientists or engineers speak about their research, followed by a Question and Answer session with journalists chaired by a member of the SMC team. The Centre genuinely is a place where journalists can meet scientists and talk openly and honestly without media management or institutional spin.

At the launch of the SMC, Baroness Greenfield emphasised that the SMC would be "unashamedly pro-science", a phrase that invited criticism from many social scientists and commentators who have welcomed a climate in which science has been subjected to more scrutiny and criticism by the public and the media. It is clear that those involved in establishing the SMC did not feel as comfortable with the growing willingness to distrust and question scientific expertise. For them the loss of trust in science, resulting from sensationalised media rows over BSE, GM and MMR, were bad for science and for society and so the SMC's core remit was identified as helping to renew public trust in science. However, observers of the SMC will testify that the Centre's interpretation of being "pro-science" is a rather more sophisticated and nuanced one than some critics have given us credit for. Broadly the concept is interpreted as the Centre's commitment to the scientific process, method and an evidence-based approach to the big issues of the day. The SMC is pro-science, but without being completely uncritical or unquestioning, and it offers up measured, evidence-based criticism on the stories of the day. As such, the SMC does not take a position on the scientific controversy of the day but facilitates and encourages scientists to enter the fray. It is common for scientists on SMC Round-Up releases to disagree fundamentally with each other, and we have run media briefings on all sides of contentious debates, such as the efficacy of Alzheimer's drugs, the effects of cannabis on the brain, and the contribution of aviation to climate change.

But the SMC has argued passionately with editors that the sacred code of journalistic balance is dangerous when it results in news programmes presenting the debates about MMR and climate change, for example, as battles between two halves, when in both cases the balance of scientific opinion is overwhelmingly on one side. I have argued that to reflect the real balance of the scientific debate on climate change, the media should run 99 interviews with mainstream climate change scientists for one with every sceptic. Similar principles apply to the SMC's own application of balance. Because the Centre is not interested in spinning or managing messages, the views that we project into the media are those held by the scientists and engineers on our database. If all of the eminent engineers and nuclear experts on that database believe that nuclear power should be part of the tool-kit for tackling climate change, then those are the sentiments that will emerge from the Centre. As a centre which openly reflects the views of mainstream science, we do not see it as our responsibility to scour the country for climate change sceptics or plant scientists who are opposed to GM simply to present a false and distorted sense of being balanced. However, when there is a genuine division of opinion within mainstream science, the SMC will reflect this and SMC Round-Up press releases regularly present a series of conflicting views from eminent scientists on a whole range of issues.

13.5 Science in the Media: Turning the Tide?

The first of many challenges to the scientific community's caution and disdain for the media's framing of certain issues came with a debate amongst leading scientists about how to respond to a series of major front page news stories announcing claims of the first cloned human being. Maverick cloners Panos Zavos and Severino Antinori and even an American religious sect called the Raelians had, over a period of 2 years, stolen headline news with their un-proven claims that they had produced the first cloned baby. None of these claims had been published in the scientific literature or even announced at scientific conferences. Instead it seemed all these people had done was book a hotel room, invite the media and make their claims. Astonishingly, despite the lack of a shred of evidence in any of these cases, each one made headline news throughout the mainstream media, including the broadsheets and more "upmarket" publications. We felt that the publicity for these mavericks was feeding the totally inaccurate view that mainstream science was in a race to clone the first human, and once the SMC was up and running, this was one of the issues that would be very much on our radar. This misinformation mattered a lot because at the time many stem cell scientists were looking for public support for their attempts to develop their work with therapeutic cloning, through which they could derive patient-specific stem cells vital for research purposes. The widespread misapprehension that scientists wanted to clone human beings was extremely damaging to these scientists. When the SMC approached the UK's leading fertility and cloning experts, we discovered that there had been a *de facto* boycott of human cloning stories for some time, with scientists so angry at the media's irresponsible reporting of these unsubstantiated claims that they refused to comment on them. It was immediately apparent that the media were once more not hearing from the right experts on the issue. This struck us as being something the SMC had to challenge. While there was an intellectual integrity to the stance that serious scientists should refuse to grace these stories with their presence, this boycott was completely invisible to the UK public who merely saw a series of unchallenged claims that scientists were in a race to clone a human. We went out of our way to kick-start a debate within the scientific community about the need to change our attitude to these stories, and argued the case that what was needed was for the scientific community to speak out every time the media covered these stories to condemn the irresponsible claims and reinforce the message that mainstream science was not engaged in cloning humans. While some continued to insist that we would lack credibility if we condemned the media for running these stories while at the same time feeding these stories with experts and information, the SMC argued that this was exactly what we needed to do. Indeed we argued for combining the two – with any experts interviewed using their time and column inches to criticise the media for running these stories with absolutely no evidence.

The first test came early one Saturday morning in 2004 when I turned on the *Today* programme on BBC Radio 4 at 6.30am to discover that the lead story was yet another announcement that the first human clone was to be revealed later that day – yet again by maverick cloner Panos Zavos. With some trepidation I called the leading cloning and fertility experts with whom we had been debating the issue, and by 10.00 am I had found six leading experts ready to give up their Saturday to go into television and radio studios to condemn Zavos and emphasise the reality that mainstream science was not interested in cloning human beings. This strategy paid off with headlines literally changing as the morning went on from "Doctor implants cloned embryo" to "Scepticism greets cloning claim" (BBC News 17 January 2004). By mid-day every channel and radio station – all of whom were running the story as headline news – had included an interview with one of the scientists the SMC put forward, and by Sunday every single national newspaper carried a headline and lead paragraph that referred to the scientific community's dismay at the story rather than Zavos' claims. This was our first piece of evidence of how scientists can overcome the culture of caution, and change what the public see and hear by engaging with the story. We now have hundred of similarly compelling cases of where a story was changed for the better through the willingness of top scientists to engage as and when the debate was happening.

Hopefully it is clear from this example that the SMC's philosophy and approach to science in the media define the way that we operate. Seeing science in the headlines as an opportunity rather than a threat, believing that engaging can change the story, encouraging openness and discouraging spin all influence the style and atmosphere at the Centre. There is another feature of the SMC that I believe has improved not only media coverage but also the scientific community's experience of engagement, and that is our attempt to give science its place in the sun. In the very early days the SMC used to approach "pro-science" spokespeople outside of science to do media work – and a glance at our early press releases will reveal patients, politicians and even Bishops in our round-ups. This practice changed very early on when we realised that the problem we were to set up to solve was precisely the ease with which journalists could access patients, campaigners and clerics, in contrast to research scientists. Added to this we discovered that if policy makers, politicians or patient campaigners sat on panels at our briefings, the whole tone of the briefings changed away from science and towards "messages" and "policy". From an early stage, and with the guidance and support of our Board of Advisors, it became a characteristic of the Centre that we would focus primarily on scientists and engineers, and consciously attempt to ensure that research science gets a much higher profile in the controversies of the day. It is now very, very rare to ever meet a politician, patient or campaign group at the SMC, and we take great pride in the fact that, when journalists attend SMC briefings, they get much more than a science story – they get a small lesson in the way science operates and develop an ever greater understanding of and sympathy with the experts carrying out the research themselves.

It is well known within science circles that science news is covered more accurately when covered by specialist science, health or environment reporters, rather than general news reporters or political editors. Academic studies of the media's coverage of GM and MMR have shown that accuracy and truth are often the first victims when the stories move from the specialist science pages to the front pages. However, one of the most positive developments in the past 5 years is that this trend has been reversed, and it is now almost certain that any major breaking science, health or environment story will stay with the specialists. Interestingly the Human Fertilisation and Embryology (HFE) Bill, which provoked huge controversy by permitting research on human-animal admixed embryos, was covered almost exclusively by science reporters, even when it was debated in the Commons and became the subject of a political row. The work of the SMC and other science press offices, and the support we give to specialist correspondents, has undoubtedly helped to strengthen their hand by ensuring that they get the best science stories in advance of non-specialists in the news room. The SMC has continued to champion specialist reporters both within the scientific community and in all our dealings with news editors. It is our strong view that they are the best allies of science in the media, and that we lose them at our peril.

13.5.1 An Example to Aspire To: The Battle for Human-Animal Embryos

As discussed in this chapter, the SMC was built on the foundations of stories widely perceived to have gone badly wrong for science. "What Went Wrong" with the media coverage of the BSE crisis, GM crops and the combined MMR vaccine has driven the work of the SMC and inspired us to support scientists in becoming more effective at engaging with the big controversial stories that hit the headlines. And now the SMC has a story of "What Went Right". The battle for human-animal admixed embryo research that raged in the media from 2006 to 2009 has finally provided the Centre with a model of how scientists *should* engage with the media. A more detailed account of the scientific community's efforts to overcome the ban on human-animal embryos has just been published.[1] However, it is worth highlighting that many of the approaches that the SMC has pioneered were applied to the media work around this contentious issue. It is a little known fact that leading stem cell

[1] See Academy of Medical Sciences et al. (2009); for an analysis of the UK newspaper coverage of this issue, see Williams et al. (2009).

scientists were the first to kick off the media debate about human-animal embryo research when they briefed science reporters at the SMC in June 2005. Stephen Minger, Robin Lovell-Badge and the late Anne McLaren did what all scientists should do: They took one of the most controversial areas in their field and gave journalists an in-depth briefing on the scientific potential as well as the ethical issues around this research; even voicing their concerns about a gaping regulatory loophole. Having generated the first UK headlines on human-animal embryos, this was followed up by a series of SMC press briefings as scientists moved towards applying for licenses to use animal eggs in their therapeutic cloning work. Then, when the subject became the most hotly debated aspect of the UK Government's HFE Bill, these scientists again made themselves easily accessible to science and health journalists from the tabloids, broadsheets, radio and television news. Along the way, the scientists involved had to cope with some scary headlines, with graphics depicting giant mutant rabbits and cow heads on human bodies, but their bravery in tackling these controversies in the full glare of the national news was rewarded by largely excellent coverage. Headlines such as "Embryology Bill: Bishop's 'Frankenstein' attack smacks of ignorance" (*Times*, 24 March 2008) also show how the "monster metaphor" in media discussions of science has developed. Once so pervasive in debates surrounding cloning, assisted reproduction, GM and other controversial areas of science, the metaphor of science as an unholy creation out of control is starting to be replaced with one in which the monster represents the misunderstanding or misrepresentation of science, which the scientists themselves step forward to correct. This is a prime example of the monster for once being tamed. More importantly, by the time the Department of Health indicated their desire to ban this research in December 2006, in the early drafts of the Government's planned fertility laws, the health and science reporters on every national news outlet already understood the basic science involved and could explain why the experts wanted to pursue it. They also trusted and respected the key scientists. The accurate and well-informed media coverage of one of the most controversial issues of our time was evident throughout this debate's twists and turns. In May 2008, our elected representatives reflected public opinion by voting in favour of allowing the research to continue, and in November later that year the bill received Royal Assent and became the HFE Act. Here, on show, was the exact opposite of the national GM debate – policymakers and the public came to an informed decision on a controversial new area of scientific research following a major national debate in which the voice of research scientists was heard loud and clear. Another issue worth noting is the huge support given to stem cell scientists by all of the UK's leading medical research and scientific bodies. None stayed silent or ran in the opposite direction, as has previously happened with issues like GM and animal research. Instead the Wellcome Trust, the Medical Research Council, the Academy of Medical Sciences and the Royal Society all backed this controversial research in the boldest and most public way, and the Association of Medical Research Charities (AMRC) and the Genetic Interest Group brilliantly galvanised over 200 patient and medical research charities to back the bill. More than anything else, this is a story of top scientific researchers taking the decision somewhere along the line that engaging with public concerns is a key part of their roles as scientists. In doing so, they made history and changed the narrative of science and the media from a negative one to one that is both positive and inspiring.

13.6 Conclusions

Since 2002 the SMC has been doing the most risky of all things – putting scientists at the eye of the media storm on some of the most controversial science stories of our times. By rights the SMC should by now have amassed hundreds of horror stories of sensationalised articles, and scientists so badly distorted that they will never do media again. In fact we have the opposite; we have thousands of examples of controversial science and engineering stories covered accurately and responsibly because some of the best experts in the world worked with some of the best journalists in the world to create something we should all be proud of. There are many problems left to solve and the fast rate of scientific change combined with developments in the media, and new media, present novel challenges almost daily. However, we are confident that the SMC's approach is and will continue to be fit for purpose.

References

Academy of Medical Sciences, Medical Research Council, Science Media Centre, Wellcome Trust (2009). *Hype, hope and hybrids: Science, policy and media perspectives of the human fertilisation and embryology bill*. Available at: http://www.acmedsci.ac.uk/publications.

Boyce, T. (2007). *Health, risk and news: The MMR vaccine and the media*. Bern: Peter Lang.

Davies, N. (2008). *Flat earth news*. London: Chatto & Windus.

House of Lords Select Committee on Science and Technology (2000). *Inquiry into Society's Relationship with Science*. Report available at: http://www.publications.parliament.uk/pa/ld199900/ldselect/ldsctech/38/3801.htm.

POST Report (1993). *On genetically modified food*. www.parliament.uk/post/pn138.pdf.

Science Media Centre Consultation Report (2002). Available at: http://www.sciencemediacentre.org/uploadDir/536adminconsultation_report.pdf.

Williams, A., S. Gajevic, J. Lewis, and J. Kitzinger (2009). *UK National Newspaper coverage of hybrid embryos: Sources, strategies and struggles. Science Media Centre* (Report). See: http://www.cf.ac.uk/jomec/research/researchgroups/riskscienceandhealth/fundedprojects/hybridembryos.html.

Part VI
Media Impact on Scholarly Communication

Chapter 14
Scientists' Blogs: Glimpses Behind the Scenes

Brian Trench

14.1 Introduction

Scientists operate in an Internet-saturated environment and their pervasive use of email and the web for professional and public communication and, in particular, the implications of the web's adoption for scholarly publishing have been the object of much professional discussion and formal analysis over more than a decade (e.g., Rzepa 1998; Peterson 2001; Dumlao and Duke 2001). But scientists' use of more conversational Internet media, specifically web logs (blogs), has not been examined as much. Much of the commentary that has been published on blogs in and about science comes from practitioners who are strongly committed to promoting this kind of communication. In this chapter, we aim to take a more dispassionate view of the extent and effects of science blogging in the context of the increasing mediatization of science and changes in professional scientific and public science communication driven by media-technological developments. We take mediatization to refer to scientists' and scientific institutions' increased and significant attention to media dimensions of their work and their increased and significant adoption of mass-media genres and platforms in their communication.[1]

We will consider the growth of science blogging, and particularly scientists' blogging, as part of the developing blogosphere and offer a characterisation of scientists' blogs, focusing in more detail on particular disciplines. In developing this characterisation we will give special attention to the insights that scientists' blogs may give on the media orientation of science, such as mediatization theory posits. We will also consider the role that scientists' blogs play in opening access to the inner workings of science; in this, we are interested to establish whether blogs as a means of personal expression facilitate public understanding of science-in-the-making.

[1] We prefer to use the term mediatization to show that this is a different, somewhat broader media concept than the medialization approach that the editors use (see Chapter 1).

B. Trench (✉)
Dublin City University
e-mail: brian.trench@dcu.ie

S. Rödder et al. (eds.), *The Sciences' Media Connection – Public Communication and its Repercussions*, Sociology of the Sciences Yearbook 28,
DOI 10.1007/978-94-007-2085-5_14,

In reviewing science blogs we were interested to establish what support, if any, could be found for the observation of nearly a decade ago that the web "opened up many aspects of scientific research previously hidden from the general public" (Peterson 2001) or for the notion that the Internet was "turning science communication inside-out" (Trench 2008). Here it was postulated that Internet communication, in opening to public view previously closed private spaces, blurs the boundaries or restructures the relations between these spheres. Blogs, with their personal, even intimate, character appear strong candidates for facilitating this "inside-out" process.

This interest in blogs as windows on previously private spaces relates to several well-known propositions on the social organisation and social relations of science that all draw implicitly or explicitly on Goffman's (1959) work on the back-stage and front-stage presentation of self. All also, to one degree or another, stress the importance of gaining access to and understanding of the back-stage preparation for the front-stage performance. These include Hilgartner's (2000) elaboration of performance, theatrical and staging metaphors in relation to scientists' participation in public debates and Latour's (1987) view of science as Janus-faced, with two faces of ready-made science and science-in-the-making. Latour focused on securing entry "through the back door" to science-in-the-making as of particular interest in understanding the social constitution of science. Durant (cited in Gregory and Miller 1998) suggested that scientific literacy could be considered as knowing many facts of science, knowing how science works, or knowing how science *really* works and focused his attention on the last of these: "What [the public] needs, surely, is a feel for the way that the social system of science actually works to deliver what is usually reliable knowledge about the natural world." In this, Durant can also be interpreted as advocating the need for public appreciation of what goes on back-stage in science.

The editors of this present collection draw on the same lexicon in their discussion of the concept of medialization (see Chapter 1, p. 5) when they ask if science's orientation to the mass media remains "limited to activities on the front stage produced just for public view or does it extend to the backstage, thus affecting the criteria of relevance in knowledge production?" Sociologists have applied ethnographic and other methods to see what is happening back stage in science. The development of Internet media, and, in particular, of blogs appears to provide a readily accessible means to look behind the scenes. In this chapter, we are interested to see if communication in this hybrid private/public space of blogs has a bearing on the conduct of science itself. In his discussion of mediatization, Valiverronen (2001) notes:

> Communicating science to the general public may influence the mechanisms of science, and not only in the level of funding, science disputes or in the public legitimation of science. Public discourse also feeds back into science-in-the-making.

The case of climate science blogging that will be discussed in this chapter offers specific answers to the editors' questions above and some confirmation of Valiverronen's view of public communication affecting the conduct of science. Pearce (2010b) opens his extended investigation of the "Climategate" affair with a chapter entitled "Windows on a closed world" and writes in his concluding chapter:

"The doors of the labs are being opened, whether scientists like it or not." As we shall explore further in part 5 of this chapter, climate science represents a special case of highly mediatized science in which blogs have played an important constitutive role.

The following sections of this chapter trace the short history of science blogs, with particular reference to blogs published by scientists (Section 14.2), consider some of the claimed impacts of scientists' blogs on the conduct and governance of science (Section 14.3), set out some general characteristics of scientists' blogs (Section 14.4), review the intense discussions of climate science in the blogosphere and the role of blogs in the "Climategate" affair (Section 14.5) and discuss the factors constraining scientists' adoption of blogs in their peer-to-peer and public communication (Section 14.6).

14.2 The Slow Growth of Science Blogging

The growth of blogging since the early 2000s has been dramatic. Total weblogs were estimated at about one million in 2003 and over four million in 2004. Another 2 years later, the Pew Internet and American Life Project (2006) reported that 8% (12 million) of 147 million adult users of the Internet in the United States kept a blog, while 39% (57 million) read one. By late 2006, the specialist web site Technorati.com was "tracking more than 57 million blogs and counting" (Sifry 2006). In 2008, the same source gave the total number of blogs as 133 million (Sifry 2008).

Free blogging software reduced the entry-cost to zero and the entry-time to minutes and helped drive blogging as a near-mass phenomenon. By 2004, the accumulated blogs were being referred to as a collective space, the blogosphere, meriting analysis in a special issue of the computing journal *Communications of the ACM* (December 2004). *Business Week* writers Baker and Green (2005) described blogs as "simply the most explosive outbreak in the information world since the internet itself" that would "shake up just about every business". Interviews with bloggers associated with Stanford University revealed that bloggers' primary motivations were to document one's life, to provide comment and opinion, to work out emotional issues and to promote conversation (Nardi et al. 2004). But bloggers were also credited with breaking major political stories in the United States (Rosenbloom 2004). Rosenbloom noted that technological research disciplines were well represented among blogging communities and, more recently, Davidson and Vaast (2009) suggested that "tech bloggers may act as an active minority within technology-focused discourse communities and, in doing so, influence social representations of ICTs within society".

A study of medical bloggers (Kovic et al. 2008) noted that survey respondents' motivations for blogging were different from the generality of bloggers: "Sharing practical knowledge and skills, as well as influencing the way other people think, were major reasons for blogging among our medical bloggers, but not among general bloggers."

Another such study (Lagu et al. 2008) concluded that medical blogs were "now part of the literature and media of medicine" which in the authors' view ranged from professional and scientific publications to medical dramatizations on television. The authors expected the importance of medical blogs to grow but they also noted that "authors of some medical blogs censor their thoughts and comments less than we expect they would in traditional public settings".

Science blogging has also attracted attention in academic and professional journals. Batts et al. (2008) described science blogs as having "carved out a small but influential niche". A report in *The Guardian* (McClellan 2004) offered an early view of blogging by academic researchers, but the cases cited were in popular culture, literature, political philosophy, informatics and cyberculture, not in the natural sciences. In the following year, a *Nature* report (Butler 2005) suggested there were "still only a few dozen scientific bloggers". Hannay (2007) described scientific blogging as "still a niche activity" and stated that "scientists have been relatively slow to fully embrace [the web's] potential [. . .] among a few million scientists worldwide, only perhaps one or two thousand are blogging, at least about science".

Batts et al. estimated the total number of science blogs at "over 1,200", drawing on a study published a year earlier that in turn quoted science blogger Bora Zivkovic as estimating the number of science blogs at 1,000–1,200 (Bonetta 2007). A more recent estimate (Mooney and Kirshenbaum 2009) was "some 1,000", though this was qualified as "undoubtedly a very conservative figure". Several science bloggers who responded to a survey by *Nature* blogger Martin Fenner (2008) said they expected there to be many more science blogs in 5 years' time as it becomes "more socially acceptable", or, according to one respondent, "so many science blogs that we have to specialise".

Because the definition of science blog or scientist blogger can never be unequivocally settled the numbers cannot be precise. The distinction between science blog and scientist blogger is of some significance, however: authors of blogs that are mainly or exclusively about science include graduate students, science journalists and science writers; qualified scientists may not be in a majority of those behind science blogs. What seems clear is that there are less science blogs and certainly far less scientists' blogs than the numbers of scientists in the world's Internet population would indicate. As has been previously observed in studies of science web sites (Massoli, 2007; Trench 2007, 2008, 2009), scientists and their institutions have tended to use the Internet mostly for professional communication and, where wider publics are in mind, for dissemination of scientific findings and for promotion of science to students, policy-makers, media, business partners and prospective employees. Blogging and other more highly interactive applications of the Internet do not fit comfortably into that set of priorities. However, these observations do not deprive science blogging of all importance. In certain sectors of science where, for example, knowledge is especially uncertain or controversial, science blogging may have a weight that is not measured in the total number of science blogs in the total blogosphere. This may also be true for sectors of science where there is a relatively high level of public, or amateur, participation. We will consider examples of such cases in later sections.

14.3 Uses and Impacts of Science Blogging

In discussion of science blogs claims have been made on how science blogging has influenced the practice of science itself. Science blogging has been reported as a means by which scientists have found collaborators for the authoring of significant papers (Batts et al. 2008) or have benefitted from "interesting perspectives" of site users' comments, even helping to generate "new research ideas" (Butler 2005). However, it is notable that examples given by commentators of significant impacts of blogging on the conduct of science tend to be repeated, suggesting there may not be very many such examples: The story of a PhD student in genetics, Reed Cartwright, who disagreed in his blog with a 2005 *Nature* paper and was then invited to be co-author of a paper for *Plant Cell*, has been told in *The Scientist* (Secko 2007), by Bonetta (2007) and by Batts et al. (2008).

Science blogging has also been presented as a means of "enabling a conversation between the science community and the general community" (Elliott 2006) as a way to interact with a wider audience of peers and public (Butler 2005), and as having a powerful "capacity to put a human face on science and related health issues by allowing scientists to discuss how these things affect them personally" (*Nature Methods* 2009). It has been claimed that the notably successful blog Pharyngula has become a "universal, interactive rallying point for understanding and discussing evolutionary development" (Batts et al. 2008), making its originator, P. Z. Myers, a "rock star of scientist bloggers" (Bonetta 2007), perhaps mainly, however, through his advocacy of science-based atheism.

Much of the analysis of science blogging has been written by enthusiasts or by observers who are also practitioners. Batts et al. (2008) set out an argument for research and academic institutions to adopt blogging actively as part of their practice. Tola (2008) considered

> the advantages of this medium are so self-evident, in terms of the possibility of gaining feedback on one's work and approaches, of finding new solutions and ideas, of meeting new colleagues and other scientists who might be contributing to the development of one's research, of starting new collaborations, even of finding new positions, that it is really difficult to imagine why a scientist, especially a young one at the beginning of her own career, should not feel like entering this collective conversation.

Wilkins, a philosopher of science and blogger (Evolving Thoughts), describes science blogging as personal, ephemeral, and "more intimate and responsive" than other forms of science communication (2007). Schmidt, a climate scientist and blogger (Realclimate.org), sees blogs as a way for scientists to talk to the public directly, casually and in depth about complicated and contested scientific topics (Gramling 2008).

This merging of professional and public spheres of communication, without the intermediation of journalists, has been represented as one of facilitating public peer review (Batts et al. 2008), or of harnessing "collective intelligence" and "wisdom of the masses" (Minol et al. 2007). By analogy with "Web 2.0" – the purported new face or phase of the Internet that is genuinely interactive and participatory, and of which blogs are a representative expression – "science 2.0" is sometimes invoked to refer to a collective, inclusive endeavour in which both citizens and experts are engaged (Waldrop 2008).

Some commentary on the Climategate affair sees in it the emergence of "extended peer review". Jerome Ravetz, who coined that phrase almost two decades ago with Silvio Funtowicz in their elaboration of "post-normal science" (Funtowicz and Ravetz 1993), said of the climate science debate that it demonstrated the need for and inevitability of such public scrutiny:

> It is hard to see how this extended peer community of the scientific blogosphere could be silenced or suppressed, once it has shown its power. Doubtless it will be vulnerable to misuse and abuse, just like democracy in the political sphere, and so it will need guidance [. . .] and courtesy (cited in Pearce 2010b).

Similarly, journalist and blogger Patrick Courrielche (2010) believes Climategate "triggered the death of unconditional trust in the scientific peer review process" and the maturing of a new movement of peer-to-peer review.

However, it is clear that the potential of science blogging to significantly affect communication among scientists and relations between scientists and lay publics depends to some degree on the level at which blogging and other more highly interactive Internet media have been adopted in science. Physicist Michael Nielsen (2009) observed that "scientists have been relatively slow to adopt online tools such as comment sites and Wikipedia". Waldrop (2008) noted, "although wikis are gaining, scientists have been strikingly slow to embrace one of the most popular Web 2.0 applications: Web logging, or blogging". He quoted Christopher Surridge, managing editor of the Web-based journal *Public Library of Science On-Line Edition*, as saying that "scientists don't blog because they get no credit" and this was echoed in the comment of *Nature's* editorial writer (*Nature* 2009) that "blogging will not help, and could even hurt, a young researcher's chances of tenure".

What the available literature indicates is that, despite strong advocacy of the merits of blogging in science and some notable instances of highly visible scientist bloggers, scientists are significantly under-represented among bloggers in general and little evidence has been reported of blogging having a tangible impact on the conduct of science.

14.4 Features of Scientists' Blogs

To form an overview of the characteristics of scientists' blogs, in early 2010 we reviewed 20 such blogs that were frequently mentioned in commentaries on science blogs or in listings of "top science blogs". The sites reviewed included well-known blogs such as Pharyngula, NASA Watch, Blog Around the Clock, Bad Astronomy, Cosmic Variance and Highly Allochthonous. As explained below, blogs primarily concerned with physics, nanotechnology and climate science were also reviewed as separate sub-groups. A study of the modes of communication in 11 science blogs (Kouper 2010) included some also among our core group of 20 and concluded that

> they provide information and explain complicated matters, but their evaluations are often trivial and they rarely provide extensive critique or articulate positions on controversial issues ... In their current multiplicity of forms and contents science blogs present a challenge rather than an opportunity for public engagement with science.

Kouper's study emphasised differences between different branches of the sciences as factors in the heterogeneity of science blogs.

Our review of scientists' blogs was guided by general considerations of good practice on the web, including those of authorship, sources, content, frequency, usability and interactivity. We found great diversity in the frequency of updates (from several times daily to less than monthly), in the types of sources used (more often general media than scientific media), and in the types of information featured (dealing more often with contexts of science such as publishing, ethics and policy, than with science content). In view of the overriding concerns of this chapter we focused on issues of interactivity – the scope and quality of exchanges between blog publishers and visitors – and of transparency of scientific process – that is, the public access provided to negotiation of different views and of uncertainty in scientific information (or "science-in-the-making").

On authorship as a criterion, we noted that blog-owners could generally be identified, along with their professional affiliations or lack of affiliation but other contributors to the sites included some who were identified by arbitrary user names. In some cases, it appeared that regular visitors to a blog have come to know each other and the identity of the person behind the user name. This can engender unbalanced exchanges, in which the identities and credentials of the participants are not universally known.

We also noted the kinds of topics that stirred most comment and any tendency for discussion to get heated and to focus on how the argument is made, and who is making it, over the substance of the issue. This criterion relates to the frequently observed tendency of Internet discussions to deteriorate to *ad hominem* argument, a tendency we assumed would be a significant deterrent to scientists' participation. We found that the topics prompting most comment tended to be well-known as controversial in broader public and media spaces, e.g., science and religion, the hacked climate science emails, alternative medicine and the Obama administration's policy on the US space programme. But it was noted that in a majority of cases, even the most-commented postings elicited less than 20 responses. The low level of discussion and the absence of debate were the most frequently made observation in relation to this criterion and that concerning the nature of debate. Where significant debate was found it was generally well-mannered. As we shall see later, this did not apply in the case of climate science or, indeed, in physicists' discussions of the activity of blogging itself.

It has been claimed that blogs open "windows into academic coffee room chatter of the sort the media is not normally privy to" (Tomlin 2007) and it has been argued that blogs, in making this possible, serve an important function because, "unlike laws and sausages, the public should see science during its manufacture" (Wilkins 2008). But we found that less than a quarter of the blogs provided even occasional looks behind the scenes of science. Several of the blogs that did so were focused on relatively abstruse areas of physics. For this reason, the review of physics blogs was extended.

Concern has been expressed in physics circles that "physicists may be getting left behind" in comparison with other researchers using blogs and wikis (Griffiths 2007), but the same author also claimed that physics blogs were "starting to have a real impact on the way researchers communicate" (Griffiths 2007). We observed fairly frequent links between blogs and formal publications on the arXiv pre-print physics publishing site. These offered relatively rare examples of blogs facilitating public view of science-in-the-making. In February 2010, for example, the arXiv blog provided access to vigorous exchanges between physicists on an arXiv paper predicting an 11-fold increase in hurricane activity with a 2-degree rise in global temperature. The commenters countered claims of "pure hype" and "tailoring the data" with claims of "good science" and "clear model".

The public visibility of this debate might be taken as an example of what the physics blog Cosmic Variance aims at, that is, "building bridges between the world of specialists and interested outsiders. Blogs offer both immediacy and unfettered access to the inner workings of mysterious vocations of all sorts, which is hard to get from more formal journalism" (Carroll 2007). In February 2010, Cosmic Variance co-publisher Sean Carroll wrote critically on a *New York Times* report of findings from Brookhaven Laboratory's Relativistic Heavy Ion Collider that were claimed to have "broken" the laws of physics, specifically the law of parity. His posting prompted dozens more that politely disagreed. Anybody browsing science blogs could observe the fact of this disagreement among physicists, if not necessarily understand its basis.

Astrophysicist Clifford X's blog, Asymptotia, drew 10,000-plus comments over its first three-and-a-half years of operation in response to his invitation to "engage in conversations with me [and] ... with other commenters". The conversations sometimes become heated, including accusations of misrepresentation, even around very abstract theoretical topics. Brazilian physicist Christine Dantas closed her blog, Background Independence, in November 2006, explaining that she was uncomfortable with the way her contributions to debate on string theory were represented (Griffiths 2007). Chad Orzel (Uncertain Principles blog) declared in 2007 that he was "fairly disgusted" with the "antics" of string theory blogs and, in February 2010, announced that he would suspend writing his own blog for Lent, declaring that "reading blogs is pissing me off to no good end". There was criticism of Orzel for his "self-aggrandizing public display" but also support, including a comment that blogs "although potentially capable of generating productive conversation, tend towards bloggers shouting their opinions on the rooftops".

The evidence here of early disaffection or fatigue with blogging mirrors the short experiment by a well-known particle physicist in another form of so-called "social media". Brian Cox, a young English physics professor with a very strong mass-media profile, started podcasts on the CERN web site in 2007 and, despite securing the co-operation of high-profile interviewees, ceased this activity in 2008.

As a field of scientific research and technological development that is still taking shape, nanotechnology appeared as a strong candidate for public discussion of its procedures and its possibilities, including its risks. However, the main content of the frequently cited Soft Machines blog comes in the form of mini-essays on the fundamentals of nanotechnology and on topics of science in society. Many other nanotechnology-focused blogs are more strongly promotional in orientation: Nanotech-Now and Nanotechbuzz are primarily vehicles for nanotechnology business news.

Nanotechnology also features strongly in blogs that present perspectives on future trends: Nanodot is maintained by the Foresight Institute and covers artificial intelligence and robotics as well as nanotechnology business stories; Singularity Hub, Accelerating Future and Next Big Future present speculations on technological futures and generally optimistic accounts of technological research and innovations.

A common feature of these nanotechnology or futures blogs is the low level of comment and the almost complete absence of critical debate on nanotechnology. These issues are aired, however, in CRNano, a blog maintained by the Centre for Responsible Nanotechnology that is dedicated to promoting awareness of the benefits and risks of nanotechnology and to promote "wise, comprehensive, and balanced plans for responsible worldwide use of this transformative technology".

The possible risks in nanotechnology were also aired in the frequently mentioned Nanobot blog, owned by journalist Howard Lovy, who shared his concerns about some of the hyperbole surrounding nanotechnology. This blog started in late 2003 and had its highest level of posts (over 500) in 2004. But in another demonstration of early disaffection with science blogging Lovy closed the blog in mid-2009, declaring his disillusion with internet communication of science: "I have already rejected Web 2.0. I am almost ready to tell Web 1.0 to get lost as well."

From these reviews we can observe that scientists' blogging presents a diversified picture but even within this diversity we find very little evidence to support the claims reported above of blogging's significant role in communicating science or its significant impacts on science. The most notable evidence of blogs playing an important innovative role in communicating science by facilitating public view of science-in-the-making was found in physics, perhaps paradoxically in aspects of that discipline where there was very little of obvious public value at stake in commercial, ethical or political terms. It was in physics, too, that some evidence was found of blogging playing a substantive constitutive role in the science itself. In both of these regards, the picture is different when we turn to climate science blogging.

14.5 The Special Case of Climate Science

Climate science is uncertain in its interpretations of historical records and prehistorical evidence, it is necessarily speculative in extrapolation to future trends, it presents knowledge that has strong political and ethical implications and its collective wisdom is represented globally on a quasi-political platform, the

Intergovernmental Panel on Climate Change (IPCC). For these and other reasons, climate science is notably attractive to media and has become especially attuned to the logics of mass media and actively concerned with ensuring appropriate media coverage. Media attention to climate science has been consistently high in recent years; it is also notably diverse and diffuse, covering the full range of older mass media – newspapers, magazines, television, radio – as well as the full range of newer digital and online media – electronic publishing, institutional and personal web sites, blogs, forums and social networking sites.

However, formal studies of media representations of climate science have tended to concentrate on "traditional" media: In their analysis of media coverage of climate change, contributors to an edited volume (Boyce and Lewis 2009) examine almost exclusively print media coverage in various countries; just one contributor (Gavin 2009) focused on the web. Carvalho (2007) examines climate change discourses in three elite British newspapers up to 2001; her study makes no reference to the Internet as a vehicle of disseminating or discussing scientific information about climate change. However, Rogers and Marres (2000) had earlier taken climate change as a case for demonstrating how a mapping technique might be applied to web debates.

Pew Internet and American Life Project (2006) reported that 20% of Americans used the Internet as their primary source of information on science but of those who stated a particular interest in climate change, 49% had received information on climate change from the web or by email. Gavin (2009) examined what information citizens had available to them on the web relating to climate change politics. He noted the large quantity of information and comment accessible on the web but he lamented the quality of what he found on blogs:

> The contributions do contain moderate exchanges of evidence and argument, but there are high numbers of controversial and uncheckable assertions ... Entries are often highly disjointed and difficult to follow – part polemic, part rant, part ramble, part squabble, and often involving people flatly contradicting or sniping at one another.

If that represented the whole story of climate blogging, it would be easily dismissed. But in late 2009 and early 2010, climate science was thrust into the public spotlight as perhaps never before and communities of bloggers played tangible roles not only in how the science was publicly received, but also in how it was constructed internally. Nearly 3 years after its publication, the Fourth Assessment Report (AR4) of the Intergovernmental Panel on Climate Change (IPCC) was scrutinised with new vigour, errors were identified and retractions made, all in the full glare of mass media and online media attention, and sometimes directly triggered by media initiatives.

These developments followed a November 17, 2009 posting by "FOIA" of a short message on the climate change sceptic blog, The Air Vent. The posting had a link to a compressed date file posted on an Internet server ostensibly based in Russia. The compressed file contained 1,073 emails and 3,587 other documents apparently hacked from the server of the Climatic Research Unit (CRU) at the University of East Anglia in England. The hack of emails appears to have arisen from a series of unmet freedom of information (FoI) requests to the CRU during 2009 for release of data on weather patterns for a long list of named countries. The campaign of FoI requests was run through the Climate Audit blog, where supporters found advice on phrasing the requests and co-ordinated their selection of countries whose climate data was requested.

The pseudonym, FOIA, was tellingly the one used by the person who posted a comment on The Air Vent on November 17, 2009. As it happened, The Air Vent's owner, Jeff ID, was away from home deer-hunting. He returned to his blogging on November 18 but did not notice until the next day that FOIA's posting linked to hundreds of emails and other internal documents from CRU that represented, in Jeff ID's words, "62 mb (megabytes) of gold".

By the evening of November 19, links to the documents had been posted on several blogs, including Watts Up With That, Climate Audit, Climate Skeptic and The Blackboard. All of these blogs represent views at odds with the IPCC on climate science. Some concentrate on technical aspects of the calculations of global warming, others on the governance of climate science through IPCC, and others on the perceived political plot behind climate change policy.

While Jeff ID took time to reflect on the propriety of making the material generally available, he was overtaken by other bloggers and soon some blogs were struggling with the load. Pearce (2010b) notes that selected emails between climatologists employed by or associated with CRU were reproduced repeatedly for their shock effect. Comments on one blog were amplified through linking to and from many others. From this hyperactive blogosphere it was a short step to online news media, starting with a New Jersey newspaper, *Essex County Conservative Examiner*, but by November 20 also reaching Fox News, *Wall Street Journal*, BBC, National Public Radio and *New Scientist*.

The debate continued in the run-up to the COP15 climate change summit that opened in Copenhagen on December 7. This meeting of political and scientific leaders took place in the media's embrace, including that of blogs and bloggers. 5,000 journalists from 180 countries were accredited to cover the meeting; the majority of these were from media outside the "mainstream". The blogs represented there included those of consultancy companies like Deloitte and Arup, energy companies like Vattenfall and The Solar Company, non-governmental organisations like Oxfam, World Wildlife Fund and 350.org, student groups from Rice University, Houston, Texas, and Chalmers University in Sweden, along with diverse media outlets such as Google, *Lloyd's Register* and BBC.

Before, during and after COP15, the work of the IPCC was under an increasingly critical spotlight, not only shone by climate-sceptic bloggers. In early December 2009, BBC News had first reported questions raised about the basis for the IPCC's assessment of the rate of disappearance of Himalayan glaciers. These questions "reverberated around climate web sites" (Black 2010). The *New Scientist* recalled on January 11, 2010 (Pearce 2010a) that in 1999 it had reported a claim similar to that contained in AR4, but referring only to the central and eastern Himalayas. On January 20, 2010, the IPCC issued a short statement that the paragraph in question was "poorly substantiated" and "the clear and well-established standards of evidence required by the IPCC procedures were not applied properly".

A week later, the IPCC spelled out in a 1,000-word statement its principles and procedures for compiling assessment reports, including the "meticulous" reassessment of the scientific information contained in AR4. The "Climategate" affair, as it has become known, could not have grown as it did were it not for the availability of blogs to quickly disseminate information and arguments. Indeed, it might never have happened at all were it not for the active presence in climate science debates of blogger Steve McIntyre (climateaudit.org) who ran the blog-based campaign of freedom of information requests for release of worldwide meteorological data used by CRU in compiling global temperature records. This campaign provided the trigger and the context for the messy process that became Climategate.

The blogosphere debates provoked by the hacked documents and the following process – again, mainly Internet-based – of refutation and subsequent retraction of claims contained in IPCC reports eventually led to the announcement by the United Nations of a review of the IPCC's work. The Inter-Academy Council's review group reported in August 2010 (Inter-Academy Council 2010), making recommendations on more open recognition of the uncertainty of climate science knowledge, on more rigorous review of information taken into IPCC reports and on improving the transparency of the IPCC's review processes. Much of the proposed reform of climate science governance and co-ordination through IPCC is due, at least indirectly, to a blog-based campaign to let public light into private, professional spaces.

What may have started as the provocative act of one technically literate individual ballooned through blogs into a global crisis for a branch of the natural sciences that involved several of the world's leading professional societies, the international leaders of the IPCC and the world's highest-placed public administrator. The involvement of the UN Secretary-General underlines the particular character of the climate science debate. But atypical cases can also serve to highlight some general features. The climate science case indicates that the importance of extra-scientific factors in a particular branch of science gave particular importance to science blogs in communicating and influencing the content and conduct of that science. Weingart (1998) noted in an early treatment of the medialization of science that "science-media coupling" tended to be stronger where political dimensions were prominent in the science. One of his case studies was climate science and over a decade later in a new digital media environment Climategate reaffirmed the significance of that case.

A detailed investigation of this affair (Pearce 2010b), completed 6 months after the initial posting of the hacked CRU documents, describes it as much more than a local disciplinary spat:

> It was also a battle for ownership of data, the building blocks of scientific theories, and a battle to open up the closed world of scientific peer review to challenge by outsiders. This was about more even than climate science; it was about the conduct of all science. A battle for the soul of science.

And it was, almost entirely, a battle fought in blogs.

14.6 Discussion

Discussions of political blogging seem certain that the blogosphere has become a significant space for political communication, not just in relatively liberal societies where various media play watchdog roles on authority but also, and in specifically different ways, in authoritarian societies where print and broadcast media are heavily restricted. Hughes and Kellmann (2009) noted that during the 2009 Iranian elections "disenfranchised Iranis have often turned to the blogosphere to engage in commentary critical of the regime". Middle Eastern activist bloggers grouped at Cyberdissidents.org have been credited with significant influence in the 'Arab Spring' revolts of late 2010 and early 2011. Sunstein noted (2007) that "political blogs are a small percentage of the total, but they are plentiful, and they seem to be having a real influence on people's beliefs and judgments". In individual cases that he draws from Hewitt (2005), he shows that political blogs have exposed lies or corrected false claims and thereby affected the careers and standing of individual US politicians. However, in a comment that resonates with the discussion above of climate change blogs he notes that debate is polarised in the blogosphere, as blog readers tend to read blogs that resonate with their own point of view, and political blogs tend strongly to link to like-minded blogs.

In Technorati.com's 2010 survey of bloggers, politics was rated the field of activity on which blogging has the most impact: 46% of the respondents ranked politics first, compared with 5% for science (Technorati 2010). Politics features prominently in discussions of blogs and their impacts, and it may be that the conditions that promote the use of blogs in political communication are precisely the conditions that constrain their use in professional and public science communication. Politics is to a high degree concerned with values, beliefs, opinions, feelings and personal experiences and while all of these are present factors in science as in any other cultural sphere, their weight is very considerably less than in politics.

The polarisation of debate to which Sunstein refers is an outcome of differences in values that may be deep-seated: Polarisation comes naturally in politics (except where it is suppressed). The particular contribution of blogs to this process is not origination but amplification. When scientists consider possible paradigm shifts and when they compete for promotion or for institutional support, differences may be expressed in strong terms but not generally in terms of fixed polarised differences.

As we have observed, the case of climate science is in many respects special: Political and ideological factors are prominently in play, including through the involvement in the public debates of "amateurs" who may have technical competence to follow the arguments but whose main motivation for becoming involved is to fight a cause. By virtue of their involvement but also for other reasons, the general atmosphere of the climate science debate has become very highly charged. This is very evident in the blog debates, where Sunstein's observation of the development of separate camps in the political blogosphere also applies. The representatives and supporters of different positions group around their poles of attraction and rarely venture directly into the opposing camp.

Prof. Judith Curry, a climatologist from Georgia Institute of Technology, is a rare example of a (qualified) defender of the IPCC who contributes to anti-IPCC blogs. She is specifically concerned that the "debate has deteriorated ... into competing lines of propaganda" (Curry 2010). Her occasional contributions to anti-IPCC blogs appear aimed at defending the possibility of rational, critical debate against the odds and have attracted strong criticism from both sides. She has argued that "ignoring sceptics from outside the field [of professional climate science] is inappropriate. Einstein did not start his research career at Princeton, but rather at a patent office" (Pearce 2010b).

Based on his detailed examination of the Climategate documents and of the responses to their release, Pearce (2010b) considers that they have shown "scientists cutting corners, playing down uncertainties in their calculations, and then covering their tracks by being secretive with data and suppressing dissent". Scientists should own up to that, he believes.

In other domains, the vehemence and vituperation of contributions to open forums have begun to turn some scientists away from Internet debate. Even a public scientist and active polemicist like Richard Dawkins has found the tone and temper of Internet discussion hard to take. The discussion forum on his web site was closed in February 2010 after he was threatened in violent manner. "There is something rotten in the internet culture," Dawkins commented (Turner 2010). Behind the cloak of anonymity or pseudonyms personal insults have been thrown, including by established professionals. But when climate scientist Roger Pielke investigated the identity of a pseudonymous blog commenter who appeared to threaten his children he found that the culprit was a professor (Turner 2010).

The tenor of Internet debates on science and religion, in which Dawkins is centrally involved, and on climate science are cautions against over-optimistic readings of the potential of science blogs to create a new public sphere, at least the kind of public sphere envisaged by Habermas (1989), in which public opinion can be formed through rational discussion. It was tempting to believe that Internet technologies could be used to create a space in which interested publics and scientists of various backgrounds and orientations could exchange views and information freely and thus engage in "co-production of knowledge", as Gibbons et al. (1994) postulate in their account of Mode-2 science. The conversations many commentators envisaged among scientists and between scientists and publics are not much in evidence, except in climate science, where such conversations easily degenerate into name-calling or focus on trivial aspects of the issue.

As noted earlier, some see possibilities for a kind of public peer review through blogs. Paradoxically, we have found indications of such possibilities in the contrasting domains of astrophysics and particle physics, on the one hand, and climate science on the other. Whereas personal value systems are major factors in the production and reception of climate science and amateurs can be significant players, neither of these applies in astrophysics and particle physics. Thus, we have cases in which blogging is a relatively significant aspect of the professional and public communication of science, where social implications have either a very high or very low presence. The case of nanotechnology compounds the paradox: In this emerging science, there are discernible social issues that have been explored through various public engagement initiatives but these appear under-represented in the blog discussions.

The short history of the Internet has been marked by waves of high expectations that it would facilitate the formation of new relationships and communities. Such expectations have been expressed even more vigorously around the developed of so-called "Web 2.0" technologies or "social media". Blogs are seen as part of this supposed new era of highly interactive Internet media based on user-generated content. Blogs on science have been presented as a means to create new relations between scientists and lay publics and to support public and peer scrutiny of new developments in science, thus "fundamentally changing the nature of science communication" (Bubela et al. 2010). In some commentaries, however, this view has also been tempered by awareness of the "dubious quality" of much of the information on science available on the web (ibid.) and awareness that "for most scientists and academics, blogs and wikis remain unattractive distractions from their real work" (MASIS Expert Group 2009).

This chapter adds to these cautions, noting the very limited and possibly declining take-up of blogs by scientists, observing the generally low levels of peer-to-peer and professional-to-public discussion on even well-established scientist blogs, but also drawing attention to cases of scientist blogs that have opened new spaces for improving scientific accountability. In this way, it suggests that the mediatization of science in respect of the newer forms of online media is very partial and very uneven. The potential of scientists' blogging to contribute to reshaping relations between sciences and publics is evident but it appears largely to be an unrealised potential. That may relate to factors in the professional culture of the sciences which deter scientists from engaging in an online medium that is inherently predisposed to personal, affective communication. It would take a different kind of exploration to establish how scientists' media adoption and adaptation are shaped by their social organisation and institutional cultures.

Acknowledgments The author acknowledges the assistance of graduate student Sean Marshall in the review of general features of science blogs.

References

Baker, S. and H. Green (2005). Blogs will change your business. *Business Week*, 39(31), 56–67.

Batts, S. A., N. J. Anthis, and T. C. Smith (2008). Advancing science through conversations: Bridging the gap between blogs and the academy. *PLoS Biology*, 6(9), 1837–1841.

Black, R. (2010). UN climate body admits 'mistake' on Himalayan glaciers, BBC News website, posted 19 January 2010 (last accessed on March 15, 2010).

Bonetta, L. (2007). Scientists enter the blogosphere. *Cell*, 129(3), 443–445.

Boyce, T. and J. Lewis (eds.) (2009). *Climate change and the media*. New York: Peter Lang.

Bubela, T., et al. (2010). Science communication reconsidered. *Nature Biotechnology*, 27(6), 514–518.

Butler, D. (2005). Joint efforts. *Nature*, 438(1 December), 548–549.

Carroll, S. (2007). Blogging for physics. *Physics World*, January, 14.

Carvalho, A. (2007). Ideological cultures and media discourses on scientific knowledge: Re-reading news on climate change. *Public Understanding of Science*, 16(2), 223–243.

Courrielche, P. (2010). Peer-to-peer review: How Climategate marks the maturing of a new science movement, posted at http://bigjournalism.com/pcourrielche/2010/01/08/peer-to-peer-review-how-climategate-marks-the-maturing-of-a-new-science-movement-part-i/ (last accessed on October 11, 2010).

Curry, J. (2010). Reflections on climategate. *People & Science*, March 13.

Davidson, E. and E. Vaast (2009). Tech talk: An investigation of blogging in technology innovation discourse. *IEEE Transactions on Professional Communication*, 52(1), 40–60.

Dumlao, R. and S. Duke (2001). The web and e-mail in science communication. *Science Communicatio*n, 24(3), 283–308.

Elliott, S. (2006). Science blogs as a vehicle for upscale ads. *The New York Times*, January 20.

Fenner, M. (2008). Why do we blog and other important questions, answered by 34 science bloggers, posted on http://blogs.nature.com/mfenner/2008/11/30/why-do-we-blog-and-other-important-questions-answered-by-34-science-bloggers (last accessed on 11 March 2010).

Funtowicz, S. and J. Ravetz (1993). Science for the post-normal age. *Futures*, 25, 739–755.

Gavin, N. (2009). The web and climate change politics: Lessons from Britain. In T. Boyce and J. Lewis (eds.), *Climate change and the media*. New York: Peter Lang, pp. 129–144.

Gibbons, M., et al. (1994). *The new production of knowledge: The dynamics of science and research in contemporary societies*. London: Sage.

Goffman, E. (1959). *The presentation of self in everyday life*. Garden City, NJ: Doubleday.

Gramling, C. (2008). Science bloggers question their role. *Geotimes*, 53(6), 47.

Gregory, J. and S. Miller (1998). *Science in public: Communication, culture and credibility*. New York: Plenum Press.

Griffiths, M. (2007). Talking physics in the social web. *Physics World*, January 24–27.

Habermas, J. (1989). *The structural transformation of the public sphere – An inquiry into a category of bourgeois society*. Cambridge, MA: MIT Press.

Hannay, T. (2007). Web 2.0 in science, *CTWatch Quarterly*, 3(3), 19–25.

Hewitt, H. (2005). *Blog: Understanding the information reformation that's changing your world*. Nashville: Thomas Nelson.

Hilgartner, S. (2000). *Science on stage: Expert advice as public drama*. Stanford, CA: Stanford University Press.

Hughes, D. and R. Kellmann (2009). Blogging's global impact and the future of blogging, posted at http://technorati.com/blogging/article/day-5-twitter-global-impact-and/page-2/ (last accessed on March 15, 2010).

Inter-Academy Council (2010). *Climate change assessments – Review of the processes and procedures of the IPCC*. Amsterdam: Inter-Academy Council.

Kouper, I. (2010). Science blogs and public engagement with science: Practices, challenges, and opportunities. *Journal of Science Communication*, 9(1), posted at http://jcom.sissa.it/archive/09/01/Jcom0901%282010%29A02/ (last accessed on December 22, 2010).

Kovic, I., I. Lulic, and G. Brumini (2008). Examining the medical blogosphere: An online survey of medical blogger. *Journal of Medical Internet Research*, 10(3), posted at http://www.jmir.org/2008/3/e28/ (last accessed on December 22, 2010).

Lagu, T., et al. (2008). Content of weblogs written by health professionals. *Journal of General Internal Medicine*, 23(10), 1642–1646.

Latour, B. (1987). *Science in action – How to follow scientists and engineers through society*. Cambridge, MA: Harvard University Press.

MASIS Expert Group (2009). *Challenging futures of science in society: Emerging trends and cutting-edge issues*. Brussels: European Commission.

Massoli, L. (2007). Science on the net: An analysis of the websites of the European public research institutions, *Journal of Science Communication*, 6(3), posted at http://jcom.sissa.it/archive/06/03/JCOM0603%282007%29A03/ (last accessed on December 22, 2010).

McClellan, J. (2004). Inside the ivory tower. *The Guardian*, Online supplement (23 September), 24.

Minol, K., et al. (2007). Portals, blogs and co.: The role of the Internet as a medium of science communication. *Biotechnology Journal*, 2(9), 1129–1140.

Mooney, C. and S. Kirschenbaum (2009). Unpopular science. *The Nation*, August 17.
Nardi, B. A., et al. (2004). Why we blog. *Communications of the ACM*, 47(12), 41–46.
Nature (2009). Editorial: Filling the void. *Nature*, 458(19 March), 260.
Nature Methods (2009). Editorial: Lines of communication. *Nature Methods*, 6(3), 181.
Nielsen, M. (2009). Doing science in the open. *Physics World*, May.
Pearce, F. (2010a). Debate heats up over IPCC melting glaciers claim, NewScientist.com, posted 11 January 2010 (last accessed on March 15, 2010).
Pearce, F. (2010b). *The Climate Files – The battle for the truth about global warming*. London: Guardian Books.
Peterson, I. (2001). Touring the scientific web. *Science Communication*, 22, 246–255.
Pew Internet and American Life Project (2006). The Internet as a resource for news and information about science, posted at http://www.pewinternet.org/Reports/2006/The-Internet-as-a-Resource-for-News-and-Information-about-Science.aspx (last accessed on March 11, 2010).
Rogers, R. and N. Marres (2000). Landscaping climate change: A mapping technique for understanding science and technology debates on the World Wide Web. *Public Understanding of Science*, 9(2), 141–163.
Rosenbloom, A. (2004). The blogosphere. *Communications of the ACM*, 47(12), 31–33.
Rzepa, H. (1998). The Internet as a medium for science communication. In E. Scanlon et al. (eds.), *Communicating science: Volume 1 – Professional contexts*. London: Routledge, pp. 141–149.
Secko, D. (2007). Scooped by a blog. *The Scientist*, 21(4), 21.
Sifry, D. (2006). *State of the blogosphere*. October 2006, posted 7 August 2006 at http://www.sifry.com/alerts/archives/000443.html (last accessed on March 15, 2010).
Sifry, D. (2008). *State of the blogosphere*. September 2008, posted 22 September 2008 at http://www.sifry.com/alerts/2008/09/technoratis-state-of-the-blogosphere-september-2008/ (last accessed on December 17, 2010).
Sunstein, C. R. (2007). *Republic.com 2.0*. Princeton: Princeton University Press.
Technorati (2010). *State of the Blogosphere 2010* at http://technorati.com/blogging/feature/state-of-the-blogosphere-2010/ (last accessed on December 17, 2010).
Tola, E. (2008). To blog or not to blog, not a real choice there. *Journal of Science Communication*, 7(2), posted at http://jcom.sissa.it/archive/07/02/Jcom0702%282008%29C01/Jcom0702%282008%29C06, (last accessed on December 22, 2010).
Tomlin, S. (2007). Blogging science. *Science & Public Affairs*, September 23.
Trench, B. (2007). How the Internet changed science journalism. In M. Bauer and M. Bucchi (eds.), *Journalism science and society*. London: Routledge, pp. 133–142.
Trench, B. (2008). Internet: Turning science communication inside-out. In M. Bucchi and B. Trench (eds.), *Handbook of public communication of science and technology*. London: Routledge, pp. 185–198.
Trench, B. (2009). Science reporting in the electronic embrace of the Internet. In R. Holliman, E. Whitelegg, E. Scanlon, S. Smidt, and J. Thomas (eds.), *Investigating science communication in the information age: Implications for public engagement and popular media*. Oxford: Oxford University Press, pp. 166–180.
Turner, A. (2010). The poisonous pro-am clash in the boffins' blogosphere. *Sunday Times*, February 28.
Valiverronen, E. (2001). From mediation to mediatization: The new politics of communicating science and technology. In U. Kivikuru and T. Savolianen (eds.), *The politics of public issues*. Helsinki: Department of Communication, University of Helsinki, pp. 157–177.
Waldrop, M. (2008). Science 2.0: Great new tool or great risk? *Scientific American*, May, 68–73.
Weingart, P. (1998). Science and the media. *Research Policy*, 27(8), 869–879.
Wilkins, J. (2008). The roles, reasons and restrictions of science blogs. *Trends in Ecology and Evolution*, 23(8), 411–413.

Nature Neuroscience (2009). Editorial: Breakdown of communication. *Nature Neuroscience*, 12, 951.

Chapter 15
Debating Troy in the Mass Media – The Catalytic Impact of Public Controversy on Academic Discourse

Susann Wagenknecht

15.1 Introduction

The mass media pays close attention to science. Scientists, in turn, and scientific organizations seem to be ever more susceptible to media coverage. This mutual interest of science and the mass media gives rise to multifarious interactions. The medialization approach questions whether and how these interactions affect the inner workings of science (see Chapter 1). It is against this background that this paper sets out to demonstrate what effects a public debate can have on academic discourse. So far, there have been few attempts to study public scientific controversies with regard to their impact on scientific knowledge production and the mechanism that creates this effect (Bucchi 1996; Lewenstein 1995; Weingart and Pansegrau 1999). This paper seeks to examine a controversy that breaks out in the mass media and is only later taken up in academic communication. A scientific controversy is here understood to encompass an inner-academic debate and – in cases such as the one under consideration – a debate in the mass media (see Section 15.2). The controversy, which will now be considered in some depth, developed around a recent German media story about Troy (2001–2005). This paper argues that the Troy controversy must be seen as a truly scientific[1] controversy, even though it is fought out in the feature sections of widely read newspapers as well as within the inner circle of academia. It will show that the mass media discourse can have a substantial impact on an inner-academic debate and should therefore be regarded as an integral part of scientific controversy.

[1] 'Science' is used here in the German meaning of '*Wissenschaft*' and, thus, is not limited to the natural sciences, but includes social sciences and humanities. In this paper, the expressions 'scientific' and 'science' are applied to disciplines such as history or philology for two reasons: First, there is no good alternative translation of '*wissenschaftlich*,' as 'academic' or 'scholarly' do not always fit. Second, the theoretical background this paper draws upon has been developed for the natural sciences. Nevertheless, at least in this case, theories about scientific communication do a good job in elucidating phenomena within the humanities.

S. Wagenknecht (✉)
Aarhus University
e-mail: su.wagen@ivs.au.dk

S. Rödder et al. (eds.), *The Sciences' Media Connection – Public Communication and its Repercussions*, Sociology of the Sciences Yearbook 28,
DOI 10.1007/978-94-007-2085-5_15,

In this paper, conclusions are drawn from a twofold empirical analysis. Section 15.3 will provide a historically oriented depiction of the controversy's overall course. The account will consider sources including academic publications, researchers' websites and newspaper articles. This will be followed by a detailed analysis of the public debate on Troy with the help of qualitative coding techniques.[2] Section 15.4 will provide a detailed account of the debate's argumentative structure. Meanwhile, in Section 15.5 it will be argued that the public debate exerts a catalytic influence on academic discussion not just by virtue of the mass media's pervasiveness, but by virtue of its discursive configuration.

15.2 Inter-specialist Controversy in the Mass Media

Scientific controversies are lasting disputes among researchers that concern knowledge claims that are generally held to be determinable within science. The literature on scientific controversies shows that there is always more at stake than seemingly irreconcilable hypotheses: Controversies arise when basic background assumptions clash (Baltas 2000: 44), when modes of knowledge production differ fundamentally (Nowotny 1975: 37) or/and when there is no consensus on how to access a certain 'reality' (Pickering 1981). Controversies are always of social relevance for the scientific community involved. Epistemic arguments gain social impact, as 'scientific authority' is concerned (McMullin 1987: 51).[3]

Scientific controversies can draw on various communication channels, among them the mass media. This, however, does not mean that the scientific discourse can appropriate the mass media. The latter certainly take an active role in shaping public debates (Mazur 1981; Biezunski 1985). It is of utmost importance to keep in mind that the mass media follow their own rationale. Whatever content the mass media transmits, it has to comply with generally accepted mass media frames. Scientific controversies can be attractive to report upon, as they satisfy the dramatic imperative of the mass media and its inclination to personalization and polarization (Bell 1991, 156ff; Galtung and Ruge 1973: 67ff; see Stichweh 2005: 104) – provided that the academic question at stake is of any interest to a larger public. Obviously, a scientific argument cannot be elaborated in as great detail within a popular frame as in strictly academic publications, which means that inter-specialist communication via the mass media may at first glance appear as an inconvenient detour. However, the mass media can offer some advantages. Apart from their ability to reach large audiences, they also allow for a quicker publication than academic journals. In addition to – so to speak – amplitude and frequency, it is also their 'genre' that might

[2] As this debate is waged in German, all quotations are translations by the author.

[3] This is not to say that scientific arguments are genuinely political or at least 'epistemically permeable', as argued within the Strong Programme (Collins 1981). Yet, it was the Strong Programme that granted controversies a prominent place in Science Studies (see Bloor 1976) and directed attention towards what one could call their 'social side'.

appeal to researchers. Mass media formats require a different style and offer a different scope for argumentation. They call for a particular framing of issues. Despite, or rather, owing to that, the mass media can serve as a platform for inter-specialist communication (Lewenstein 1995). They address lay and expert audiences simultaneously. How these audiences receive the information will, of course, differ, but this does *not* imply that the former take scientific controversies in the mass media seriously, whereas the latter perceive them as mere façade. The debate on Troy exemplifies that (a) it can be genuinely scientific knowledge claims that are discussed by researchers in the mass media and that (b) the public discourse can have a decisive impact on the discourse within an esoteric circle of researchers. One should note here that the Troy debate in German newspapers is essentially composed of *researchers'* contributions. It would not be accurate to describe the Troy debate as *journalistic* coverage. Of course, the newspaper editors function as gate keepers who decide on whom to admit to the floor, but they – at least in this case – do not uphold a bias in coverage for long. Besides, it is basically the researchers who decide on what to say. Therefore, the fact that a public debate has an impact on inner-academic discourse is not a threat to the autonomy of science.

The public Troy debate is primarily an instance of inter-specialist communication in a specific space: the *Feuilleton*, the highly intellectual feature pages found in all renowned German newspapers. In order to describe the communication of scientific knowledge claims, it is important to develop sensitivity towards the situatedness of science communication – the specific ecological niches in which science communication might thrive, such as the *Feuilleton*. The *Feuilleton* is the section in German newspapers that provides socio-cultural coverage of all kinds of occurrences. These pages cultivate a reflexive, distanced commentary that is to some extent independent from everyday breaking news (Steinfeld 2004: 20f). The *Feuilleton* differs from the newspapers' science sections in its tendency to reflect *on* science and is in this respect not a medium for usual science popularization (see Kohring 2005: 288). What distinguishes the *Feuilleton* from news reporting is the authors' poetic style and their subjective coverage of the issue (Stöber 2005: 183ff). In this vein, the *Feuilleton* positions itself in a sense in opposition to academic publishing, even though its authors and subjects usually show great affinity towards the scientific community and scientific knowledge (Blödorn and Langer 2004: 162f). Researchers and other intellectuals who are not perceived as journalists are often invited to publish articles in the *Feuilleton* section.

Usually scientific knowledge claims are first diffused within academia before being transmitted to a wider audience, so there is a general movement from 'esoteric' to more and more 'exoteric' audiences (Fleck 1935) and, consequently, a passing through different 'expository genres' (Cloître and Shinn 1985). However, there are exceptions to the rule. In cases of 'deviation' – that are in fact perceived as such by the researchers concerned – knowledge is displayed in a more exoteric genre before being formulated within more esoteric frames. According to Cloître and Shinn, this happens because one genre might be more appropriate for one kind of content than another: 'In deviation, cognition is intentionally shifted from one expository genre to another with the idea in mind that expository standards and criteria of the alternative category are more felicitous for the growth of the idea at hand' (ibid.: 51). Bucchi applies their notion of 'deviation' to scientific controversies played out in the mass media. He analyzes cases where scientists deviate from the habitual path of science popularization by initiating a debate in the print press. Bucchi shows for a number of cases that the public discourse about controversial scientific issues comprises several layers of 'boundary work'[4] (Bucchi 1996: 382ff). He characterizes public scientific controversies as multi-layer conflict negotiation – a negotiation that is not about consensus building, but that is driven by both sides' efforts to mark out their terrain. Bucchi describes three superposed boundaries at stake between conflicting parties:

i. a boundary between conflicting research paradigms
ii. a boundary between scientific communities, usually disciplines
iii. a boundary between science and non-science

The latter boundary, Bucchi points out, is the most dominant one in public terms. Bucchi observes a tendency towards generalization,

> a tendency to transform [...] [internal, SW] tensions into more general matters of demarcation between disciplines, and eventually into matters of demarcation between orthodoxy (science) and deviance (non-science), which can be resolved by the public degradation and expulsion of 'heretical scientists' from the scientific arena (ibid.: 383).

One could possibly add a fourth boundary that is negotiated in publicly displayed controversies, since Bucchi notes that science communication itself becomes an issue. In addition to the above mentioned boundaries, adversaries fight over the delineation between communication practices perceived as standard and those practices perceived as deviant (ibid.: 387). All these multiple intertwined levels of demarcation make controversial science communication in public a complex matter worthy of fine-grained analysis. A case in point is the above mentioned heated media debate about research on Troy that developed in German newspapers between 2001 and 2005.

15.3 The Troy Controversy 2001–2005

To cut a long story short, the Troy controversy deals with the interpretation of excavation findings in a place called Hissarlik, situated in Northwest Turkey, within a bigger picture of the late Bronze Age. It was Heinrich Schliemann who first excavated remnants of ancient settlements in Hissarlik in the late nineteenth century. He claimed to have found the Homeric Troy, the battleground of the very Trojan War that paved the way for the ascending Greek – and hence, at least from a classical

---

[4] Bucchi here refers to Gieryn's concept of 'boundary work', which is originally meant to describe social mechanisms of exclusion mobilizing analytical distinctions on a conceptual layer (Gieryn 1983).

perspective – European civilization. Already during Schliemann's lifetime scholars were very skeptical about identifying the settlements below today's Hissarlik as the 'Homeric Troy,' the urban scenery described in Homer's *Iliad.* Yet, the equation of the excavated with the Homeric Troy has been of great appeal to a larger public ever since.

A century later, Manfred Korfmann, one of Schliemann's successors in Hissarlik, found himself amidst a polemic public debate, because he did not explicitly rule out that equation. Since the 1980s, Korfmann had been excavating different layers of classical and pre-classical settlements beneath the very slope where Schliemann claimed to have found the Homeric Troy. Korfmann received a chair in prehistoric archeology at Tübingen in 1982. In 2001/02 he curated a Troy exposition that brought his life's work on Troy to a wider public. Having worked on Turkish soil for many years, Korfmann maintained a special relationship to Turkish academia and politics. In 2002 he became the first foreign honorary member of the Turkish archeologists' association and was awarded a medal of merit by the Turkish Secretary of State. Shortly before his death in 2005 he accepted the Turkish nationality (Schuder 2005: II.1826; Beleke 2005: 745f). Korfmann's research in Hissarlik was funded to a significant extent by the Gottlieb Daimler and Karl Benz Foundation.[5]

Apart from Korfmann, it is his one-time adversary Frank Kolb who figures most prominently in the debate under scrutiny. When the controversy unfolded, they both held prestigious tenure positions at the University of Tübingen. In the 1990s Korfmann and Kolb collaborated in a graduate school on Anatolian history at the University of Tübingen where Kolb had been professor for ancient history since 1986. He had gained archeological experience from field survey activities in Western Asia Minor. Kolb's monograph *Die Stadt im Altertum* (1984) is regarded as a touchstone in literature on settlements in ancient and pre-ancient times. In this text he states that the Bronze Age Troy was not an important city, but a rural settlement.

Korfmann and his team, in contrast, interpret the excavated remnants in Hissarlik as evidence of a significant Bronze Age settlement – a settlement that once was a rich commercial town and served as residence of potentates, with a fortified outer city of an urban character below the king's castle. It was clear, therefore, long before the controversy broke out that Korfmann's Troy interpretation is at odds with Kolb's position. In a conference related journal article, Korfmann objects to Kolb, referring to him without providing the reference:

> [...] new findings oblige us to revise our understanding of Troia's economic and political status during the late Bronze Age. In fact, it was far from being the 'poor little settlement' with 'no claim to be called a city', which until recently some [sic!] scholars supposed (Korfmann 1995: 179f).

Korfmann's image of the archaeologically reconstructible Troy matches the Troy as described in Homer's *Iliad* and, as a matter of fact, the Homeric Troy is more or less explicitly present in his writings at least since the mid-1990s: 'If we compare

[5] See Projekt Troia (2007), Projektfinanzierung. http://www.uni-tuebingen.de/troia/deu/finanz.html (last accessed on June 29, 2010).

the settlement exposed by Schliemann, Dörpfeld and Blegen with the picture to be gained from Homer, the walls have until now always seemed to enclose an unexpectedly small area' (Korfmann 1995: 173).

In the 1990s Korfmann and his team claimed to have found even more evidence to support the hypothesis that the excavated Troy strongly resembles the Troy depicted by Homer. Resemblance suggests identity as the scholar of ancient Greek culture, Joachim Latacz, argues. He links Korfmann's archaeological findings to his philological argument which is elaborated in detail in his monograph *Troia und Homer* (Latacz 2001a). Latacz suggests that there is an immediate historical truth to the Troy in Homer's epic. According to Latacz, the essential narrative material in Homer's *Iliad* dates back to the late Bronze Age, which is exactly the time when today's Hissarlik was, at least according to Korfmann, a rich residential settlement. Apart from that, Latacz believes that he can show the historical existence of Homer's Troy in Hittite sources from the Bronze Age that have been created independently from the *Iliad* and the *Odyssey*. One of the main pillars of his line of argumentation is to identify the Greek terms for Troy as found in the *Iliad* with Hittite geographical names. There is, however, no consensus at all in Hittitology (see Heinhold-Kramer 2003). In a short review, fellow scholar Wolfgang Kullmann judges Latacz' book inadequate, since it does not properly represent the state of art in research on Homer (Kullmann 2001: 663).

Although Kolb, Kullmann and a number of other scholars do not agree with the way Korfmann interprets his Hissarlik excavations with respect to the Homeric Troy, critique is astonishingly scarce before 2001 (Weber 2006: 13). Critique, if expressed at all, is not formulated in monographs or any other comprehensive way, but in a popular science book (Hertel 2001) and sporadic short reviews (Kullmann 2001; Cobet 2001). Kolb himself publishes no journal article opposing Korfmann. It is only in June 2001 that Kolb verifiably utters critique within an academic setting,[6] when he delivers a speech at a symposium related to the Troy exhibition in Braunschweig, that is later published as a paper in 2003 (Kolb 2003). Although there is clearly a clash even before the exhibition, there is definitely no controversy about Korfmann's Troy interpretation before that time – at least no debate that left any retraceable record, such as journal articles (Table 15.1).

The exhibition, entitled *Troy – Dream and Reality*, brought an end to the silent tolerance of Korfmann's Troy interpretation in the scientific community. The exhibition opened on March 17, 2001, and was shown in Stuttgart, Braunschweig and Bonn until April 2002. In Braunschweig alone, the exhibition attracted more than 300,000 visitors.[7] The state presidents of Germany and Turkey acted as joint patrons

[6] It is only in 2001 that he communicates in accessible written form that he has been critical about Korfmann's hypotheses for some time and that he has uttered his critique verbally on various occasions before the public debate took place. See http://www.uni-tuebingen.de/fakultaeten/fakultaet-fuer-philosophie-und-geschichte/seminare/historisches-seminar/abteilungeninstitute/alte-geschichte/personal/prof-dr-f-kolb/troia-ii.html (last accessed on June 29, 2010).

[7] See Troia – Traum und Wirklichkeit (2001). http://www.troia.de/pages/portal.php (last accessed on June 29, 2010).

Table 15.1 Chronology of events

March 17, 2001	Inauguration of the Troy exhibition in Stuttgart
June 8 and 9, 2001	Symposium on Troy in Braunschweig
July 24, 2001	The Däniken accusation in the *Schwäbische Tageblatt*
February 15 and 16, 2002	Symposium on Troy at the University of Tübingen
April 1, 2002	End of Troy exhibition
September 10–13, 2002	*Deutscher Historikertag* in Halle/Saale
August 11, 2005	Korfmann's death

for the exhibition for which Korfmann in collaboration with Latacz and others acted as curator. Thus, the exhibition followed Korfmann's line of argumentation.[8] It was in fact a wooden model of late Bronze Age Troy that eventually incited Korfmann's critics to protest in the mass media. The model showed a large, densely tilled lower city below the castle, surrounded by a well-fortified wall. This model was a manifestation of Korfmann's hypothesis that Troy at this time could have had up to 10,000 inhabitants. It is this model that Kolb primarily had in mind when accusing Korfmann of deceiving lay audiences (Kolb 2001a, b; see Cobet and Gehrke 2002). He argued that there was no archaeological evidence for such a dense population below the castle. Korfmann seemed to accept this point, as he later partially distanced himself from the model. In an interview given in October 2001, he said that it was the model maker's decision to visualize excavation findings in a way that showed a possible scenario, although the scenario was not completely substantiated by research results (Korfmann 2001: 18).

When the exhibition opened in March 2001 there was a lot of media attention, but only Kolb's polemic comment on Korfmann's work in a regional newspaper triggered the actual controversy. On July 24, the *Schwäbisches Tageblatt* quoted Kolb calling his colleague Korfmann a 'Däniken of Archeology' (Bachmann 2001).[9] A week earlier, on July 17, Kolb had given an interview to the *Berliner Morgenpost* in which he accused Korfmann of 'deceiving' the public (Kolb 2001a). In this article, Kolb criticized Korfmann for interpreting archeological findings wrongly and for presenting his interpretation to a wider public without accounting for the critique of scientific peers. In addition, Kolb accused the *Frankfurter Allgemeine Zeitung* (*FAZ*) of biased coverage, as the editorial office had earlier refused to publish a critical comment on Korfmann's exhibition.[10] In fact, a member of its editorial staff, the science journalist Michael Siebler, was known to have good relations with Korfmann. During the debate Korfmann himself rarely appeared as author or

[8] See Troia – Traum und Wirklichkeit (2001) Konzeption. http://www.troia.de/pages/konzeption.html (last accessed on June 29, 2010).

[9] Erich von Däniken, notorious for original claims about extraterrestrial influences on early civilizations around the world, is a popular science phenomenon in its own right. See Däniken, E. (2010). http://www.daniken.com/e/index.html (last accessed on June 29, 2010).

[10] This is all meticulously documented on Kolb's webpage. See footnote 6. See also Korfmann's webpage for his perspective on the debate: Projekt Troja (2003). http://www.uni-tuebingen.de/troia/ (last accessed on June 29, 2010).

interviewee in the mass media, but Siebler came across as one of his most eager supporters. He published a range of articles in favor of Korfmann in the *FAZ* – along with other researchers, such as the philologist Joachim Latacz and the archeologist Wolf-Dietrich Niemeyer.

Kolb, too, was backed by various colleagues, among them philologists, archeologists such as Dieter Hertel *and* fellow historians, namely Hartwin Brandt and Wolfgang Schuller. Hertel, Schuller and four other professors wrote an open letter[11] to the *FAZ* editorial office, complaining about the newspaper's unambiguous support for Korfmann's position. The editors, however, refused to publish their statement, and it was not until September 2001 that the *FAZ* published one of Kolb's supporter's articles in its feature section (Schuller 2001). The coverage policy of the *FAZ* is an important issue for the researchers concerned, as the newspaper's feature pages are regarded as a crucial platform for popularization – and inter-specialist communication, too. The *FAZ* is one of the leading intellectual newspapers in Germany. Its readership profile encompasses the well-educated middle classes.[12] The public debate on Troy was fought out intensively in several German newspapers between 2001 and 2003; in the *FAZ*, however, the debate continued until Korfmann's death in 2005.

The public debate had a substantial impact on the academic discussion. Once the public debate had been triggered, one can observe that more and more people engaged in the controversy about Troy – in the *Feuilleton* debate as well as in truly academic settings. A symposium tied into the exhibition was organized in Braunschweig in June 2001. Then in February 2002, the rectorate of the University of Tübingen tried to pour oil on the troubled waters by organizing a conference with the hope of dispersing some of the resentment.[13] Half a year later, in September 2002, the Troy issue was broached in a discussion in the *Deutscher Historikertag*.[14] Direct encounters between adversaries were more or less limited to those occasions and although publications were spawned by these conferences, the adversaries did not publish together (e.g., Ulf 2003). Korfmann's critics did, however, formulate their position in extensive academic publications (ibid; Hertel 2008). Although Korfmann and his supporters phrased their hypotheses more carefully after 2002 (see Korfmann 2006), they never did a turnaround and no compromise was found.

---

[11] The letter is published online. See Funke, P., Gehrke, H.-J., Hertel, D., Hesberg, H. v., Hölkeskamp, K.-J. and W. Schuller (2001) Anlage 1. http://www.uni-tuebingen.de/fileadmin/Uni_Tuebingen/Fakultaeten/PhiloGeschichte/Historisches_Seminar/Alte_Geschichte/Troia/troia1anlage1.pdf (last accessed on June 29, 2010).

[12] For readership statistics, see Frankfurter Allgemeine Zeitung (2009) Leseranalyse Entscheidungsträger in Wirtschaft und Verwaltung. http://www.m-cloud.de/lae2009/titelportraits/Site/default.html (last accessed on June 29, 2010). These statistics serve marketing activities. For the year 2001 these data are unfortunately not available.

[13] See Eberhard Karls Universität Tübingen (2010) Troia Symposium Abschlussdiskussion. http://timms.uni-tuebingen.de/List/List01.aspx (last accessed on June 29, 2010).

[14] As for the program, see Deutscher Historikertag (2010) Programm. http://www.historikertag2002.uni-halle.de/programm/7.shtml (last accessed on June 29, 2010).

15.4 Debating Troy in the Media

From the above description it becomes apparent that the controversy about Troy does not unfold before the *Feuilleton* comes to function as its arena. Therefore, the question arises by what mechanism the public debate exerts influence on the discussion within academia. This question is best approached through studying the particular characteristics of discursive architecture that distinguish the public debate.[15]

The controversial media debate cannot be regarded as mere popularization of an inner-academic conflict. It is different: It is *both* more *and* less than a genuinely scientific debate. It deals with arguments of genuine scientific interest, but in a truncated way that makes it impossible for non-experts to assess points in detail. One could say that the inter-specialist discourse in the mass media cites arguments elaborated in academic publications in order to deal with them in a broader context. Seven central issues can be identified: The first four of them concern genuine research issues; the latter three deal with epistemic and ethical standards.

1. Archeological interpretation of excavation findings
2. Philological interpretation of written sources
3. Historicity of Homer's *Iliad*
4. Possibilities for merging conflicting viewpoints
5. Formulation of hypotheses for eso- and exoteric publics
6. Academic authority
7. Quality of the public debate to which they contribute

As for the first four, proposals for consensus building may be considered rhetoric devices for hedging positions rather than reconciliation offers. Both parties see their position as superior and attack their adversaries in a harsh tone. Neither side accepts the critique as worthy of consideration. Korfmann, Latacz and their collaborators sketch a possible historical scenario by merging the excavated Troy, the philologically corroborated Troy and the Homeric Troy. Their aspirations for a

[15] The debate about Troy in the German newspapers has been studied in detail with the help of qualitative content analysis (Mayring 2003, 2007), thereby guaranteeing a symmetrical approach (see Bloor 1976) to the controversy without favoring one party over another. Qualitative coding techniques have been employed to find out *how* researchers actually discuss their issues in public. The material which underpins this question is encompassed in all the articles that were published on the Troy issue in the *Frankfurter Allgemeine Zeitung* after the Däniken accusation from June 24, 2001 until 2005, including the appraisal of Korfmann's lifework after his death. The *FAZ* covers the debate with unrivalled intensity and persistency – partly due to it trying to save face when confronted with the accusation of being partial. Indeed it is not before September 2001 that a pronounced Korfmann critic is admitted to the floor. For details on data selection and analysis, see my master's thesis (Wagenknecht 2010).

global interpretation are most obvious in – if not inextricably tied to – exoteric communication, such as popular science books (e.g., Korfmann and Mannsperger 1998) and notably the Troy exhibition.

There is an apparent asymmetry in argumentation: Whereas Korfmann and his supporters put forward an all-encompassing vision of history with great public appeal, their adversaries rather limit themselves to a fragmented patchwork of moderately, often negatively formulated hypotheses. They argue that, given the scarce evidence from the late Bronze Age, Korfmann's Troy interpretation is not justified and that his single disputable hypothesis does not add up to a well corroborated overall interpretation. Instead, 'partial truths add up to a whole falsity' (Schuller 2001). His adversaries call Korfmann's Troy interpretation highly speculative and trimmed for 'selling' to a broad non-academic public. In short, Korfmann is criticized for popularizing premature guesswork. His supporters, instead, stress the overall consistency of his interpretation of excavation findings. They point to the sparse evidence available in prehistory – a circumstance which puts, according to them, an inevitable limit to the certainty of knowledge from that time (Latacz 2002a). Latacz accuses Korfmann's critics of 'excessive skepticism' (Latacz 2001b). The underlying question negotiated here is when a hypothesis is to be regarded as justified and which degree of justification is necessary for a hypothesis to allow for popularization to a lay public.

Apparently, questions of genuine scientific interest are complemented by issues of more general epistemological and ethical interest in the public debate, such as justification standards for hypotheses, authority and the quality of researchers' communication within science and beyond. These issues are not broached from a detached stance, but are weapons in the adversaries' argumentative arsenal. Whereas Korfmann's supporters claim that the public debate on Korfmann's Troy research is unnecessary and even damaging for science's public image (Latacz 2002a, b), his opponents see the *Feuilleton* debate as necessary as long as it triggers an academic discussion (Blum 2002; Brandt 2002b). Both parties complain about their adversaries' impertinence and the polemical tone that characterizes the *Feuilleton* and the later evolving debate within academic science.

Negotiating academic authority means negotiating the legitimacy to utter an opinion on Troy research. Legitimacy is tied to scientific expertise, which, in turn, is defined differently by the debating parties. Those supporting Korfmann argue that personal experience in terms of an immediate – one could say 'tactile' – contact with the material research object is vital for academic authority, i.e., only those who take part in excavating Troy have the authority to come up with an overall interpretation of their archaeological findings. Others only 'disgrace themselves' if they engage in a discussion about archaeological results (Niemeier 2002). One should note here that Korfmann and his team hold exclusive excavation rights for Hissarlik granted by the Turkish government. Moreover, Korfmann's supporters deny authority to their critics by saying that the latter would not keep up with current research results – if they did, they would acknowledge Korfmann's Troy interpretation as plausible, since recent excavation findings undeniably corroborate his view (Latacz 2002b). However, it is exactly those findings that his critics call into question.

Still another criterion for ascribing authority is disciplinary membership. Primarily Korfmann's supporters use disciplinary boundaries for claiming archeology's authority vis-à-vis ancient history and for de-legitimizing their critics as non-experts of pre-classical history (Latacz 2002a). At the same time, however, they present Korfmann's Troy interpretation as an interdisciplinary endeavor, yet claim autonomy from classical history (Latacz 2001b). Korfmann's critics, too, make use of their disciplinary membership to justify their authority to speak out on Troy research. They argue that all research on past events has to comply with history's disciplinary standards of considerate hermeneutics and critical source analysis (Brandt 2004) – especially when it comes to publicly displaying a research field for which laymen would hold historians responsible.

The distinction between disciplinary fellows and foreigners is one key dichotomy displayed in the *Feuilleton* articles, while the delineation between adherence to academic standards and their violation is another one, similar albeit with different roots. Whereas disciplinary membership is presented as a descriptive distinction, the distinction between 'scientific' and 'non-scientific' is palpably normative and applies in principle to all issues discussed in the debate. From the practitioners' perspective, 'scientific' is what supposedly fosters solid epistemic progress. Both parties accuse each other of violating the ethical and epistemic standards of their profession, but differ in what they perceive as questionable practice. Whereas Korfmann and his supporters stress the immediate demonstrative quality of their excavation findings (Niemeier 2002), their critics highlight precision in terminology, especially when it comes to classifying the late Bronze Age Troy as 'rural' or 'urban settlement' as was Korfmann's stance (Brandt 2002a).

The underlying pattern is simple: Both parties refer to themselves as fulfilling science's ethical and methodological requirements, whereas their adversaries do not. Both Korfmann's supporters and his critics hold that the impertinent and simplistic manner in which the public debate is fought – by their respective adversaries, of course – is unacceptable and destructive. Latacz, for instance, characterizes the publicly held symposium in Tübingen to which a large number of journalists was admitted as 'continuous fire of apodictic pretensions whose simplicity amazed the attacked time and again' (Latacz 2002a). Besides that, however, they differ noticeably in what they label as unscientific. Korfmann's supporters criticize their adversaries first and foremost for initiating a polemic public debate – in other words, for deviation from the usual path of science communication and popularization. Korfmann's critics, in turn, argue that Korfmann and his team violated standards of good scientific practice twice. First, regarding academic knowledge production, their hypotheses are formulated imprecisely and based on speculation rather than on thoroughly considered empirical evidence. Second, regarding popularization and especially the Troy exhibition in 2001/02, they allegedly popularize badly corroborated hypotheses in a suggestive and simplified way, yielding an implicit political message. Korfmann is accused of deliberately presenting a picture of the past that locates Europe's cultural origins in Asian Turkey by claiming that there was indeed a mighty Troy and probably also a Trojan War won by the Greeks.

In fact, Bucchi's three levels of conflict flagging can be retrieved quite easily from the *Feuilleton* debate on Troy. The split between conflicting parties is generalized in a way that almost eclipses their conflict regarding the excavation findings and their interpretation. The conflict is reframed in a way that further adds to the split between Korfmann's supporters and his critics. In so doing, the public debate amplifies the relevance of the conflict for the scientific communities concerned. Even researchers who do not deal with Troy research immediately feel the need to either back or oppose one side in the conflict. As a result, the debate picks up more and more scholars as it develops.

15.5 Conclusions: Catalytic Dynamics in Public Debates

General conclusions should be drawn with caution. The Troy controversy has its peculiarities. The case at hand is not only idiosyncratic, but unusual within literature on science communication. First of all, it is primarily researchers that argue with each other. Journalists act as gatekeepers and adherents, but in contrast to other public controversies such as the Goldhagen debate (Weingart and Pansegrau 1999), they do not constitute a major opposition to the academic mainstream. In addition, it is worth noting that the controversy takes place among humanities scholars, not natural scientists. Humanists and social scientists may be generally more prone to engage with the mass media, especially in the *Feuilleton,* than natural scientists.[16]

The Troy case suggests that publicly debating science is not just an ordinary academic discussion by other means, but it is not detached from the academic discourse either. Academic and public discourses are interrelated. It is useful to distinguish mass media communication from academic communication, but one should refrain from overstating disparity at the expense of empirical adequacy. Specific mass media formats, such as the *Feuilleton,* allow scientific hypotheses to be considered in an appropriate and truthful manner. The *Feuilleton* is a privileged ecological niche in the mass media with a particular affinity to academia. It offers a modus for displaying scientific knowledge claims in a truncated, polemic and accentuated way while simultaneously reflecting on them from a wider angle. The *Feuilleton* debate on Troy has an impact on science. This impact is more than an amplifying effect (see Weingart 2001: 259ff); it is indeed a catalytic one because the public debate instigates a discussion in academic settings that has *not* been led – at least not re-traceably – *before.* In reaction to the public debate, academic symposia are held and comprehensive publications are edited. Journal articles and edited books subsequently take an explicit stand with respect to the conflicting positions. In addition, one can see that Korfmann and his supporters phrase their Troy interpretation ever more carefully (see Section 15.3). The public debate has a generating effect on the academic discourse.

---

[16] For social scientists and their engagement in mass media, see Osrecki (Chapter 16).

Regarding the Troy debate, Bucchi is right in pointing out that clashing research paradigms are presented as conflicts between disciplines and even as conflict between proper and 'false science' in the mass media. Still, the debate in the mass media refers to genuine scientific knowledge claims (see Section 15.4). It seems to be precisely this particular aspect that invests the public debate with catalytic power. For this catalytic effect to unfold it was, first, necessary for the debate to 'enter' the public stage. The exhibition paved the way for a broader audience and interested the newspaper editors. Second, it is the particular argumentative configuration of controversies in the mass media that facilitated the catalytic effect upon the academic discourse, since the conflict is broadened and reframed.

The involvement of the mass media has two effects that are of particular significance to the scholars engaged in the controversy. It leads to a temporarily heightened attention to Korfmann's Troy research and its possible shortcomings in a broader academic audience. In addition to that, it leads to a sharpened disciplinary delineation due to the mass media's affinity for clear cut conflicts. The first effect is in the interest of Korfmann's critics. They want Korfmann and his team to finally consider their critique by creating public attention. However, it would be wrong to regard this effect as an outcome of their deliberate actions, since Kolb and his supporters, in the beginning, found it difficult to publish in high profile newspapers and were highly dependent on journalistic gatekeepers. It was Korfmann's popularization efforts that initially made a broader audience interested in Troy research. The latter effect – a pronounced delineation of disciplines – serves Korfmann and his supporters in fending off critique.

It is important not to confuse discursive framing with actual social structure. Adversaries use disciplinary membership as means for distinguishing 'us' and 'them' and for denying 'them' authority and the legitimacy to have their say. This does not mean that the Troy controversy is indeed a conflict between two disciplines, i.e., history and archeology. The way the controversy is described in the mass media – namely, in terms of a conflict between disciplines – is not consistent with the disciplinary membership as defined by job positions of the scientists involved. The closing discussion at the symposium in Tübingen shows that historians, archeologists and philologists can be found on both sides.[17] In fact, in the case of the Troy debate, the conflict line does not run along disciplinary boundaries. This constitutes a main difference to Bucchi's analysis of public debates on Cold Fusion, the Big Bang and Louis Pasteur's public experiments on anthrax (Bucchi 1996).

So, put differently, the case at hand is not a public clash between two communities which are usually kept apart by disciplinary boundaries. Instead, there is no disciplinary boundary that would seriously prevent any outright discussion on easily noticeable differences in research results. The Troy controversy illustrates how discursive disciplinary framing inflates a controversy in the mass media such that it is

[17] The discussion was organized by the rector's office of Tübingen University. A video record is available here: Eberhard Karls Universität Tübingen (2010) Troia Symposium Abschlussdiskussion. http://timms.uni-tuebingen.de/List/List01.aspx (last accessed on June 29, 2010).

continued within science. By reframing the original conflict in broader terms, stakes are increased. In this manner, the conflict becomes laden with significance not only for those directly concerned with research on Troy or other Bronze Age settlements in Asia Minor, but more generally, hence more and more scientists find themselves engaged in the public debate. A promising conjecture here would be that a scientific controversy needs a 'critical mass' of participants and that just those scholars who are immediately dealing with Troy research do not muster enough momentum to nourish a controversial discussion within academia. The dynamics gained in this way in the first place get lost when disciplinary framing is employed in a self-referential and less adversarial way. Actually, scholars not involved in Troy research tend to engage in the debate with a more detached perspective that allows them to use the controversy as an instance for assuring themselves of their disciplinary self-concept (e.g., Cobet and Gehrke 2002; Weber 2006). When the controversy serves as an occasion for strengthening disciplinary identities, the discussion loses conflict potential and, hence, momentum. The controversy is not solved explicitly but fizzles out.

References

Bachmann, A. (2001, July 24). Dem Mythos Homer erlegen. *Schwäbisches Tageblatt*, 168, 15.

Baltas, A. (2000). Classifying scientific controversies. In P. Machamer, M. Pera, and A. Baltas (eds.), *Scientific controversies. Philosophical and historical perspectives*. New York: Oxford University Press, pp. 40–49.

Beleke, N. (2005). *Wer ist Wer? Das deutsche Who is Who? 2005*. Schmidt-Römhild: Lübeck.

Bell, A. (1991). *The language of news media*. Oxford: Blackwell.

Biezunski, M. (1985). Popularization and scientific controversy. In T. Shinn and R. Whitley (eds.), *Expository science: Forms and functions of popularization*. Dordrecht: Kluwer, pp. 183–193.

Blödorn, A. and D. Langer (2004). Feuilleton und Universität. In T. Steinfeld (ed.), *Was vom Tage bleibt. Das Feuilleton und die Zukunft der kritischen Öffentlichkeit in Deutschland*. Frankfurt/M.: Fischer, pp. 161–164.

Bloor, D. (1976). *Knowledge and social imagery*. London: Routledge.

Blum, H. (2002, April 6). Letter to the editor. *Frankfurter Allgemeine Zeitung*, 80, 50.

Brandt, H. (2002a, February 26). Troianische Phantasien. Zur modernen Irrfahrt eines Recken aus Tübingen. *Frankfurter Allgemeine Zeitung*, 48, 51.

Brandt, H. (2002b, April 8). Letter to the editor. *Frankfurter Allgemeine Zeitung*, 81, 8.

Brandt, H. (2004, February 13). Verweigerung gilt nicht! *Frankfurter Allgemeine Zeitung*, 37, 39.

Bucchi, M. (1996). When Scientists turn to the public: Alternative routes in science communication. *Public Understanding of Science*, 7(5), 375–394.

Cloître, M. and T. Shinn (1985). Expository practice. Social cognitive and epistemological linkage. In T. Shinn and R. Whitley (eds.), *Expository science: Forms and functions of popularization*. Dordrecht: Kluwer, pp. 31–60.

Cobet, J. (2001). Manfred Korfmann/Dietrich Mannsperger, Troia. Ein historischer Überblick und Rundgang. *Historische Zeitschrift*, 272, 422–424.

Cobet, J. and H.-J. Gehrke (2002). Warum um Troia immer wieder streiten? *Geschichte in Wissenschaft und Unterricht*, 5(53), 290–325.

Collins, H. (1981). Stages in the empirical programme of relativism. *Social Studies of Science*, 11(1), 3–10.

Fleck, L. (1935). *Entstehung und Entwicklung einer wissenschaftlichen Tatsache. Einführung in die Lehre vom Denkstil und Denkkollektiv*. Basel: Schwabe.

Galtung, J. and M. Ruge (1973). Structuring and selecting news. In S. Cohen and J. Young (eds.), *The manufacture of news. Social problems, deviance and the mass media*. London: Constable, pp. 62–72.

Gieryn, T. (1983). Boundary work as the demarcation from science to non-science: Strains and Interests in professional ideologies of scientists. *American Sociological Review*, 48(6), 781–795.

Heinhold-Kramer, S. (2003). Zur Gleichsetzung der Namen Ilios-Wilusa und Troia-Taruisa. In C. Ulf (ed.), *Der neue Streit um Troia. Eine Bilanz*. München: Beck, pp. 146–168.

Hertel, D. (2001). *Die Mauern von Troia. Mythos und Geschichte im antiken Ilion*. München: Beck.

Hertel, D. (2008). *Das frühe Ilion. Die Besiedlung Troias durch die Griechen (1020-650/25 v. Chr.)*. Zetemata, Monographien zur Klassischen Altertumswissenschaft, Heft 130. München: Beck.

Kohring, M. (2005). *Wissenschaftsjournalismus. Forschungsüberblick und Theorieentwurf*. Konstanz: UVK.

Kolb, F. (1984). *Die Stadt im Altertum*. München: Beck.

Kolb, F. (2001a, July 17). Traumgebilde. Interview. *Berliner Morgenpost*, 192, 15.

Kolb, F. (2001b, September 25). Wider dem troianischen Populismus. Warum die Debatte um die Bedeutung Troias nicht früher geführt wurde. *Schwäbisches Tageblatt*, 222, 21.

Kolb, F. (2003). Ein neuer Troia-Mythos? Traum und Wirklichkeit auf dem Grabungshügel von Hisarlik. In H.-J. Behr, G. Biegel, and H. Castritius (eds.), *Troia – Traum und Wirklichkeit. Ein Mythos in Geschichte und Rezeption*. Braunschweigisches Landesmuseum, pp. 9–39.

Korfmann, M. (1995). Troia: A residential and trading city at the Dardanelles. In R. Laffineur and W.-D. Niemeier (eds.), *Politeia – Society and State in the Aegean Bronze Age. Proceedings of the 5th International Aegean Conference*, University of Heidelberg, pp. 173–184.

Korfmann, M. (2001). Mit Homer halte ich es wie mit Edgar Wallace. Interview. *Literaturen*, 10, 18–23.

Korfmann, M. (2006). Troia – Archäologie eines Siedlungshügels und seiner Landschaft, Einleitung. In M. Korfmann (ed.), *Troia. Archäologie eines Siedlungshügels und seiner Landschaft*, Mainz: Zabern, pp. 1–14.

Korfmann, M. and D. Mannsperger (1998). *Troia – Ein historischer Überblick und Rundgang*. Darmstadt: Wissenschaftliche Buchgesellschaft Darmstadt.

Kullmann, W. (2001). Latacz, Troia und Homer. *Gnomon*, 73, 657–663.

Latacz, J. (2001a). *Troia und Homer. Der Weg zur Lösung eines alten Rätsels*. München: Köhler und Amelang.

Latacz, J. (2001b, October 9). Der Mond auf seinen zarten grünen Matten. Nur erst als kleine Wolke schwebt! – Zu der Gefahr, sich in der Troia-Frage selbst im Licht zu stehen. *Frankfurter Allgemeine Zeitung*, 234, 54.

Latacz, J. (2002a, Februray 22). Moderner Krieg um Troia. Beim Tübinger Symposion kämpft mancher Recke unfair. *Frankfurter Allgemeine Zeitung*, 45, 49.

Latacz, J. (2002b, March 6). Letter to the editor. *Frankfurter Allgemeine Zeitung*, 56, 8.

Lewenstein, B. V. (1995). From fax to facts: Science communication in the cold fusion saga. *Social Studies of Science*, 25(3), 403–436.

Mayring, P. (2003). *Qualitative Inhaltsanalyse. Grundlagen und Techniken*. Weinheim und Basel: Beltz.

Mayring, P. (2007). Qualitative Inhaltsanalyse. In U. Flick, E. Kardorff, and I. Steinke (eds.), *Qualitative Forschung. Ein Handbuch*. Reinbeck: Rowohlt, pp. 468–475.

Mazur, A. (1981). Media coverage and public opinion on scientific controversies. *Journal of communication*, 31, 106–115.

McMullin, E. (1987). Scientific controversy and its termination. In T. Engelhardt and A. Caplan (eds.), *Scientific controversies. Case studies in the resolution and closure of disputes in science*. Cambridge: Cambridge University Press, pp. 49–92.

Niemeier, W. D. (2002, March 16). Viel Rauch, wenig Braten. Wer was wie wahrnimmt: Noch eine Antwort im Troia-Streit. *Frankfurter Allgemeine Zeitung*, 64, 52.

Nowotny, H. (1975). Controversies in science: Remarks on the different modes of production of knowledge and their use. *Zeitschrift für Soziologie*, 4(1), 34–45.

Pickering, A. (1981). Constraints on controversy: The case of the magnetic monopole. *Social Studies of Science*, 11(1), 63–93.

Schuder, W. (2005). *Kürschers Deutscher Gelehrten-Kalender 2005. Biobibliographisches Verzeichnis deutschsprachiger Wissenschaftler der Gegenwart*. München: Saur.

Schuller, W. (2001, September 12). Der Hügel, wo sie wandeln, liegt im Schatten. Indes er drüben schon im Lichte webt? Zur Kontroverse um die Ausgrabungen in Troia. *Frankfurter Allgemeine Zeitung*, 212, 52.

Steinfeld, T. (2004). Was vom Tage bleibt. Das Feuilleton und die Zukunft der kritischen Öffentlichkeit in Deutschland. In T. Steinfeld (ed.), *Was vom Tage bleibt. Das Feuilleton und die Zukunft der kritischen Öffentlichkeit in Deutschland*. Frankfurt/M.: Fischer, pp. 19–24.

Stichweh, R. (2005). *Inklusionen und Exkklusionen. Studien zur Gesellschaftstheorie*. Bielefeld: Transcript.

Stöber, R. (2005). *Deutsche Pressegeschichte. Einführung, Systematik, Glossar*. Konstanz: UVK.

Ulf, C. (2003). Wozu eine Bilanz? In C. Ulf (ed.), *Der neue Streit um Troia. Eine Bilanz*. München: Beck, pp. 9–15.

Wagenknecht, S. (2010). *Die Rolle des Feuilleton in wissenschaftlichen Kontroversen am Beispiel der Troia-Debatte*. Unpublished Master's thesis, University of Bielefeld.

Weber, G. (2006). Neue Kämpfe um Troia. Genese, Entwicklung und Hintergründe einer Kontroverse. *Klio*, 88(1), 7–33.

Weingart, P. (2001). *Die Stunde der Wahrheit? Zum Verhältnis der Wissenschaft zu Politik, Wirtschaft und Medien in der Wissensgesellschaft*. Weilerswist: Velbrueck.

Weingart, P. and P. Pansegrau (1999). Reputation in science and prominence in the media: The Goldhagen Debate. *Public Understanding of Science*, 8(1), 1–16.

Chapter 16
Diagnosing the Present: Towards a Sociology of Medialized Social Science

Fran Osrecki

16.1 Introduction: A Sociology of Sociological Knowledge

As sociologists of science we are trained in analyzing "hard" sciences. Over the last 30 years, we developed powerful theoretical and methodological tools to deconstruct those sciences which, at first sight, seem objective, free of ideological speculation and self-interest. However, we still lack a corresponding model for social sciences. A bundle of reasons could explain this fact. First of all, social sciences seem to be regarded as an "easy" case for science studies, since they often not even try to present their theories and procedures as superior to other forms of public debate. If one of the central aims of science studies is to show how notions of objectivity are constructed and strategically used, social sciences are certainly not the most obvious example that comes to mind. Another reason is that scholars of science are mostly social scientists themselves. When they talk about social sciences, they cannot act as if they were talking about an external object. They rather use social scientific concepts as *tools* of explanation. Only very few science scholars will treat social scientific approaches as "interesting cases." Instead, most of us will ask whether actor-network theory (ANT) or differentiation theory or symbolic interactionism or science policy analysis (to name a few approaches) is the more *adequate* or *valid* explanation for what happens in the lab or in the scientific journal. When social scientists analyze social sciences they act in the role of the critical peer. The lack of research on social sciences can therefore be explained by the fact that social scientists cannot keep their distance when it comes to the concepts they use or the concepts they want to criticize. If we want to develop a model for analyzing social sciences, we therefore have to make a radical step and relate the core of social scientific knowledge production to social structures.

The approach I chose to develop such a model has two main sources. The first is *classical sociology of knowledge*, a line of thought broadly associated with Karl

F. Osrecki (✉)
Osnabrück University
e-mail: fran.osrecki@uni-osnabrueck.de

S. Rödder et al. (eds.), *The Sciences' Media Connection – Public Communication and its Repercussions*, Sociology of the Sciences Yearbook 28,
DOI 10.1007/978-94-007-2085-5_16, © Springer Science+Business Media B.V. 2012

Marx and Friedrich Engels ([1845] 1947), Max Scheler ([1926] 1980), and, most prominently, Karl Mannheim ([1929] 1936). It is important to keep in mind that this version of sociology of knowledge should not be confused with the sociological analysis of everyday knowledge as proposed in *The Social Construction of Reality* by Berger and Luckmann (1966). Rather than asking how everyday knowledge is constituted, classical sociology of knowledge develops sociological tools for investigating elaborate social scientific and/or philosophical arguments. My approach combines this "classical" version of sociology of knowledge with differentiation theory. The basic idea here is to ask how elaborate knowledge is developed and processed under conditions of functional differentiation. Originally developed by Niklas Luhmann (e.g., Luhmann 1980),[1] this question was further explicated and applied to sociological knowledge by André Kieserling (2004). Kieserling's suggestion is that in a functionally differentiated society, sociological knowledge also becomes differentiated according to the functional system where it is processed. One important line of differentiation is the one between academic and mass media publics. In the course of the twentieth century, so the argument is unfolded, we became witnesses of a differentiation of sociological knowledge leading to diverse *genres* of sociological argumentation. On the one hand, there emerged *general sociological theories*, almost exclusively discussed in academia, thus meeting hardly any response from the public sphere. On the other hand, there emerged sociological *diagnoses of the present state of society*, which strive for mass media attention while to a large extent disregarding standards of academic rigor.

So far, this suggestion has not been tested empirically. To do so, I will argue from the perspective of role theory and propose a model for analyzing medialized sociology. In the course of the chapter, I will illustrate the argument in the following steps:

1. The concept of medialization rests upon the notion of an autonomous core of knowledge production. The role set corresponding to medialized science is the media expert whose research, even if presented in mass media, can be evaluated according to academic standards. In cases where this is not possible, scientific norms have been violated.
2. In social sciences, it is more difficult to draw a clear line between the autonomous core of knowledge production and public deliberation. Accordingly, social scientists address mass media in at least three different and overlapping roles: as media experts, as public intellectuals, and as diagnosticians of the present state of society.
3. As diagnosticians of the present, sociologists can address both an academic and a mass media public by making social change newsworthy. At first sight, diagnosticians of the present state of society act in a hybrid role, bridging the gap between academic and mass media communication.

[1] For an introduction to Niklas Luhmann's sociology of knowledge, see, e.g., Staeheli (1997).

4. On a textual level it can be shown, however, that diagnoses of the present, as a genre of sociology, have to neglect certain academic standards to turn social change into newsworthy events. To make this point clear, I will exemplify the specific features of diagnoses of the present as opposed to general sociological theories by comparing the models of social change in David Riesman's *Lonely Crowd* and Jürgen Habermas's *Theory of Communicative Action*.
5. Two overall conclusion are drawn from this argument. First, although in sociology there seems to be a fluent continuum between academic and publicly accessible debates, the distinction between general theories and diagnoses of the present allows a distinction between an autonomous core of knowledge production and medialized social science. Second, in line with the differentiation of sociological genres, it is also possible to formulate an analytical role differentiation.

16.2 Medialization, Popularization, and Autonomous Science

As Franzen et al. have pointed out in their introduction to this volume, analyzing science from the view of differentiation theory means basically distinguishing modes of communication. It does not imply that science is by any means exterior to society, nor does it imply any idea of superiority in comparison to other forms of communication. Following Luhmann, we should rather work with the concept of a functionally differentiated society without a hegemonic center of command – be that economy, politics, mass media, law, religion, education, or science (Luhmann 1982). Luhmann's functional systems, and for that matter Bourdieu's societal fields, are not hermetically sealed monads. They are highly inter-dependent and can perform their autonomous societal function only in accordance with their neighboring, coequal functional systems. Inter-system communication – e.g., between science and mass media – is insofar a very common phenomenon. Functional differentiation, however, implies that the meaning, i.e., the interpretation of the very same communicative action, takes different courses depending on where it is processed. The *popularization of science* concept as proposed by Whitley (1985) or Stichweh (2003) is to a large extent informed by this fundamental idea. As already stressed by the editors in the introduction, we can use the idea of popularization only if we can, however marginally, distinguish between knowledge constitution and its broader diffusion. From this point of view, popularization means that the communication of scientific knowledge cannot cross the boundary between the esoteric production center and an exoteric audience without changing its meaning.

The *medialization of science* approach is a consequent elaboration of this concept. It too starts from an autonomous, esoteric production center which has to deal with various forms of popularization. From intra-scientific communication among colleagues, over interdisciplinary communication up to "handbook-science" and academic teaching: Specialized knowledge gets simplified and apodictic or is transformed according to the special needs of local collaborations (see, e.g., Galison

1997, 1999). At some point, however, scientific communication is addressed to a general public which is represented by mass media. When experts present their research in newspapers, on TV or in widely read web-based information platforms, they have to simplify their results and also have to keep in mind the selection criteria of mass media communication, i.e., what media studies call "news values" for a classical study (see, e.g., Galtung and Ruge 1965). Media attention can cause public support and therefore is highly welcomed by most researchers. The medialization debate, on the other hand, shows that the tighter coupling between the two fields can also affect the self-steering mechanisms of science and, eventually, even undermine the autonomy of its esoteric core.

In the model of medialized science, the autonomy of scientific communication does not imply power over other parts of society. It describes exactly the opposite constellation: By leaving the lab, science loses its control over the mode of scientific knowledge production and hands it over to a public sphere that has its own rules of dealing with it. The differentiation model allows the public to speak back in its own words – the words of mass media.[2] In this context, a kind of spherically stratified concept of science can be used. Medialization can be pictured as a continuous or gradual process where scientific communication is staged to fit the selection criteria of mass media. Metaphors can be used, juicy human interest stories can be highlighted in favor of dull experimental procedures, sensational novelty can be claimed where actually miniature puzzle solving happened. Medialization can even go further and some studies show that even high prestige journals like *Science* or *Nature* select articles according to their newsworthiness (Franzen 2009). At a certain point, however, scientific communication reaches the outer shell of the medialization continuum and merges into mass media. The scientists' fall from grace is exemplified when the distinction of truth and false gets eclipsed by the distinctions "news or not news," "spectacular or boring." In other words: Striving for mass media attention can, in extreme cases, foster fraudulent practice.

As a matter of course, it is not always easy to tell where popularized science ends and the violation of scientific norms starts. One obvious reason is that there exists no clear "gold standard" for good science, simply because the levels of scientific rigor and creativity are endlessly enhanceable. Another reason is the evident time lag between the publication of a scientific novelty in broad mass media and the process of peer review. It takes a while to check whether the sensational finding which caused the media attention complies to academic standards. In cases where new facts are diffused publicly prior to extensive academic scrutiny, their reliability can only be determined retrospectively. However, as long as the media experts in question act as academic professionals, the process of peer review subsequently enables

[2] Although some sociologists might argue that mass media produce only distorted versions of public deliberation, it is very hard to imagine how modern societies could create a general, i.e., *socially inclusive* public without relying on mass media. The only historical example would perhaps be the eighteenth century *salon*, where publics could still be based on direct face-to-face interactions. A model suitable only for small bourgeois circles, surrounded by pauperized, illiterate masses (see, e.g., Habermas 1989).

the scientific community to draw lines between popularized (but overall sound) and poor scientific performance. This was the case with Hwang Woo-suk's research on stem cells. Here, originally sound looking scientific breakthroughs partly proved to be made up to gain mass media attention (e.g., Chong and Normile 2006; Rusnak 2006).

Although I share most of the views held by the concept of medialization, I doubt that it can be used as a general model of the science-media relationship. It is mainly a model for analyzing "hard" sciences, because it presupposes an autonomous core of knowledge production. Even if medialization is a matter of degree, it can only be measured if there is a non-medialized, purely academic sphere of reasoning. When poor scientific practice enters the core of knowledge production via the detour of mass media, scientific norms have obviously been violated. In social sciences, it is far more difficult to locate such an autonomous sphere and, consequently, to work with the medialization concept. Not because there exist no clear and rigorous social scientific procedures of validation and certification. Social sciences can indeed distinguish between true and false. Yet, unlike in the case of "hard" sciences, social scientists address the publics of mass media in largely *overlapping* roles. The first one is the *media expert*, whose role-specific actions can adequately be grasped by the model of medialized science. The other is the *public intellectual*. Here, we have a role setting which cannot be studied by analyzing media expertise, simply because public intellectuals are not experts in the narrow sense.

16.3 Media Experts and Public Intellectuals: A Delicate Distinction for the Social Sciences

The relation between mass media and social sciences is still not a widely discussed area of research in science studies. Only a few scattered papers have dealt with the question in recent years (Gans 1997; Schwartz 1998; Felt 2000; Revers 2009). An at least loosely connected discussion, however, did not emerge. The main characteristic of social scientific expertise seems to be that it is ill-represented in the mass media when compared to "hard sciences" (Felt 2000). Yet, social scientific experts do appear in mass media, mostly as academic researchers discussing and interpreting data on social problems. Their invisibility can partly be explained by the fact that in this case media attention seems to imply concealing disciplinary background. Revers (2009), for example, argues that sociologists tend to present themselves as "pension experts," "demographers" or similar when addressing a non-academic public. Obviously, low public standing of some social sciences means that enhancing public visibility as an expert goes hand in hand with hiding professional heritage. While this might be a special feature of certain social sciences in the mass media, we can nevertheless assume that medialization effects will not be completely different from what has already been described for the "hard sciences." We can expect simplified or apodictic presentations, quick solutions for highly complex social problems, connections to actual events and a lot of old wine in new skins. But above all, we can also expect fraud. Data sets can be manipulated, non-existing interview material can be staged, field research might not have happened at all. It goes without saying that, due to low media coverage, exposing such practices would not create a Hwang-style scandal. Social scientific experts are an exotic species in mass media after all. And, as a matter of fact, we do not have a gold standard for ultimately good social science. But as long as we can distinguish modes of communication, we can tell when medialization turns into questionable scientific practice.

We can distinguish the role of the media expert (social scientific or other) from one specific role in which social scientists often address broad publics. It is the role of the public intellectual. This is one of the few cases where the knowledge production of academic personnel cannot clearly be separated from public discourse. So let us first look at a very broad and inclusive definition of the role[3]:

> In short [...] the intellectual writes for the general public, or at least for a broader than merely academic or specialist audience, on 'public affairs' – on political matters in the broadest sense of that word, a sense that includes cultural matters when they are viewed under the aspect of ideology, ethics, or politics (which may all be the same thing). The intellectual is more 'applied,' contemporary, and 'result-oriented' than the scholar, but broader than the technician. Approximate synonyms for 'intellectual' in this sense are 'social critic' and 'political intellectual.' The intellectual, so defined, *is* the public intellectual [...] (Posner 2001: 23).

To what extent do we need to distinguish this role from the role of the media expert? First of all, public intellectuals, unlike media experts, need not popularize existing academic research to gain public attention. They might be highly prestigious experts and, in fact, most of them are employed at universities today (Coser 1965; Jacoby [1987] 2000; Michael 2000; Joffe 2003). But their role is to discuss issues which, due to their novelty, broadness and public accessibility, do not fit into the structure of specialized research programs. This is the case with the description of novel social, cultural, or political trends, yet to be studied in detail by the (social) scientific community (Posner 2001: 45f).

One of the pivotal traits of modern knowledge production is its high degree of differentiation. Scientific expertise is a role model for this process. When public intellectuals discuss "public issues," they not only cease to be experts but also enter a field for which expert knowledge does not exist yet in modern society. There is no scientific specialty for "brand new" matters of general public interest. Therefore public intellectual commentary cannot be evaluated by applying scientific standards. It is important to emphasize that scientific evaluation standards need not be entirely disciplinary. Interdisciplinary and transdisciplinary research can indeed broaden the scope of interest while keeping the boundary between knowledge production and knowledge distribution intact. The same holds true for policy proposals based on expertise. In all those cases we can ask whether the idea behind the translation process was sound, whether the line between "popular" and "bad" has been crossed (Posner 2001: 36ff).

---

[3] For further reading on defining public intellectuals, see, e.g., Etzioni (2006).

For the natural sciences the role of the public intellectual does not create much confusion. The expressive genres usually associated with this role – general political commentary, social critique, literary criticism – will not be taken as contributions to natural scientific knowledge production. In fact, numerous natural scientists have played the role successfully without producing much entanglement on where to draw the line between their specialized area of expertise and public commentary. Albert Einstein is as much of an example as are contemporary natural scientists turned public intellectuals like Steven Weinberg, Steven Hawking or Stephen Jay Gould. In social sciences, however, those two realms still cannot be separated clearly. Not only are the social scientists with the highest scientific reputation almost simultaneously public intellectuals; but public intellectual commentary is also regarded as potentially improving or enriching the "core" of social scientific knowledge production. For large parts of the social scientific community it is a desideratum to maintain a high degree of permeability between academic discourse, public deliberation and political activism (see, e.g., Clawson et al. 2007; Nichols 2007).

As long as the roles of the public intellectual and the social scientist cannot be discerned unambiguously, the sociology of medialized social science should not be restricted to the analysis of media expertise. In the following section, I will show that in the course of the twentieth century the relationship between social science and mass media was marked by two interrelated, but opposed processes: the one was the development of a strictly academic terminology and corresponding research programs. The effect was a squeezing-out of public intellectual commentary from academic social science. The second process was basically a reaction to the first and is represented by the ongoing demand for bridging the gap between social scientific academic discourses and public deliberation. In the description of this process, I will concentrate on the case of sociology.

16.4 From Intellectual Discourse to Diagnoses of the Present

The term "intellectual" has a long tradition. It first became widely used in Tsarist Russia during the 1860s, describing political opposition of the educated strata. As a noun, identifying an ideal type or a role set, it became a matter of broad discussion during the Dreyfus affair in the 1890s (Lipset 2000: 1156). The role quickly evolved as one of the major concerns of early sociology. The discussion centered around questions of the class character of intellectuals. Between the 1900s and the 1920s, three lines of interpretation were developed which, until today, dominate the sociology of intellectuals (Kurzman and Owens 2002). The historically earliest was proposed by Julien Benda, who claimed that intellectuals formed a *class in themselves*, pursuing specific interests which could neither be subsumed under the interest of the bourgeoisie nor the working classes. A radical reply was developed by the Marxist thinker Antonio Gramsci who rejected Benda's "class in themselves" approach by claiming that each social class produced its own intellectuals whose main job was to elaborate particularistic *class-bound* interests. Finally, the most influential concept was developed by Karl Mannheim and his idea of "free-floating" intellectuals who were neither advocates of their own class interests nor the educated forefront of another class. Instead, so the argument goes, the social inclusiveness of the university brings receptive persons from different social strata in contact with each other, thereby enforcing mutual adoptions of class-specific perspectives. Intellectuals who were trained in such an institution would be able to expand their biased worldviews and could rather represent the *interests of society at large* than any other social class.

In either approach, the role was understood very comprehensively. Intellectuals could be poets, novelists, journalists, philosophers, natural and social scientists, artists or spiritual leaders – in short, everyone who "by the presentation of symbols to be appreciated" can "elicit, guide, and form the expressive dispositions within a society" (Shils 1958: 7). In a similar vein, Parsons defined intellectuals as individuals specializing in cultural concerns by applying universalistic standards like objectivity and interest in theoretical and empirical analysis regardless of their relevance for immediate action (Parsons 1969). From this perspective, it is very hard, if not impossible, to distinguish intellectuals from scientists and to analyze the relation of the two.

Only in recent years, the sociology of intellectuals began to ask whether narrower and more specific definitions should be favored to emphasize the differences between roles which must not be confused with the role of the intellectual. Oevermann (2003), for example, analyzes the intellectual as a *transitory practice* between science, politics and public deliberation. Intellectuals share some features of politicians, scientists and journalists without being absorbed by either of them. With politicians, intellectuals share the commitment to value laden argumentation, yet unlike politicians, they are not supposed to make decisions. Their influence does not rest on pragmatic compromise, but exclusively on the convincing logic of the better argument. This is what they share with scientists. What separates them from scientific practice is that they are expected to defend certain values openly and that they have to address a non-specialist, *general public*. In this sense, their major concern is not methodologically sound critique, but rhetorically effective, practical criticism which defends certain values against others. Addressing a general public is what they have in common with journalists. Oevermann (2003: 17ff), however, argues that intellectual and journalist practice can be distinguished on two levels. First, intellectuals deal with topics which concern an interested public at large. Usually, their commentary is not related to report on daily events or issues touching persons in their specific roles. Intellectual commentary deals either with societal macro-structures or syndromes which are seen as relevant for every interested person, regardless of his or her profession, political affiliation, social status, etc. Second, intellectuals are not just retailers of news. They are in permanent contact with a general public held together by a shared set of values and a common base of knowledge. The relationship between the public and the intellectual is one of reciprocity – it is not a process of top-down information, but of mutual deliberating critique. The role of the intellectual is best performed in what Oevermann (2003: 48) calls "anticipatory" or "prophylactic" diagnosis of crises. In this respect, intellectuals are either interpreters or creators of crises which emerge when a general public encounters the breakup of routinized macro-societal structures.

Admittedly, this is a very narrow definition of the intellectual role set. But as an ideal type it serves its function to analyze why and to what extent social structures deviate from the clear case. Keeping in mind that Oevermann's typology is not an empirical induction, it is possible to describe the development of sociology as a process of *gradual detachment* from the ideal-type role of the intellectual. Until the late nineteenth century it was very hard to tell apart sociology and public intellectual deliberation. Sociologists were simultaneously social critics, journalists, political commentators and public intellectuals in the sense stated above: They diagnosed the present state of society in terms of crises and could expect direct or quasi-direct response from a learned, though neither specialized nor university-trained bourgeoisie. An anecdotal proof for this highly permeable role set is that the masterpieces of early sociology (and other social sciences) were written without a specialist jargon. Simmel's famous *Sociology* as well as Durkheim's *Division of Labor in Society* can easily be read by anyone with a general education in philosophy and history. This, of course, was only true for a tiny minority, but to read one of those "classics" of sociology, a specialist sociological training was neither necessary nor was it available at universities.

Insofar, early sociology comes very close to the model of public deliberation between intellectuals and a small but universally trained public. At this stage it was not only hard to distinguish sociology from public intellectual deliberation, but also to define clear boundaries between sociology, other social sciences, humanities, aesthetic reasoning and broad political statements. Intellectual discourses could be all at once. The best example for such "universal" intellectual discourses are theories of cultural decline which spread between the fin de siècle and the inter-war period. A famous example surely is *The Education of Henry Adams* (Adams [1917] 1999): auto-biography, political commentary, historical philosophy and social science in one piece. Even extremely voluminous and highly demanding elaborations like Spengler's *Decline of the West* (Spengler 1918) and Arnold Toynbee's *A Study of History* (Toynbee 1935) were not written for exclusively academic audiences. Concepts like Spengler's diagnosis of an ending "faustian era" indicate that the presumed auditory should, above all, be familiar with classical literature and arts. Roughly speaking, the prerequisite for participating in intellectual/social scientific debates was the possession of a private library.

The situation began to change slowly during the last decade of the nineteenth century, first in the US. Here, the relatively independent university system facilitated the institutionalization of specialized sociological departments and sociological journals. This process was followed by successful attempts of keeping non-academic and hybrid audiences out of the business. The first "victims" were social reform movements and the so-called "Christian sociology" (Evans 2009). Although Evans tries to show that this was a process of deliberate boundary-work, the following decades showed that, *despite numerous efforts*, sociology failed in re-establishing ties to extra-academic audiences by providing an applied version of professional practice. The discipline, for a good part *unwillingly*, became institutionalized as a "staff-oriented profession" – its only audience were other sociologists and students (Janowitz 1972). In such an environment, theories and methods could be developed whose understanding took years of specialized training. Even a high degree of general education in humanities was not sufficient to understand cutting-edge developments in sociology.

> In short, from its very beginning, sociology had a departmental basis; it was more than the effort of a single professor. Moreover, *it could not be thought of as a collection of men who were only part of a learned society*. They were academic professionals, since professionalization for them meant control over their research and academic freedom (Janowitz 1972: 110, emphasis added).

From this perspective, the institutionalization of sociology was more than boundary-work. By the mid twentieth century, the discipline could exclude non-academic audiences not only by strategic action, but simply by the level of abstraction and detail which were imposed by rules of academic rigor. A private library was not enough: You had to study sociology if you wanted to participate in sociological discussions. In Western Europe, a similar development could be observed, although with a certain time-lag, for reasons analyzed elsewhere (e.g., Shils 1970).

Not later than in the 1950s, the unity of sociology and public intellectual deliberation was disrupted by a combination of elaborated empirical methods and Parsonian structural functionalism (Turner and Turner 1990). By that time, sociology could be presented as a mature science, with a coherent body of knowledge, autonomous theoretical devices and clear-cut methods – the development of sociology seemed to be a story of scientific progress (Lipset and Smelser 1961). This, however, does not imply that the whole discipline was at any time thoroughly dominated by structural functionalism and quantitative methods. Critique came quickly, and it was fierce. Yet, various forms of criticism had to be formulated in a way which could keep up to an extremely high level of theoretical coherence and empirical detail.

Despite this trends towards academization, sociology never succeeded in expelling a publicly accessible alternative from its core of knowledge production. Throughout the second half of the twentieth century, the discipline provided a great deal of approaches which were treated as contributions to academic discussions and were simultaneously read by a non-academic public. Recent developments in sociology indicate, however, that this kind of hybrid mixture of sociology and public discourse has to be treated as a special fraction of an otherwise professionalized academic discipline. However, unlike in the "hard" sciences, engaging in purely academic, esoteric discourses still is an option not preferred by all sociologists. In sociology, impetuous debates periodically arise around the question if the academization and detachment from public deliberation has gone too far.

In the US, the most recent discussion about the discipline's public accessibility was initiated by Michael Burawoy. As the newly elected president of the American Sociological Association, he called for a revival of what he called *public sociology* (Burawoy 2005). Two anthologies followed and turned Burawoy's proposal into a heated debate (see Clawson et al. 2007; Nichols 2007). For Burawoy, the history of sociology, too, is a process of differentiation which at the beginning of the twenty-first century figured out four distinct, but interrelated types of sociology: professional, policy, critical, and public sociology. Without going deeper into the taxonomy, public sociology for Burawoy is marked by its relation to *extra-academic* audiences and by providing *reflexive* knowledge. By doing public sociology, professionals should strive for consensual dialogue with this non-academic audience under the conditions of what Habermas calls "communicative action" (Burawoy 2005: 9). In this role, the professional sociologist should turn into an "organic" public intellectual by defending the value orientations of civil society, i.e., voluntary associations, social movements, political parties, trade unions, etc. Notably, public sociology should not replace the other types of sociology. What Burawoy calls for is strengthening public sociology within a fractioned discipline which, on the one hand, is dependent on professionalized academic knowledge production but, on the other hand, should not be dominated by it exclusively. Communicating with a wider than academic public, sociology could return to the original mission to tackle social problems for which an over-professionalized sociology has lost its sensitivity.

Apart from critiques concerning Burawoy's taxonomy, the discussion soon turned into a dispute about the *value freedom* of sociology. Obviously, Burawoy's call for public sociology could not make anybody happy. For the one side, his proposal granted too much autonomy for professional sociology, thereby only consolidating its questionable dominance over social criticism and political commitment. For the other side, value-laden argumentation, consensual solidarity with activists and the role of the pubic intellectual were simply incommensurable with social science. Burawoy's Marxist inclination might, among other reasons, explain this turn of the debate. A satisfying solution was not suggested. Sociology could either be done to solve sociological puzzles, or to change the world – each variety ignores the other and both have their place under the academic roof. Eventually, the plan for a revival of public sociology just created incentives to draw an even sharper line between politicized and esoteric knowledge production (e.g., Turner 2005). Notwithstanding, some commentators tried to transcend this inextricable deadlock by claiming that political commitment was not the essential problem of public sociology. Boyns and Fletcher (2005), for example, raised concerns about Burawoy's idealized images of the public intellectual's role in modern public opinion formation. From their point of view, a general, i.e., heterogeneous public can only be addressed with the mass media as an intermediary. A public sociology which tries to appeal not only to small circles of already politicized activists runs the risk of becoming yet another aspect of popular culture and turning the public sociologist/intellectual into a media celebrity (Boyns and Fletcher 2005: 13f). Sociology's potential of critical reflection would be seriously harmed if it surrendered to the selection criteria of mass media communication.

Here we get to the central issue of this paper: The role of the public intellectual seems to be a vital competence of sociology and goes beyond popularizing scientific results. At least some fractions of the discipline imagine that addressing non-academic publics can be part of sociological knowledge production and that in the role of the public intellectual sociologists can overcome an academization "gone mad." Now if being a public sociological intellectual today supposedly implies embracing mass media, than in sociology we would have a *medialized core* of knowledge production alongside esoteric contributions which presuppose disciplinary training and circulate only within expert circles. The dispute about public sociology supports this view insofar, as there seem to be different degrees of public accessibility in sociology. Where to draw an exact line between sociology as a science and sociology as public discourse turns out to be extremely difficult and contested. Apparently, we have to deal with blurred boundaries.

Yet, differentiation theory advises us to dare a closer look. Especially scholars influenced by the writings of Jürgen Habermas and Niklas Luhmann have argued that public deliberation and sociology do not form a continuum, but rather distinct *genres*[4] with clear references to either mass media or science. The term used to describe the sociological genre with close connections to mass media is "diagnosis of our time" or "diagnoses of the present."[5] Here, sociology strives for media attention neither by political critique nor by popularizing research, but rather by presenting social change and novel social trends in a newsworthy style. For Peters (2008), this genre deals with general social, historical and cultural changes which are diagnosed under two aspects: an extensive social crisis with potentially hazardous effects and/or a fundamental social transformation leading to a new historical stage and a new type of society in a foreseeable future. Either way, social change is depicted in a dramatized, moralized, excited, and hyperbolic fashion. According to Peters, this genre can easily be digested by broad publics, since it allows mass media to present social transformation by applying the same "news values" as they come to play in daily news coverage. The argument was further elaborated by Kieserling (2004), who links the specific argumentative structures of the genre to general developments in sociology. For him, the formation of this medialized genre indicates that general sociological theory and public intellectual commentary ceased to form a unity, despite the fact that both have a common topic: macro-scale social transformation. From this perspective, diagnoses of the present are not only marked by a proximity to mass media, but also by a distance to academic macro-theories.

Instead of accepting somehow blurred boundaries between the two, we should rather be aware of diverging system references and ask how the distinct genres interact. Both Kieserling's and Peters's elaborations stayed on a rather theoretical and explorative level. By using the example of a widely read diagnosis of the present, I will suggest how the argumentative structures of this genre can be distinguished empirically from academic general theories in sociology. In line with the

---

[4] When using the term "genre," I rely on the concept of "communicative genres" as proposed by Bergmann and Luckmann (1995: 289): "Thus, 'genre' belongs to a level which is located between the linguistic, coda-related, and the institutional, social structure-related determination of communicative processes. It is characterized by social modeling of the key features of communicative acts." Originally developed for analyzing genres of oral communication, the concept is broad enough for interpreting written communication.

[5] The term was originally coined by Karl Mannheim (1947). The German expression is usually *Zeitdiagnose* or *Gegenwartsdiagnose*. For further reading (in German), see, e.g., Schimank (2000), Reese-Schäfer (1996) or Lichtblau (1991).

central argument developed above, the main task is to show that the medialization of sociology is not only a question of popularization, but one of knowledge production. In this respect, it is not enough to show that a certain sociological study was well received by mass media – the closeness to extra-scientific selection criteria has to be found on the primary level of argumentation, i.e., in the text itself.

16.5 Making Social Change Newsworthy

So far we encountered three roles in which sociologists aim at a broad, non-academic public. The first was the media expert who popularizes existing, mostly specialized research. The second role, still largely neglected by the sociology of science, was the public intellectual who provides politically commentary in close contact with activists or an politically committed reading and viewing public. Neither of those two roles create much confusion with academic contributions. Popularization and medialization might be a question of degree, but fraudulent practice is simply deviant behavior, even if it causes temporary media attention. Value laden solidarity, on the other hand, might be appreciated by some sociologists, but at no time was it ever possible to pledge the whole discipline to one political preference which could then be presented as good science without evoking dissent.

The sociologist as a diagnostician of the present state of society is the third, and certainly most complicated role. In this hybrid role, sociologists can apparently bridge the gap between the roles of the academic and the public intellectual. By providing sociological insights into the present state of society to a broad public, they can act as true "public sociologists" (see Burawoy 2005: 7ff). Now if we want to examine whether such a hybrid role setting can be performed smoothly, i.e., without thoroughly succumbing to either pure academic discourse or mass media communication, we have to make a further step and ask how the genre associated with this role is structured. In other words: Can the genre performed by public sociologists meet the expectations associated with this role? In the following section, I will argue that the argumentative structures of public sociological diagnoses show a proximity to the selection criteria of mass media, eventually hampering a sound analysis of social transformation processes. The case I want to present here will illustrate this point.

The example is David Riesman's *Lonely Crowd* (Riesman [1950] 1953). It was widely read by non-academic audiences, enabled the author to play the role of the public intellectual, described fundamental social change and was nonetheless regarded as an important contribution to academic sociology. We can think of other examples of the genre: Daniel Bell's *The Coming of Post-industrial Society* (Bell 1973), Ronald Inglehart's *Silent Revolution* (Inglehart 1977), Neil Postman's *Amusing Ourselves to Death* (Postman 1985), Ulrich Beck's *Risk Society* (Beck 1986), or Richard Sennett's *The Corrosion of Character* (Sennett 1998). However, my aim is to introduce Riesman's book as an ideal type of the genre. *The Lonely Crowd* probably is the clearest case of a diagnosis which could please both broad publics and the academic world. Furthermore, it is a comparatively well discussed approach. Until today it remains the discipline's undisputed number one bestseller (Gans 1997). The book's popularity went so far that Riesman's portrait appeared on a cover of *Time Magazine* in the early 1950s (see *Time Magazine* 1954). On the other hand, his well written study also played a significant role for empirical research and theoretical elaborations (e.g., Lipset 1961; Parsons and White 1961; Centers 1962; Greenstein 1964; Berghorn and Steere 1966). Although Riesman's diagnosis began to appear dated around the mid 1970s (Zussman 2001), it still can be regarded as a role model for contemporary diagnoses of the present.

The most important characteristic of *The Lonely Crowd* is that it describes social change in terms of *discontinuity,* favoring a view which considers social transformation to be clearly marked by unprecedented breaks or thresholds. Riesman's main argument was that each society by means of early childhood socialization produces a "social character" which fits its demographic growth potential. Societies with high death and high birth rates generate a social character which is dominated by *tradition* and obedience. Individuals have their position within society according to their ancestral role set which is hardly ever questioned. As populations grow, very diverse people are forced to live together under conditions of scarcity. Social control cannot be exerted externally by traditional roles, but is installed as an internal self-disciplining mechanism, or *inner direction.* Traditional forms of social control are replaced by generalized ideals members of society strive for without being forced to do so. The ambitious Puritan serves Riesman as an illustration of this social character, which is kept on course by a moral "gyroscope." Now under conditions of affluence and low birth rates, this form of self-control is substituted for yet another type of social character, the so called *other-directedness.* Especially among upper-middle classes in urban USA, inner-direction loses its significance and is thus replaced by a peer culture – people lose their commitment to internalized norms and act predominantly according to what they think their peers would value.

In Riesman's description, nuances between the historical phases and their corresponding social characters disappear. In other words, the three phases described in Riesman's typology are *mutually exclusive* – one sort of social character is bound to one historical phase only. The change from one historical phase to another appears to be an instant break. Actually, Riesman himself states that his intention was to describe both a society and its typical individuals with a "minimum of scaffolding" (Riesman [1950] 1953: 48), a minimum of cross-over types. Still, Riesman is aware of the problem that it is simply impossible to ascribe one character type to entire societies or to entire historical epochs. In fact, he directly responds to this criticism by admitting that all members of society are in a sense simultaneously inner-directed, other-directed, and traditional – depending on the situation they find themselves in. Social characters are thus "abstractions" or "ideal types" which may overlap (Riesman [1950] 1953: 46f). But how does he solve this dilemma? Is a given historical phase entirely dominated by a social character type or do social characters mix? His answer is that diverse social characters can indeed appear in a given society simultaneously, but each phase has dominant and subordinate social characters. Unable to claim absolute exclusiveness, Riesman introduces a repairing device: Diverse social characters cannot cohabit peacefully. His diagnosis thus becomes a conflict theory of social characters. The mutual exclusiveness of social characters is further fortified by ascribing one single character type to geographical regions. The dominant types are placed on the forefront of a given society. Other-directedness flourishes in coastal metropolitan areas, while traditional behavior survives only in rural parts of the country and, of course, is imported by immigrants (Riesman [1950] 1953: 49ff). The message is: Although we still have diverse social characters, the foreseeable future belongs to those who are best adapted to a phase of population decline – the other-directed. Other types are reduced to "outcroppings of submerged types here and there" (Riesman [1950] 1953: 49). In *The Lonely Crowd* the discontinuity of social change is constructed by an ongoing shift from one dominant character type to another.

From a historical and anthropological point of view, critics claimed that there was no evidence for a correlation between social character type and demographic growth potential (Mead 1951; Aydelotte 1953; Heberle 1956). This well-founded critique in a way misses the crucial point about Riesman's diagnosis. *The Lonely Crowd* is certainly not convincing as a historically sound sociological model, but rather as an argumentative construction of social macro-change. With the connection of character type and demographic change, Riesman tells us two things. First of all, it's not only about individuals and new styles of socialization, but about the emergence of a *new era*. The new character type begins to infiltrate all parts of society: Education and schooling concentrate on installing group conformity, politics disregards morale, work begins to be dominated by an ideology of human relations, leisure time is increasingly transformed from escapism and conspicuous consumption to an exchange of tastes. What at first sight seems to be a bunch of disparate trends, is in fact held together by a common denominator. The character type is a kind of *pars-pro-toto* device: What happens in a psychological nutshell of early childhood socialization stands for society in general. The effects can literally be encountered everywhere. In this respect, social change practically has to appear as discontinuous when it is reduced to one factor only. The correlation between character type and demographic change fulfills another function. If a society produces the character type needed for coping with changing birth rates, then new trends and syndromes associated with other-directedness are more than highbrow fashions of the chattering classes. They represent what society at large increasingly needs. This allows Riesman to treat the new character type as representing a new historical stage despite the fact that other-directed persons are not the most numerous fraction of the US population. They still form a wealthy minority (Riesman [1950] 1953: 36), but their lifestyle is a kind of avant garde. What happens in urban, coastal upper middle-class families will soon be the fate of all of us. With this argumentative device, a society-wide significance can be claimed for those trends which due to their radical novelty cannot be mass phenomena. They can be portrayed as the *rise* of a new society or a move *towards* a new historical era.

Finally, there is a third argumentative device which helps Riesman to deal with social change in terms of discontinuity. Broadly speaking, it consists of keeping historical phases unambiguous. Ongoing social change can be presented as an unprecedented break or threshold only if the recent past is reduced to a clear type which is sharply contrasted with what is happening presently. This was the focal point of those critics who attacked Riesman for an over-simplified view on inner-directedness. Berghorn and Steere's (1966) content analysis of American education guidebooks, for example, could not find an increase of other-directed values in the period between 1865 and 1929. More specifically, both inner- and other-direction values served as guidelines for education without a discernible historical tendency. Greenstein's (1964) secondary examination of an old data set, on the other hand, shows that teenagers' values towards politics were already "other-directed" around 1890. It seems as if Riesman's diagnosis of a rising other-directed era is based on a reification of inner-directedness which never was a historical reality. When describing inner-directedness, he obviously has in mind an ideal type Puritan, personified in Max Weber's famous example of *Christian*, the main character in John Bunyan's *Pilgrim's Progress* (Weber [1904/05] 2001: 62ff). Both are kept on course by internalized moral standards, both keep diaries, both adhere to diverse forms of strict self-discipline, and both rely on the same work ethics. The *idle talk* resented by the Puritan, today reappears in the inner-directed worker's contempt for the new wordy schmooze of human relations (Riesman [1950] 1953: 135ff). So the third way of creating discontinuous social change is to juxtapose the past and the present so clearly, that new trends automatically become *inversions* of old social structures. In a commentary on *The Lonely Crowd*, Margaret Mead makes exactly this point:

> In this presentation the two 'earlier' types of character are backdrops for the descriptions of the 'other-directed' man with which Riesman's major field work has been concerned, and it had better be regarded as such (Mead 1951: 496).

We can now see that the media attention Riesman's diagnosis received was certainly not an accident. As mass media form an autonomous system by filtering all communication according to newsworthiness, we can expect that they will be interested in social change only if it is reportable like daily news. Insofar, Riesman gave the mass media exactly the social change they wanted. We can argue with Luhmann (1996: 25ff) that mass media select information according to the level of potential surprise. Novelties have to break with existing expectations and, above all, the *item* of information has to be new. Insofar, *The Lonely Crowd* was newsworthy exactly because it described new items: a new social character and a new society, which break with routinized expectations and norms. Another important aspect of newsworthiness is the *local relevance* of news. This does not imply that mass media do not cover global events – they obviously do. However, the local relevance of novelties is relatively important in mass media, simply because consumers believe to be fairly well-informed about what happens in their neighborhood – every additional information is thus highly valued (Luhmann 1996: 29). Riesman sure is a master of creating local relevance. The syndromes discussed in *The Lonely Crowd* are tailored to find other-directedness at your own kitchen table, your children's schools, the magazines you read, and in your office. And, most importantly, you can read the book as a kind of elaborated personality test which eventually reveals whether you are other-directed yourself.

Now one could argue that describing new social trends and using clear-cut ideal types is sociology's daily business. From this angle, *The Lonely Crowd* is nothing more than a well-written piece of standard social science. Sure enough, the prerequisite of every scientific contribution is that it allows new and unfamiliar insights. But compared to news values in mass media, the scientific interest in new insights is extremely generalized. The most radical example of this generalized interest in novelty are scientific revolutions as described by Thomas Kuhn (1962). Scientific revolutions or paradigm shifts do not occur when new facts are simply added to old ones, but when whole worldviews change in the light of new knowledge. In such a situation a whole stock of existing knowledge is turned upside down (Kuhn 1962: 111ff). A crucial consequence is that in the course of a scientific revolution, the main task is to reassess *old* facts. Certainly, not every scientific contribution has to initiate a scientific revolution. The example just demonstrates that scientific reasoning and its interest for new insights is not restricted to factually new items. Occasionally, science can indeed detect objects which are new in the sense that their existence was hitherto unknown – new plants and animals, new asteroids and galaxies, new tribes in the mountain chains of New Guinea. Those are possible but highly exceptional events. Everyday science is mainly concerned with checking what has already been said by other scientists. In science, new insights can be, and often must be, related to objects which are not new at all. A generalized interest in new insights can therefore by applied to both factually new and already known objects, in the latter case by proposing an unfamiliar perspective on common knowledge.

Having said that, it becomes clear that a diagnosis of the present like *The Lonely Crowd* favors a limited or *particularistic* view on novelties. In this case, the adoption of extra-scientific selection criteria can definitely foster popularity, but limits social change to factually new trends. This is a very high price to be paid for public acceptance. In Riesman's case, restricting social change to newsworthy novelties entails an uncritical acceptance of old knowledge. The construction of an inner-directed past is a good example. At first sight, the inner-directed Puritan seems to be a usual ideal type and Riesman himself emphasizes this point clearly (Riesman [1950] 1953: 48). Weberian ideal types, however, are not empirical abstractions; they are meant to be analytical tools for highlighting the difference between the type and the empirical reality. Insofar, they never exist in reality and Weber's suggestion even was to construct ideal types in sharp contrast with observable facts (Weber [1922] 1980: 10). The empirical adequacy of an ideal type thus cannot change over time. But this is exactly what Riesman tells us. In his view, character types are abstractions, but in the sense of empirically induced generalizations. While Weber introduced the morally over-steered Puritan as an ideal type, Riesman treats him as a description of a *past reality.* The restriction to factually new trends is thus based on the *reification* of other theories. *The Lonely Crowd* can only deal with social change if it is factually new; the description of the past is left to other theories which are not treated with a skeptical distance. Ideal types are merely turned into actual types with diminishing empirical relevance as society moves towards the present. A possible term for this kind of argumentation could be *retrospective realism.* Social change can appear as an instantaneous break, if the past is described in terms of other theories which are not critically reassessed, but conventionalized into quasi-empirical observations. Highly contested theories are staged as adequate historical accounts. Weber used to be right – until now.

We can contrast such a diagnosis with general social theories which can describe macro social change without relying on argumentative moves highlighting discontinuity. Habermas's *Theory of Communicative Action* is a possible example (Habermas 1981a, b). He treats modernity as an increasing differentiation of "system" and "life-world." Habermas's view is certainly novel insofar as this theoretical framework was never used before. But the item or object he describes is not new at all. In fact, the twin terms "system" and "life-world" can be used to describe tribal societies, ancient empires, the medieval feudal order, early nation states, industrial capitalism and the modern welfare state (see Habermas 1981b: 234ff). Habermas can describe the entire social history in new terms without claiming that the present is marked by a discontinuous break with old patterns. His diagnosis of recent trends – the pathological "colonization of the life-world by the system" (Habermas 1981b: 293f) – is a consequent elaboration of the very same differentiation process he believes to be at work since hundreds of years. Furthermore, Habermas's generalized interest in developing a new perspective on social evolution forces *The Theory of Communicative Action* to reassess old knowledge. Habermas, for example, neither treats Weber's theory of rationalization, nor Marxism as adequate descriptions of a past social reality. From his perspective, both were wrong at the time they were formulated (see, e.g., Habermas 1981a: 384ff; Habermas 1981b: 499ff). To make a long story short: A general theory like Habermas's approach can analyze both common knowledge about society and factually new trends with the same pair of *new* terms – the theory comes close to Kuhn's idea of a scientific revolution.

If we take Riesman's *Lonely Crowd* and Habermas's *Theory of Communicative Action* as ideal types of different genres, a clear distinction between the two can be made regarding the level of generalization which is employed to analyze the novelty of social change. This distinction is more than a question of style. It also explains which system reference is chosen. A particularistic view on new social trends creates sharp discontinuities which can easily be transferred into mass media. On the other hand, it provokes intra-scientific critique which, as in Riesman's case, can be radical to the point where the scientific status of the diagnosis is questioned (e.g., Bain 1951; Calhoun 1951; Goldsen 1951; Francis 1953). In comparison, a generalized theoretical interest in social change will mostly reinterpret existing knowledge and disregard factually new trends.

Two clarifications have to be added to this distinction. First of all, the argumentative reference to mass media does not explain the success of a discontinuous diagnosis. It only explains one of many *prerequisites* of mass media attention. Other factors can be personal networks between the author and influential publishers, the cultural climate which makes a diagnosis plausible for general audiences, specific national traditions of intellectual elites, the author's ties to prestigious institutions of knowledge production, and, of course, the author's personal marketability (McLaughlin 1998). Furthermore, diagnoses of the present and general theories form *interacting* genres which can perform mutual functions for each other. Diagnoses of the present can, for example, refer to general theories in order to claim the status of a scientific contribution. Even if general theories are reduced to retrospective realist backdrops for current social change, quoting Marx, Durkheim, Parsons or Habermas will, at least potentially, call for intra-scientific commentary. Vice versa, general theories can use diagnoses of the present as powerful sensors for social change. If a generalized interest in novelties forces them to rewrite existing knowledge, then factually new trends can easily be overseen. Habermas's *Theory of Communicative Action* actually shows that a general theory of society runs the risk of becoming a critique of other sociological approaches with very little space for trends which have not already bothered other sociologists. Diagnoses of the present can help general theories to take new trends seriously without relying on the blunt alarmism of mass media.

We can think of various examples where general theories referred to diagnoses of the present in order to incorporate novel social trends into their general approaches. In this vein, Talcott Parsons commented on *The Lonely Crowd* (Parsons and White 1961) as he did on C. Wright Mills's *Power Elite* (Parsons 1957). Niklas Luhmann, as a prototype general theorist, occasionally took diagnoses of the present as a point of departure for applying his theory on recent developments in modern society – from environmentalism (Luhmann 1989) to post-modernism (Luhmann 1997: 1143ff). Put more generally, it is highly unlikely that a single oeuvre or even a single monograph can be subsumed under a single genre. Distinguishing genres means distinguishing *modes* of argumentation which can, and mostly do, appear alongside each other.

16.6 Conclusion: Medialized Sociology and the Public Intellectual

We can analyze the differentiation of sociological genres as a consequence of the professionalization of the discipline. The high level of theoretical abstraction and empirical detail made it impossible for general audiences to participate in cutting-edge knowledge production in sociology. This perspective forces us to distinguish at least two interrelated genres which share the same topic – macro social change – but are separated by different system references: science or mass media. Diagnoses of the present thus appeal to another presumed auditory than general theories. The yardstick for drawing that line is the level of generalization and historical detail which is applied for discussing novel social trends.

What conclusions does this argument provide for the concept of medialization of science? First of all, that it is difficult to apply the concept of medialization if the discipline in question produces a continuum between public debates and academic discourse. However difficult, this distinction can be drawn for sociology, at least in those cases where an autonomous core of knowledge production exists. From this perspective, diagnoses of the present can be treated as examples of medialized sociology, yet only insofar as there exist general theories of social change which operate as exclusively academic debates. This constellation is rather new in sociology. At least until the 1950s the continuum of purely academic vs. public intellectual debates was too permeable to demarcate and unambiguously define the two. In this respect, sociology is, with a certain time lag, undergoing a similar development as did the "hard" sciences in the nineteenth century: from gentleman science to autonomous, self-steering, academic practice (e.g., Stichweh 1990). The medialization of sociology, and diagnoses of the present as the corresponding genre, thus are effects of academic professionalization. Despite this trend, diagnoses of the present still are so widely and willingly discussed in sociology that their genre-specific traits are not yet self-evident for large parts of the academic community. As a genre, they produce direct repercussions in academic sociology, i.e., the autonomous core of sociological knowledge production. This makes them a rather good example of medialized social science. To put it differently: The easier it is to demarcate an autonomous core of sociological knowledge production, the easier it is to apply the concept of medialization.

The second lesson we can learn for the argument above is that the differentiation of sociological genres is triggered by the professionalization of the discipline which, eventually, creates differentiated role sets. Certainly, role set and genre need not be congruent. Diagnosticians of the present can and indeed do engage in purely academic debates. The very fact that most of them are employed at universities makes it impossible to withdraw from professional academic engagement entirely. Actually, there are numerous cases where a diagnosis of the present provoked highly abstract academic discussions (for the Riesman case, see, e.g., Lipset and Lowenthal 1961). Vice versa, general theorists can produce diagnoses of the present and often act as intellectual celebrities. Jürgen Habermas, Anthony Giddens, Michel Foucault, and Pierre Bourdieu certainly are prominent examples of how to switch between those two role arrangements.

Subsuming a whole approach under a single genre would be as over-simplified as claiming that a single person performs a single role throughout her or his whole career. Yet, the two roles can be distinguished analytically. Their role specific actions are related to different audiences and hence have to argue on different levels of abstraction and detail. As a consequence, they are not to the same degree dependent on peer review. Diagnosticians of the present can be prominent public figures even if their insights are, at least from an academic perspective, based on exaggerations and reductionist descriptions of the past. In other words: The success of general sociological theories depends solely on the judgment of peers, while diagnoses of the present can – as in Riesman's case – be made famous primarily by mass media coverage.

The possibility of distinguishing roles and genres in sociology is, on the one hand, an effect of professionalization. On the other hand, the differentiation of genres and roles in sociology can also be analyzed from the angle of a structural transformation of publics. I want to discuss this final aspect just briefly here. The main argument is that a unity of intellectual discourse and sociological debate was only possible under conditions of mutual critique between public intellectuals and a generally educated bourgeoisie – a non-specialized *learned society*. The massive expansion of education systems which can be observed since the 1950s changed this situation completely. On the one hand, huge parts of the population are given the chance to obtain higher education degrees and to consume culture on an unprecedentedly high level of abstraction. On the other hand, the culturally most receptive parts of modern society are specialists, with no relevant amount of knowledge about domains which were not part of their education (Shils 1964). Public intellectuals therefore cannot rely on a common knowledge base shared with the broad publics they want to address. Specialized publics may still stay interested in affairs like the general course of society, but they lack substantive criteria for evaluating intellectual discourses about such issues. In a society ruled by specialists there is no expertise for topics of general interest. The only criteria large, college-educated middle classes have to judge intellectual commentary are "appeal qualities." Addressing well-educated, but heterogeneous publics leaves the intellectual no other choice but to constantly produce shock, frisson and brilliant novelty. This development was described in Lewis Coser's concept of the *celebrity intellectual* (Coser [1973] 2006). From this perspective, diagnoses of the present are not only a consequence of a professionalized discipline, but also the product of a society which can deal with public intellectualism only if it is articulated in the language of mass media. Recent developments in the sociology of mass communication furthermore indicate that this process is reinforced by the strict rules of the attention market on the internet. The democratization of intellectual commentary initiated by the blog sphere does not seem to foster nuanced public deliberation (Freese 2009).

Commenting on an earlier version of this chapter, a reviewer remarked that a sharp distinction between diagnoses of the present and general theories runs the risk of applying double standards. Surely, medialized sociological diagnoses like *The Lonely Crowd* were and still are widely discussed in social sciences and the humanities. Being accessible to a wider than academic public does not automatically produce "bad" science. However, as long as parts of sociology claim that there should not be an autonomous core of knowledge production (see the discussion about "public sociology"), the discipline is inclined to take diagnoses of the present seriously not because they are scientifically sound but because they are discussed in the public sphere – i.e., mass media. In such cases, being aware of genre-specific distinctions can prevent the discipline from falling into Hwang-style medialization traps. On the other hand, it is also absolutely correct that engaging in autonomous knowledge production does not guarantee "good" science per se. Quite on the contrary: Academic contributions increase the possibility of being heavily criticized by informed peers. Therefore, it is hardly surprising that dozens of books and hundreds of papers were published to criticize approaches like *The Theory of Communicative Action*. In short, the distinction between medialized diagnoses of the present and general theories should not be equated with distinguishing good and bad social science. It only designates which functional system predominantly decides about the success of a given approach. Both bad scientific practice and brilliant innovations can come from within the autonomous core or, indirectly, via the channel of mass media. But I do think that a discipline which still has to delineate what counts as the autonomous knowledge production core, is more endangered by taking its medialized genres too seriously than by over-professionalization.

My aim was to show how a sociology of medialized social science can look like. Broader studies, comparing more cases can provide additional empirical evidence for the distinction between general theories, medialized sociological diagnoses and the corresponding role sets (for a broader analysis, see, e.g., Osrecki 2011). But even with the few examples given above, we can see that addressing broad publics is an endeavor with high costs. When leaving academia, sociologists do not enter a functionally neutral public sphere where they can act as free-floating social critics, without the restrictions imposed by professionalized science. Outside the university, they will just encounter other strict rules of argumentation which they cannot control autonomously. The argumentative structures of diagnoses of the present suggest that we have to refuse notions of a hybrid role set in sociology, allowing public sociologists to smoothly shift between academic and mass media publics. Or to quote Richard Posner:

> The modern academic intellectual usually cannot, as earlier generations of intellectuals could and did, pitch his writing at a level accessible to a general audience yet does not strike the author's peers as lacking in rigor – he needs two styles of writing, one for the public and one for his peers (Posner 2001: 53).

From a methodological point of view, a sociology of sociological knowledge should search for ways of connecting argumentative structures with extra-textual factors. Such an approach can neither be based only on comparing styles of argumentation, nor can it reduce its scope to external social factors. My suggestion was to link the structure of genre-specific sociological argumentation to differentiated role sets, caused by the professionalization of the discipline. This model shows that social sciences are all but easy cases for science studies.

References

Adams, H. ([1917] 1999). *The education of Henry Adams*. Oxford: Oxford University Press.

Aydelotte, W. O. (1953). Review of: The lonely crowd: A study of the changing American character, by David Riesman et al. *The Journal of Modern History*, 25, 181–183.

Bain, R. (1951). Review of: The lonely crowd: A study of the changing American character, by David Riesman. *American Sociological Review*, 16(2), 269–271.

Beck, U. (1986). *Die Risikogesellschaft: Auf dem Weg in eine andere Moderne*. Frankfurt/M.: Suhrkamp.

Bell, D. (1973). *The coming of post-industrial society: A venture in social forecasting*. New York: Basic Books.

Berger, P. L. and T. Luckmann (1966). *The social construction of reality: A Treatise in the sociology of knowledge*. Garden City: Anchor Books.

Berghorn, F. J. and G. H. Steere (1966). Are American Values changing? The problem of inner- or other-direction. *American Quarterly*, 18(1), 52–62.

Bergmann J. R. and T. Luckmann (1995). Reconstructive genres of everyday communication. In U. Quasthoff (ed.), *Aspects of oral communication*. Berlin and New York: Walter de Gruyter, pp. 289–303.

Boyns, D. and J. Fletcher (2005). Reflections on public sociology: Public relations, disciplinary identity, and the strong program in professional sociology. *The American Sociologist*, 36(3–4), 5–26.

Burawoy, M. (2005). 2004 Presidential address: For public sociology, *American Sociological Review*, 70, 4–28.

Calhoun, D. W. (1951). Review of: The lonely crowd: A study of the changing American character, by David Riesman et al. *Social Forces*, 30(1), 113–114.

Centers, R. (1962). An examination of the Riesman social character typology: A metropolitan survey. *Sociometry*, 25(3), 231–240.

Chong, S. and D. Normile (2006, January 6). How young Korean researchers helped unearth a scandal. *Science*, 311, 22–25.

Clawson, D., et al. (2007). *Public sociology: Fifteen eminent sociologists debate politics and the profession in the twenty-first century*. Berkeley et al.: University of California Press.

Coser, L. A. (1965). *Men of ideas: A sociologist's view*. New York: Free Press.

Coser, L. A. ([1973] 2006). The intellectual as celebrity. In A. Etzioni (ed.), *Public intellectuals: An endangered species?* Lanham et al.: Rowman & Littleeld Publishers, pp. 227–240.

Etzioni, A. (2006). Introduction: Are public intellectuals an endangered species? In A. Etzioni (ed.), *Public intellectuals: An endangered species?* Lanham et al.: Rowman & Littlefield Publishers, pp. 1–27.

Evans, M. S. (2009). Defining the public, defining sociology: Hybrid science-public relations and boundary-work in early American sociology. *Public Understanding of Science*, 18(1), 5–22.

Felt, U. (2000). Die unsichtbaren Sozialwissenschaften: Zur Problematik der Positionierung sozialwissenschaftlichen Wissens im öffentlichen Raum. *Österreichische Zeitschrift für Soziologie*, Sonderband 5, 177–212.

Francis, E. K. (1953). Review of: the lonely crowd, by David Riesman. *The Review of Politics*, 15(4), 534–537.

Franzen, M. (2009). Torwächter der Wissenschaft oder Einfallstor für die Massenmedien? Zur Rolle von *Science* und *Nature* an der Schnittstelle zwischen Wissenschaft und medialer Öffentlichkeit. In S. Stöckel et al. (eds.), *Das Medium Wissenschaftszeitschrift seit dem 19. Jahrhundert. Verwissenschaftlichung der Gesellschaft – Vergesellschaftung von Wissenschaft*. Stuttgart: Franz Steiner Verlag, pp. 229–252.

Freese, J. (2009). Blogs and the attention market for public intellectuals. *Society*, 46, 45–48.

Galison, P. (1997). *Image & logic: A material culture of microphysics*. Chicago and London: University of Chicago Press.

Galison, P. (1999). Trading zone: Coordinating action and belief. In M. Biagoli (ed.), *The science studies reader*. New York: Routledge, pp. 137–160.

Galtung, J. and M. H. Ruge (1965). The structure of foreign news: The presentation of Congo, Cuba and Cyprus crises in four Norwegian newspapers. *Journal of Peace Research*, 2, 64–91.

Gans, H. J. (1997). Best-sellers by sociologists: An exploratory study. *Contemporary Sociology*, 26(2), 131–135.

Goldsen, J. M. (1951). Review of: The lonely crowd: A study of the changing American character, by David Riesman. *The Public Opinion Quarterly*, 15(1), 158–161.

Greenstein, F. I. (1964). New light on changing American values: A forgotten body of survey data. *Social Forces*, 42(4), 441–450.

Habermas, J. (1981a). *Theorie des kommunikativen Handelns. Band 1: Handlungsrationalität und gesellschaftliche Rationalisierung*. Frankfurt/M.: Suhrkamp.

Habermas, J. (1981b). *Theorie des kommunikativen Handelns. Band 2: Zur Kritik der funktionalistischen Vernunft*. Frankfurt/M.: Suhrkamp.

Habermas, J. (1989). *The structural transformation of the public sphere*. Cambridge: Polity Press.

Heberle, R. (1956). A note on Riesman's The Lonely Crowd. *The American Journal of Sociology*, 62(1), 34–36.

Inglehart, R. (1977). *The silent revolution: Changing values and politics among Western publics*. Princeton: Princeton University Press.

Jacoby, R. ([1987] 2000). *The last intellectuals: American culture in the age of academe*. New York: Basic Books.

Janowitz, M. (1972). Professionalization of sociology. *The American Journal of Sociology*, 78(1), 105–135.

Joffe, J. (2003). The decline of the public intellectual and the rise of the Pundit. In J. Weinberger et al. (eds.), *The public intellectual: Between philosophy and politics*. Lenham: Rowman & Littlefield Publishers, pp. 109–122.

Kieserling, A. (2004). *Selbstbeschreibung und Fremdbeschreibung: Beiträge zur Soziologie soziologischen Wissens*. Frankfurt/M.: Suhrkamp.

Kuhn, T. S. (1962). *The structure of scientific revolutions*. Chicago: The University of Chicago Press.

Kurzman, C. and L. Owens (2002). The sociology of intellectuals. *Annual Review of Sociology*, 28, 63–90.

Lichtblau, K. (1991). Soziologie und Zeitdiagnose. Oder: Die Moderne im Selbstbezug. In S. Müller-Doohm (ed.), *Jenseits der Utopie: Theoriekritik der Gegenwart*. Frankfurt/M.: Suhrkamp, pp. 15–47.

Lipset, S. M. (1961). A changing American character? In S. M. Lipset and L. Lowenthal (eds.), *Culture and social character: The work of David Riesman reviewed*. Glencoe: Free Press, pp. 136–171.

Lipset, S. M. (2000). Intellectuals. In E. F. Borgatella and R. J. V. Montgomery (eds.), *Encyclopedia of sociology. Volume 2*. New York et al.: Macmillan, pp. 1355–1359.

Lipset, S. M. and L. Lowenthal (1961). *Culture and social character: The work of David Riesman reviewed*. Glencoe: Free Press.

Lipset, S. M. and N. J. Smelser (1961). *Sociology: The progress of a decade*. Englewood Cliffs: Prentice-Hall.

Luhmann, N. (1980). *Gesellschaftsstruktur und Semantik. Studien zur Wissenssoziologie der modernen Gesellschaft. Band 1*. Frankfurt/M.: Suhrkamp.

Luhmann, N. (1982). *The differentiation of society*. New York: Columbia University Press.

Luhmann, N. (1989). *Ecological communication*. Chicago: The University of Chicago Press.

Luhmann, N. (1996). *The reality of mass media*. Stanford: Stanford University Press.

Luhmann, N. (1997). *Die Gesellschaft der Gesellschaft. Band 2*. Frankfurt/M.: Suhrkamp.

Mannheim, K. ([1929] 1936). *Ideology and utopia*. London: Routledge.

Mannheim, K. (1947). *Diagnosis of our time: Wartime essays of a sociologist*. London: Kegan Paul, Trench, Trubner & Co.

Marx, K. and F. Engels ([1845] 1947). *The German ideology*. New York: International Publishers.

McLaughlin, N. (1998). How to become a forgotten intellectual: Intellectual movements and the rise and fall of Erich Fromm. *Sociological Forum*, 13(2), 215–246.

Mead, M. (1951). Review of: The lonely crowd: A study of the changing American character, by David Riesman. *The American Journal of Sociology*, 56(5), 495–497.

Michael, J. (2000). *Anxious intellects: Academic professionals, public intellectuals, and enlightenment values*. Durham and London: Duke University Press.

Nichols, L. T. (ed.) (2007). *Public sociology: The contemporary debate*. New Brunswick and New Jersey: Transaction Publishers.

Oevermann, U. (2003). Der Intellektuelle – Soziologische Strukturbestimmung des Komplementär von Öffentlichkeit. In A. Franzmann, S. Liebermann, and J. Tykwer (eds.), *Die Macht des Geistes. Soziologische Fallanalysen zum Strukturtyp des Intellektuellen*. Frankfurt/M.: Humanities Online, pp. 13–75.

Osrecki, F. (2011). *Die Diagnosegesellschaft: Zeitdiagnostik zwischen Soziologie und medialer Popularität*. Bielefeld: Transcript.

Parsons, T. (1957). The distribution of power in American society: Review of: The power elite, by C. Wright Mills. *World Politics*, 10(1), 123–143.

Parsons, T. (1969). The intellectual: A social role category. In P. Rieff (ed.), *On intellectuals: Theoretical studies, case studies*. Garden City: Anchor, pp. 3–24.

Parsons, T. and W. White (1961). The link between character and society. In S. M. Lipset and L. Lowenthal (eds.), *Culture and social character: The work of David Riesman reviewed*. Glencoe: Free Press.

Peters, B. (2008). *Public deliberation and public culture: The writings of Bernhard Peters, 1993–2005*. Foreword by Jürgen Habermas, edited by Hartmut Wessler. Basingstoke (Hampshire): Palgrave.

Posner, R. A. (2001). *Public intellectuals: A study of decline*, Cambridge, MA and London: Harvard University Press.

Postman, N. (1985). *Amusing ourselves to death: Public discourse in the age of show business*. London: Methuen.

Reese-Schäfer, W. (1996). Zeitdiagnose als wissenschaftliche Aufgabe. *Berliner Journal für Soziologie*, 6, 377–390.

Revers, M. (2009). Sociologists in the press. *The American Sociologist*. doi: 10.1007/s12108-009-9079-9 (published online 23 September 2009).

Riesman, D. ([1950] 1953). *The lonely crowd: A study of the changing American character*. With Reuel Denney and Nathan Glazer. Abridged by the authors. Garden City: Doubleday.

Rusnak, A. J. and A. Chudley (2006). Stem cell research: Cloning, therapy and scientific fraud. *Clinical Genetics*, 70, 302–305.

Scheler, M. ([1926] 1980). *Problems of a sociology of knowledge*. London: Routledge & Kegan Paul.

Schimank, U. (2000). Soziologische Gegenwartsdiagnosen – Zur Einführung. In U. Schimank and U. Volkmann (eds.), *Soziologische Gegenwartsdiagnosen 1: Eine Bestandsaufnahme*. Opladen: Leske + Budrich, pp. 9–22.

Schwartz, P. (1998). Stage fright or death wish: Sociology in the mass media. *Contemporary Sociology*, 27(5), 439–445.

Sennett, R. (1998). *The corrosion of character: The personal consequences of work in the new capitalism*. New York: Norton.

Shils, E. (1958). The intellectuals and the powers: Some perspectives for comparative analysis. *Comparative Studies in Society and History*, 1(1), 5–22.

Shils, E. (1964). Mass society and its culture. In N. Jacobs (ed.), *Culture for the millions? Mass media in modern society*. Boston: Beacon Press, pp. 1–27.

Shils, E. (1970). Tradition, ecology and institution in the history of sociology. *Daedalus*, 99(4), 760–826.

Spengler, O. (1918). *Der Untergang des Abendlandes. Band 1: Gestalt und Wirklichkeit*. Wien: Braumüller.

Staeheli, U. (1997). Exorcising the 'popular' seriously: Luhmann's concept of semantics. *International Review of Sociology*, 7(1), 127–145.

Stichweh, R. (1990). Self-Organization and autopoiesis in the development of modern science. In W. Krohn et al. (eds.), *Selforganization: Portrait of a scientific revolution*. Dordrecht: Kluwer, pp. 195–207.

Stichweh, R. (2003). The multiple publics of science: Inclusion and popularization. *Soziale Systeme*, 9, 210–220.

Time Magazine (1954). Social scientist David Riesman: What is the American character? *The Time Magazine*, 64(13), Cover page.

Toynbee, A. J. (1935). *A study of history*. London et al.: Oxford University Press.

Turner, J. (2005). Is public sociology such a good idea? *The American Sociologist*, 36(3–4), 27–45.

Turner, S. and J. Turner (1990). *The impossible science: An institutional analysis of American sociology*. London: Sage.

Weber, M. ([1922] 1980). *Wirtschaft und Gesellschaft: Grundriss der Verstehenden Soziologie*. J.C.B. Mohr: Tübingen.

Weber, M. ([1904/05] 2001). *The protestant ethic and the spirit of capitalism*. New Translation and Introduction by Stephen Kalberg. Los Angeles: Roxbury.

Whitley, R. (1985). Knowledge producers and knowledge acquirers. In T. Shinn et al. (eds.), *Expository science. Forms and functions of popularisation. Sociology of the Sciences Yearbook.* Dordrecht: Kluwer, pp. 3–28.

Zussman, R. (2001). Still lonely after all these years. Review of: The lonely crowd: A study of the changing American character, by David Riesman. *Sociological Forum*, 16(1), 157–166.

Chapter 17
Making Science News: The Press Relations of Scientific Journals and Implications for Scholarly Communication

Martina Franzen

17.1 Introduction

The concept of the medialization of science postulates that there is an increasing orientation of science towards the media by adapting to news values in order to get public support (Weingart 2001; Chapter 1). Since social systems reproduce themselves through communication, it seems promising to locate any changes at the core of science, i.e., the scientific paper. To address the thesis of a "medialization of science," I will refer to the organizational level of scientific publishing and focus, first, on the editorial selection process of manuscripts of scholarly journals and, second, on the impact of single papers on science and the media. The argument to be developed is that at the organizational level of high-impact journals the dual programming towards scientific significance, on the one hand, and general interest, on the other, can result in a media conflict in science.

Scholarly journals primarily address the scientific community but a few journals are regularly used as a source for issue selection in science journalism that widens the scope from scholarly peers to a broader public. A professional press service allows for broad news coverage of published new scientific findings. The press relation efforts of journals in the biomedical field are the most pronounced that correlate with public interest in health issues. However, biomedical journals are often criticized for sensationalizing scientific findings (e.g., Schwartz et al. 2002). Moreover, the two most prestigious basic research journals *Science* and *Nature* are frequently alleged to favor the newsworthy with their editorial decisions (Kennedy 2002; Abbott 2006a). Observations like these point to divergent scientific and media-related criteria in scientific publishing. Because of their impact on science and the mass media alike, it comes as a surprise that the medialization discourse has ignored the role of scholarly journals. This chapter tries to fill this gap by investigating the journals' operation modes concerning the science-media

M. Franzen (✉)
Bielefeld University
e-mail: martina.franzen@uni-bielefeld.de

S. Rödder et al. (eds.), *The Sciences' Media Connection – Public Communication and its Repercussions*, Sociology of the Sciences Yearbook 28,
DOI 10.1007/978-94-007-2085-5_17,

coupling. Empirical insights are gathered by focusing on a field that is supposed to be medialized (stem cell research).

The chapter begins by introducing the scientific paper as the key element of science (Section 17.2). Scientific journals constitute a medium that forms scientific communication. Editorial guidelines shape the content and the style of presentation of scientific knowledge. Professional press relations of scholarly journals allow for a broad coverage of new scientific findings. However, only a few journals are regularly monitored by science journalists, depending on their scientific prestige and their press service (Section 17.3). While certain journals like *Science* and *Nature* are considered the most relevant ones from both perspectives, science and the media, it can simultaneously be observed that they have the highest article retraction rates (see Nath et al. 2006). This calls for further exploration: The central question is whether a media orientation of these journals, as expressed in the editorial programs favoring general interest, can create conflicting expectations of scientific robustness and newsworthiness on the part of the authors (Section 17.4). The results of a qualitative analysis of three prototypical cases in stem cell science reveal various modes of orientation towards the media with different effects on the self-reproduction mechanism of science (Section 17.5). The contextualization of single papers suggests that the general interest journals play a relevant part in the medialization of science but they also have to cope with the undesired side effects to the reliability of sensational findings. In their role as gate keepers of science, journal editors are forced to (re-)establish trust inside and outside of science; a recent implementation of new data control mechanisms ahead of print is supposed to ensure the credibility of new findings – however, this points to a growing mistrust towards scientific authors (Section 17.6).

17.2 The Scientific Publication System

While scholars in the sociology of science and Science and Technology Studies (STS) discuss a range of ideas on what constitutes science, they tend to agree that publication plays a relevant part. The scientific paper is regarded as the science-specific form of communication (Stichweh 1987), the product and aim of all scientific endeavors (Knorr-Cetina 1981; Latour and Woolgar 1986 [1979]) or just "the basic unit" of science communication (Goffman 1981). The foundation of scientific journals in the seventeenth century was insofar a prerequisite of modern science as journals made a broad diffusion of scientific findings possible and institutionalized one of the main characteristics of science, organized scepticism through peer review. Scientific journals flourished in the nineteenth century (Garfield 1987: 164) and became the primary medium of scholarly communication.[1] In contrast to the monograph, which often spanned a life's work of scientific endeavors, scientific periodicals offered their authors to publish much smaller pieces of research.

[1] In the humanities, however, monographs are still dominant to communicate new findings.

The genre of the experimental report emerged historically (Bazerman 1988) and produced catalytic effects for scientific development (Zuckerman and Merton 1971). By the diffusion and archival storage of scientific findings, it became possible to distinguish between new and old knowledge. The semantics of novelty and the aim to conduct original research can be regarded as a side effect of printing technology (Luhmann 1997: 996; Esposito 2005: 167), codified in journals' editorial guidelines favoring new and original research results.[2]

Only findings that are written down and published can diffuse into the global communication of science to create a network of scientific knowledge. As Weingart (2003: 187) puts it: "Discoveries which are not published in an article or any other form do not exist for the scientific community, which means they have not been certified." However, it is worth to consider that the presentation of knowledge is always a selective reconstruction of perception and observation activities that govern the production process of knowledge. The dependence on technical diffusion media (print or electronic) opens up a space for a variety of presentation modes of knowledge extending from the inner circle to a wider public. From this perspective, science is structured to a large extent by requirements that also govern the communication process of the mass media.[3] This should be kept in mind when reflecting the science-media relationship. Limitations of space in the mass media as well as in scientific journals restrict the amount of information to be presented, although electronic publishing has recently softened this criterion. With regard to formal scientific publications, author instructions clearly define the article format and shape the way findings are presented in relation to the journal's scope.

For scholarly communication, a specific style of presentation has developed over time. The characteristic feature of scientific papers is the formal structure, i.e., the division into four parts: Introduction, Methods, Results and Discussion (IMRAD). "This so-called 'IMRAD' structure," according to the guidelines of the International Committee of Medical Journal Editors (IMCJE 2007), is "not simply an arbitrary publication format, but rather a direct reflection of the process of scientific discovery."[4] Editorial programs of scientific journals specify the style of presentation in relation to disciplinary standards following the journals' readership that is mostly constituted by the small community of specialists in a particular field rather than the general public addressed by the mass media. How far the presentation of scientific knowledge is editorially prescribed is illustrated by the following example of *Science's Brevia*, a peer reviewed format of original research manuscripts developed in 2001:

---

[2] The preference for novelty is expressed in the so-called Ingelfinger rule.

[3] Luhmann conceives technical distribution media as a prerequisite and, at the same time, as a definitive characteristic of the social system of the mass media (1996: 10f) – a standpoint that created some theoretical confusion (see, e.g., Göbel 2006).

[4] However, the argument of a direct reflection of the process of discovery is disputable, given the media specific discrepancy of the production and presentation mode of science.

> *Brevia* are brief contributions of *600 to 800 words* of text (including authors' names, affiliations, and references) accompanied by *one illustration* (a maximum of four panels) or table and *not more than six references*. Titles should be *no more than eight words* in length. Authors should also submit an *abstract of 100 words* or less, which will appear online only. *Supporting online material is allowed* (no more than one illustration or table or 500 words of text) (*Science* 2007, my emphasis).

This example shows that, due to limitations of space, scientific results are occasionally pressed into very rigid formats. However, the requirements for manuscripts to be published in multidisciplinary journals differ from those of specialty journals. The research topics chosen for *Science* in the category *Brevia* should be eligible for a broad scientific public and beyond:

> *Brevia* present research results on subject matter attractive to, and understandable by, scientists from a wide range of fields. Interdisciplinary work, or experiments or analyses that produce a *result of general interest*, are especially appropriate for this section (*Science* 2007, my emphasis).

Along with the focus on topics of general interest, the author instructions for this special format also include stylistic advice for writing a paper:

> Authors should *avoid highly technical presentations* and *jargon* specific to particular disciplines (*Science* 2007, my emphasis).

On the surface, these author instructions seem to be quite similar to those for science journalists: Choose a topic of general interest, avoid technical jargon and limit the writing to one page. In shaping the kind of knowledge presentation, the "mediality of the journal" comes to the fore (Franzen 2011). The mode of selection and presentation of scientific results exercised by *Science's Brevia* facilitates a broader diffusion via mass media. One scientist delineates the media response on her *Brevia* article as follows: "Wow, mad day – I have done about 14 interviews and have had photos taken. Two camera crews are coming and I'm going on five radio programs!" (cit. in AAAS n.d.). This kind of intermediary format, *Science's Brevia,* is indeed an exception and not the rule in scientific publishing. But it mirrors the difficulty to draw a demarcation line between science and the mass media just on the basis of textual products, which is in line with Hilgartner (1990: 528, Fig. 2) who concludes: "Popularization is a matter of degree." What actually qualifies communication as being scientific is judged by the internal control mechanism of peer review, the integral part of scientific publishing. The question therefore is: on what basis are publishing decisions on submissions eventually reached?

17.3 Publishing Decisions at Scientific Journals

The editorial criteria of what to publish vary among scholarly journals: "[T]he decision to publish or not to publish is not a judicial act but a necessarily subjective expression of interest (which is why it is in the best interests of science that there should be a great many scientific journals)," as John Maddox, *Nature*'s former Editor-in-Chief, has put it (Maddox 1969: 129).

At *Science,* for example, editorial decisions are made according to the paper's expected broad impact on science and society.[5] "*Science*, like other high-profile journals, aggressively seeks firsts: papers that generate publicity and awe in the scientific community and beyond," as *Science* writer Jennifer Couzin (2006: 23) explains. The selection process at *Science* is triaged[6]: (1) On the first level, there is a pre-selection of manuscripts by editors and board members based on a "mixture of novelty, originality, and trendiness" (Denis Duboule, cit. in ibid.: 24). At this stage, about 80 percent of submissions are usually rejected without external peer review. (2) At the level of peer review, two or more specialists in the field look at its technical rigor and novelty[7] and are asked to evaluate whether it constitutes a "*Science* paper" (Lawrence 2003: 260). The professional editorial staff members are advised by a Board of Reviewing Editors that comprises 140 scientists. (3) The final decision is made by the staff editors based on the reviews and the editorial selection criteria mentioned above. The acceptance rate today is less than 8 percent.[8] Particularly interesting or topical manuscripts are published ahead-of-print in *Science Express*.

It becomes apparent that editorial decisions in highly selective multidisciplinary journals are not only based on papers' scientific merits safeguarded by peer review, but also on criteria that reminds us of the selection criteria of the mass media (i.e., trendiness). Considering the theoretically drawn boundaries between science and the mass media (see Chapters 1 and 2), on the level of organization with regard to the particular editorial programs of scientific journals, the distinction between truth-oriented and news-values-oriented communication seem to get blurred. The question arises if the operation modes of scientific journals towards public interests can have any effects on the self-steering mechanism of science.

To foster or just to enable media coverage of research findings, a lot of prestigious journals maintain professionalized press relations.[9] In the journalist's view, the knowledge presented in scholarly journals is certified through the internal control mechanism of peer review: "[I]f their story subsequently turned out to be flawed the blame would fall on the experts, the editors, and the referees and not on them" (Entwistle 1995: 922). In prior-to-publication press releases, editors alert journalists to original articles they think newsworthy.

Journals' press releases are usually distributed one week prior to the paper's publication with a press embargo to allow for further journalistic investigations. Whereas the journals strictly adhere to the embargo by emphasizing "that the embargo serves scientists, authors, journalists and the public," a quote by the Nature

[5] Nearly the same procedure can be applied for *Science's* competitor *Nature*. See Franzen (2009) for more details.

[6] Information is gathered from the Science Contributors FAQ at http://www.sciencemag.org/about/authors/faq/index.dtl (last accessed on December 1, 2010).

[7] Science's Information for Reviewers of Research Articles. http://www.sciencemag.org/about/authors/RAinstr.pdf (last accessed on January 26, 2009).

[8] The same figure can be applied for other top journals. See McCook (2006).

[9] Successful media relations serve in general the goal to legitimate an organization (see Peters et al. 2008 and Chapter 11).

Publishing group, it also draws criticism. Vincent Kiernan is among the pundits and in his opinion: "It seems likely that a major motivation for journal editors in offering embargoed information is maximizing the publicity that the journal receives from the popular press" (Kiernan 1997: 316). Under the current institutional and economic constraints in news organizations, the embargo system fosters so-called "pack journalism", i.e., reporters feel obliged to cover a story because they think their competitors will (Williams and Clifford 2009). The result is "a steady stream of largely uncritical [science] stories" (Macilwain 2010).

Summing up the literature on the role of journals in international science journalism, one can conclude that a few journals to a great extent determine the information pool from which research stories are selected (Weigold 2001; Siegfried 2006). In the biomedical field, only a handful of journals are routinely scanned by science journalists (Blöbaum et al. 2004; Wormer 2008). Among the basic research journals which have the greatest media appeal are *Science* and *Nature*. Both in particular strive to get public attention by publishing "firsts" and by regularly announcing scientific breakthroughs in form of press releases and editorial material. Pahl (1998) reported that it is not just the impact factor but also commentaries on original papers like news or review articles that influence media coverage. Others have shown that media reports on research papers are largely based on press releases issued by the journals (Entwistle 1995; De Semir et al. 1998; Bartlett et al. 2002).

Which papers actually appear in the press is pre-selected by the journals in two ways: first, the editorial decisions on which scientific topics are eventually published and, second, the selection of articles highlighted by commentaries and press releases. The Nature Publishing Group, servicing a bulk of journals in different subfields, promises their authors to "receive maximum exposure for their work in the world's most important newspapers, magazines, radio and television channels."[10] Gaining publicity can be one incentive for authors where to publish. But what actually counts is scientific merit, and this is first of all assessed by the peers. Peers assess not only if manuscripts can be published, but they give credit to authors by citing their papers.

17.4 A Media Conflict in Science?

Reputation in science rests on publications. This has led to the well-known publish-or-perish dynamics in modern science. Eugene Garfield explored the origin of the phrase "publish or perish" and brought to light that the issue was first tagged in a 1942 monograph by Logan Wilson, a later student of Robert K. Merton. He wrote: "The prevailing pragmatism forced upon the academic group is that one must write something and get it into print. Situational imperatives dictate a 'publish or perish'

[10] http://www.nature.com/authors/author_resources/why_publish_with_npg.html (last accessed on December 7, 2010).

credo within the ranks" (idem, cit. in Garfield 1996). As a result, publication lists tend to be inflated by a publication strategy towards least publishable units.

While the credo prevailed at least for the past 70 years, it was recently modified. Not only the quantity of publications is measured but also their quality. That is where the journals come into play. "Publish in high-impact journals or perish" seems to be the name of the game today in highly competitive fields like biomedical research. Since 1975, the Institute for Scientific Information (ISI)[11] annually publishes the Journal Citation Reports which provide a ranking of journals according to the impact factor, a quantitative tool based on citation frequencies to measure the relative importance of a journal within its field. With the introduction of the impact factor, a clear center-periphery structure of scientific journals became manifest with a few journals at the top that promise the highest reputational gains for authors. For basic research in molecular biology, *Science* and *Nature* are the highest ranked journals regarding the 5-Year Impact Factor, followed by the journal *Cell* in the Journal Citation Reports from 2007. Beyond their editorial focus on molecular biology, the two multidisciplinary journals are considered the most important journals in various scientific subfields (see DFG 2005 and the report by the Alexander von Humboldt Stiftung 2008).

The submission rates in high-impact journals like *Science*, *Nature* or *The New England Journal of Medicine* have increased considerably in the past decades. *Science*, for example, receives 12,000 submissions p.a. "at a rate of growth rivalling the rate of Chinese economic growth," as former Editor-in-Chief Donald Kennedy ironically mentioned (McCook 2006: 26). The increase in submissions for the top journals can best be explained by the relevance of publications for individual scientific career paths, guided by research performance evaluations based on the impact factor. The molecular biologist Peter A. Lawrence, representative for many other scientists, reflects this trend critically: "We now consider the journal to be more important than the scientific message. If we publish in a top journal we have arrived, if we don't we haven't" (Lawrence 2003: 259). High-impact journals are not only a career incentive to get tenure; in some countries, even financial benefits are paid to authors for contributions in highly ranked journals (see Adam 2002).

The rivalry between scholarly journals to publish the most intriguing papers has been intensified since new journals, some of them open access, entered the market in the past decade. However, the monopoly of interpretation of scientific quality and career advancement by only a few scholarly journals in the biomedical field has recently encountered resistance in the scientific community. The tenor is always the same: appreciate the scientific message more than the journal that published it (e.g., Rosenbaum 2008). The journal's impact factor misleads to judge the actual importance of single studies and is often misused as a measure for research quality, as its inventor, Eugene Garfield, (1998) has critically noticed. Due to its importance in research assessments, the vulnerability of the impact factor

[11] ISI was acquired by Thomson in 1992 and is now part of the Thomson Reuters Corporation.

for editorial manipulation strategies is widely discussed[12] and has provoked its provider Thomson Reuters to control for instance the journals' self-citation rates (Testa 2008). Chew et al. (2007) conducted a series of interviews with Editors-in-Chief of seven biomedical journals on the impact factor development. One strategy, they found, is to boost the impact factor via media promotion. "I consider which articles will get media coverage in making publication decisions" one interviewee states (cit. in Chew et al. 2007: 146). It is rational for scientific journals to anticipate media related criteria because there is a correlation between media coverage and citation rates of articles. Phillips et al. (1991) found that the coverage of scientific articles in the press induces an increase of citations for the original article. They argue that scientists not only become aware of scientific findings through the media, but once a finding is widely covered, it seems to become more relevant to get cited. However, courting publicity is, from the perspective of publication ethics, denounced as "massaging the impact factor" (COPE 2005). In particular *Science* and *Nature* are charged for merely selecting "flashy papers" (Kennedy 2002). On these allegations, *Nature* news editor Alison Abbott comments as follows:

> Many scientists are convinced that one selection criterion is to get *Nature* in the newspapers. But actually it is not! Articles published in *Nature* get press coverage because journalists are interested in issues of major scientific significance – we do not need to be artificially selecting for 'sexier' stories. The most important thing is that there has to be a big step such that a result means something for science in general – rather than just being incredibly good only within one discipline (Abbott 2006a: 302).

If media-related criteria like these are (supposed to be) implemented in editorial programs and the decision-making processes of scientific journals, it may create divergent expectations among authors. In line with Giancarlo Corsi (2005), who has claimed a *media conflict in science* due to the dependence on technical distribution media, my assumption is: The more media attention a scientific journal attracts, the more expectations of scientific and news relevance are conflicting on the level of issue selection. That may, in turn, irritate the scientific communication process.

From the media perspective, it is obvious that not every single research paper is of general interest to be widely covered. Here, the media logic that is oriented on news values comes into play. Badenschier and Wormer (see Chapter 4) have developed a special news value theory for science journalism. They identified 14 news factors for science journalism such as scientific, political and economical relevance, astonishment and unexpectedness. The news value "scientific relevance" is attributed to the reputation and name recognition of a scientific journal (ibid.: 79). This is in line with the empirical observation that media coverage of research news is oriented at certain journals (see Section 17.3).

The reputation of a scientific journal (e.g., the impact factor, see above) also structures the scientific awareness of new findings. However, beyond the attribution

---

[12] The *PLoS* Medicine Editors (2006) for instance highlight: "This is not a theoretical question; it is well-known that editors at many journals plan and implement strategies to massage their impact factors."

to a journal the scientific significance of *single articles* is more difficult to assess. The peer review system allows for a critical assessment of what is worth publishing in scholarly journals. If, for example, the editorial decisions what to publish deviate from the peers' votes, publication decisions are often legitimized through the articles' citation numbers, as the following example illustrates:

> [R]eferees criticized a molecular-biology paper for a lack of mechanistic insight and expressed reservations about the appropriateness of some of the techniques the authors used. But the editors felt that the therapeutic implications of the paper merited publication and, after resolving the technical issues raised by the referees, pushed ahead with publication of what turned out to be a highly cited development (Nature 2006).

But what does citation actually mean with regard to the scientific significance of single papers? Citation analysis is generally used as a quantitative tool in bibliometrics to identify communication networks, topical issues, the field size and the impact of journals. Why authors cite is an open question. Citation motives are diverse (Hornbostel 1997: 289ff; Cronin 1998) and a citation theory is not in sight.[13] The definition of citation thresholds[14] for different areas of research can inform about *highly-cited papers* of a field in annual cohorts. However, it is widely discussed whether citations are correlated with scientific significance. Cole and Cole concluded from a range of data that the number of citations is a useful "indicator of the scientific significance of published work" (Cole and Cole 1971: 25). Others argued against the appropriateness of citations as a proxy for scientific quality: "Technically speaking, the number of citations reflects popularity, not necessarily quality" (Bentley 2007).[15] The latter argument comes close to the observation by Phillips et al. (1991) that citation numbers are influenced by media coverage of scientific findings and, thus, are more or less an indicator for the science-media coupling and its resonance.

As long as "scientific significance" is difficult to measure with quantitative tools alone, it might be easier to ask the other way around: What disqualifies a scientific paper? A proxy for the *lack of scientific quality* of single articles is an ex-post correction or even a retraction. Journals[16] have established a policy of handling papers that include errors or fraud. Due to electronic archival storage, these articles are marked as corrected or even retracted in order to clean up the scientific record. In examining all retractions of biomedical articles that were indexed in the database MEDLINE between 1982 and 2002, Nath et al. (2006) found that the three journals

---

[13] See for a discussion on the utility of a citation theory the special issue of *Scientometrics* 1998, *43*(1).

[14] *Science Watch* regularly calculates the citation thresholds for highly cited papers, so-called top papers with a citation rate that comprise the top 1% of papers in a field according to the year of publication.

[15] It is moreover the popularity of the journal, its impact factor, that influences the citation rates of papers which can be described in the notion of Merton's Matthew effect (Larivière and Gingras 2010).

[16] Although an editorial policy of handling retractions is demanded in publication ethics, only a small percentage of biomedical journals have one (see Atlas 2004).

with the highest numbers of retractions were the multidisciplinary journals *Science*, *Nature* and the *Proceedings of the National Academy of Sciences* (PNAS).[17] But how can these findings be interpreted? Does the elevated error rate, in particular for the high-impact journals *Science* and *Nature*, point to a lack of scientific scrutiny in the editorial decision-making? Does it indicate that conflicting expectations on scientific and media relevance come into operation, so that news values eclipse scientific rigor in the high profile journals, as the thesis of a media conflict in science suggests? Or – vice versa – is it rather the high visibility of these journals that encourages ex-post scrutiny, so that errors are more likely to be detected?

To further explore the role of scientific journals in the medialization of science and to ask for any repercussions on science, we need to look more carefully at the very nature of publication events. A contextualization of three ex-post corrected articles and their editorial handling should shed light on divergent expectations of scientific significance and media-related criteria of attention in scholarly communication.

17.5 Characteristics of a Media Conflict: A Case Study on Stem Cell Research

To explore the thesis of a media conflict in science I conducted a case study in stem cell research – a field commonly seen as highly medialized in terms of extensive media coverage (Weingart et al. 2008; Schäfer 2009). Stem cell research is a field under great pressure in regard to clinical outcomes, economic interests and political regulations concerning the ethics of embryonic research and derivation techniques (for an overview, see, e.g., Hauskeller 2002). With the derivation of human embryonic stem cells in 1998, the scientific output has increased steadily (Camargo and Winterhager 2006; Guhr et al. 2006; Owen-Smith and McCormick 2006) but so has the scepticism around highly-publicized stem cell findings. Jeanne Loring, a stem cell scientist from the Scripps Institute in La Jolla (California) states: "When a topic is of great interest – not just to scientists but also to the public – it's certainly more likely that someone will get ahead of themselves and not be as scrupulous as scientists have to be" (cit. in Aldhous and Reich 2008). This statement points first of all to the anticipation of extra-scientific criteria that might violate scientific standards. By distinguishing between the two spheres of science and the media, I will further explore whether conflicting expectations can be reconstructed on the level of single papers.

An in-depth analysis of three ex-post corrected papers was conducted based on accompanying source material like press release articles, reviews and news articles, correspondences, explanatory statements for corrections and corresponding press

[17] A concentration of retractions in *Science* and *Nature* was also identified in a sample of stem cell papers that were each highlighted as important milestones in science policy reports (Franzen 2011).

articles in the *New York Times* and the *Washington Post*.[18] The results of the extensive analysis can only be summarized here.[19] It indicates that the three prototypic cases have a lot in common:

(1) Each original article published in *Science* and *Nature* was framed as an important step in stem cell science and was strongly promoted. They were all highly-cited, i.e., the citation numbers were beyond the annual thresholds for *top papers* in the research area of "molecular biology and genetics."[20] The articles were published online ahead of print – an editorial handling for a selection of particularly topical manuscripts described above (see Section 17.3). They were press released, discussed in a news article and two of them highlighted on the cover and by a review article in the same issue. One paper was finally accompanied by a podcast, a feature that was introduced after the other two papers had already been published. I want to define these editorial pieces as *amplifications* of the original message that govern the attention of both scientists and journalists. These amplifications of single publications distinguish *Science* and *Nature* from most other periodicals.[21] As we know from science journalism studies, the provision of context information can help to get papers reported in the press (see Pahl 1998). The media coverage of the scientific papers appeared simultaneously to the ahead-of-print publication date.
(2) Each of the three original articles immediately fuelled the political debate in the U.S. Congress. The particular findings served either the proponents or opponents of embryonic stem cell research and were used as such to strengthen the relative position in upcoming decisions on stem cell research and cloning issues. The authority given by the most eminent peer reviewed journals added weight to political statements. It is likely that the announcement of a "scientific breakthrough," for example a cloning success, before a Senate hearing will influence the political outcome and can be decisive for publication schedules, notwithstanding general editorial constraints raised by *Nature* editor Natalie DeWitt: "It was very difficult to publish any stem cell paper at a time when there is no political debate in the United States" (cit. in Wade 2002).
(3) The revelation of inconsistencies in each of the three papers was managed by the mass media. This gives empirical evidence of the watchdog role of the mass media for science (Schnabel 2004; Weingart 2004; Stollorz 2008) and illustrates

[18] These two are considered as agenda-setters for international science journalism (Nisbet et al. 2003).

[19] For a detailed version, see Franzen (2011).

[20] The field of "molecular biology and genetics" contains the most conservative numbers for "top papers." The citation thresholds used here were updated in July 2008 on the Science Watch website provided by Thomson Reuters (www.sciencewatch.com).

[21] However, other journals like *Cell* started to adapt such amplifications, introducing the "Leading Edge" section in 2005 or *PLoS Biology* with its "primers" and "synopses," "serve as mini science education modules and are designed to enhance public understanding of a key area of biology" (Chapter 18, p. 354).

the effects of the medialization process. In one case, even the *Nature* news department was the leading actor in detecting unethical research practises that resulted in the first of a series of corrections of the respective *Science* paper (Cyranoski 2004). Public scrutiny led to formal investigations in each case. Such forms of investigative journalism illustrate that the "context speaks back" (Nowotny et al. 2001) and may come close to the "extended peer review" that Funtowicz and Ravetz (1993) have called for. However, the verification of the allegations falls again to the peers. Independent research teams were assigned to replicate the experiments and to prove the results. Based on the committee reports, the journals decided on a case-by-case basis what kind of action was to be taken to correct the scientific record.

Besides the similarities given, the prototypical analysis disclosed structural differences. From the science's perspective, three types of reception of medialized papers can be distinguished: (1) The selection of the positive value "true"; (2) The selection of the negative value "not true" and (3) A code value cannot be assigned. In the following contextualization of the three prototypic papers the implications for scholarly communication will be discussed.

In the *first case*, the published findings were finally marked as false and fabricated. It was about the first therapeutic cloning success (Hwang et al. 2004) that was highly promoted by the journal *Science*. The extensive media promotion by *Science* with a publication schedule geared to the annual AAAS-meeting[22] together with the editorial framing as an important step towards cell replacement therapies to treat diabetes and other diseases is, however, inconsistent with the objections that were raised in the peer review process on the validity of the results. The report of the investigation committee *Science* had assigned to evaluate the editorial handling of the Hwang paper states:

> It is clear from the correspondence that the editors were aware of a major potential flaw [...], namely the possible occurrence of parthenogenesis. This possibility should have been pursued and eliminated, since the central contention of the 2004 paper was that a human stem cell line had been generated by nuclear transplant. This central fact was not established (Science Committee Report 2006).[23]

In this case, an orientation towards the mass media is evident on the organizational level of editorial decision-making: If providing more evidence for a bold claim is neglected in favor of an accelerated and widely noticed publication, mass-media related criteria are conflicting with scientific rigor. It can be assumed that the temporal dimension superimposes the factual dimension as one effect of the medialization of science. But what does the publication of findings that have high news value but turn out to be "not true" mean for the science system? That hypotheses are

---

[22] Meetings of the AAAS as the "World's Largest General Scientific Society" are probably the most important event worldwide in bringing together scientists and science journalists.

[23] It later turned out that the human stem cell line derived by Hwang and colleagues was in fact of parthenogenetic origin (Kim et al. 2007).

finally rejected due to new empirical findings is common practise in science and – regardless of an intention to deceive – even useful for scientific progress. The data fabrication was compared to many other incidents of fraud promptly discovered and, hence, the scientific communication process was not enduringly disturbed. After the revelation, stem cell scientists regarded the race as open again (Snyder and Loring 2006). Nonetheless, this fraudulent event fuelled mistrust among scientists and the public. The public image of stem cell science had been tarnished.

In the *second case*, the situation is more complex. The paper claims the derivation of human embryonic stem cells without the destruction of embryos (Klimanskaya et al. 2006). It was not the empirical data that were questioned later. They could stand as a re-review confirmed (Wade 2006). The problem with the paper was that relevant details in the experimental protocol were left out in order to make a political claim that was promoted by the respective journal with various editorial amplifications such as a news article, a podcast and a press release. In the paper itself, false images were used that strengthen the argument of an ethically acceptable technique. From the nature of presentation, it does not become clear that none of the embryos actually "survived": The authors merely established a proof-of-principle to be tested by further studies for its practical use. Due to early public objections raised by a German journalist, the accompanying press release and the podcast were corrected (Abbott 2006b). The findings were published three months later in a completely revised form. The first version gave a wrong impression of its political impact. However, the initial announcement of an "ethically sound" derivation procedure provokes not only a direct statement from the White House (cit. in Weiss 2006), it also shows effects on the stock market: The stock prices of the involved company leapt five-fold in just 10 h (see Abbott 2006b). The enhancement of science's public from the peers to a general audience, implicitly or explicitly addressed in scientific publications, can be regarded as another medialization effect. Whitley (1985: 19) has pointed to the fact that the more important an outside public is for science, the more it is necessary to adapt the communication to that audience's concerns. An orientation towards political and economical desires, however, bears the risk of losing a firm footing. The ultimate attribution of the positive scientific value "true," albeit decelerated, to this second type of a medialized communication conceals the fact that the mismatch between the production and presentation of findings hampered the evaluation of the technology's efficiency, as some protagonists had criticized (cit. in Abbott 2006b) and thus, any replication efforts.

The *third paper* under investigation was about the discovery of a rare stem cell population in the bone marrow that shares characteristics of embryonic stem cells (Jiang et al. 2002). The results were astonishing because they challenged the fundamental dogma of developmental biology that cellular differentiation is linear and irreversible. This made them newsworthy. From the political point of view, the potential therapeutic use of adult stem cells without the ethical burden associated with embryonic stem cells was highly appreciated and served the political opponents as an argument to ban human cloning (Wade and Stolberg 2002). Several research teams jumped on the bandwagon but failed to reproduce the results (Giles 2006). Five years after publication, the experimental protocol was in some minor aspects corrected due to journalistic allegations by the *New Scientist* (Aldhous and Reich 2007), but until now it is unclear if the politically desired findings are an artefact or rather extremely difficult to obtain. For this case, the scientific code of "true" or "not true" is not applicable, although it is the relatively most-cited item in the sample, 7 times higher than the *top paper* threshold value. The absolute citation rate of 1,664 (accessed in July 2008) conceals the fact that the paper's scientific relevance is still in question. This gives reason to doubt the usefulness of citation metrics to judge scientific relevance. It furthermore indicates that the publication of controversial findings attracts many citations and can thus be a rationale in making editorial decisions: "Well. If it is right, it is interesting. If it is wrong, it will at least stimulate a lot of discussions" (Vogel 2006: 326, referring to editorial practise at *Science*). In terms of scientific advance, the once sensational discovery appears like a fruitless undertaking.[24] Cases like these point to a media conflict in science in which the systems specific connectivity is disturbed.

17.6 Discussion: The Medialization Effects on Scholarly Communication

What lessons can be learned from the change of perspective towards scientific journals for the medialization of science? The in-depth-analysis of publication events revealed the discrepancy between the actual scientific value and the newsworthiness of single papers; favoring the latter can irritate scholarly communication up to the point that the code of "truth" is temporarily losing its orienting function. Instead of the interpretation of a de-differentiation process of society (Nowotny et al. 2001), the results rather point to irritations in the self-direction of science when editorial programs addressing different publics come into operation. In order to gain reputation for publishing in high-ranked journals, authors need to anticipate the underlying selection criteria. If journals of the highest rank are likely or believed to prefer papers that find media coverage, authors need to address this issue in order to get their paper published. To serve the media's demand for astonishing results, scientists themselves tend to overstate the societal implications of their findings.

It was not in the realm of the qualitative analysis but there is additional empirical evidence that scholarly communication is generally shifting towards attention seeking, the proper characteristic of mass media communication. Key word analyses of scientific papers show a trend towards more "biased words," i.e., "claims of 'crucial', 'critical' or 'unique' events as well as 'important' or 'original' discoveries" that are suspected to bias the interpretation of the research article (Fraser and Martin 2009: np.). Other studies show a shift in scientific language of research

[24] Cole and Cole (1971: 25) have argued: "It is unlikely that any work which is wrong without being a 'fruitful error' will ever accumulate many citations." This case is a counter-example, however. Thus, more qualitative studies on citation counts are needed to address questions like these.

articles towards the characterization of *new* and *unexpected* findings: "One might think that academic machismo or realism would cause scientists to downplay their surprise, but, on the other hand, overstating the level of astonishment may occur when striving for media attention" (Jasienski 2006). Notwithstanding that "novelty" is a main characteristic of science and the mass media, one can assume that the shift in scientific style towards the media rules of attention mirrors first of all the editorial guidelines of scientific journals and the competition around the most intriguing papers (see Section 17.2). Besides the observation of strategies to adapt to mass media related criteria in the presentation of knowledge, the crucial question remains: what does medialization mean for knowledge production?

The case study on stem cell science indicates that the medialization of science comes with undesirable side effects to the reliability of published findings. If scientific rigor is eclipsed by criteria of news values, political relevance or economic interests, the credibility of science as a social enterprise is jeopardized. It therefore reveals the downside of a "strongly contextualized knowledge production" (Nowotny et al. 2001). High expectations towards scientific output for technical advances, therapeutic applications and ethical solutions can lead to what Luhmann (1990: 237ff, 622f) has denoted as the inflation of the science-specific medium "truth." In times of inflation, the empirical verification and the preciseness of terms and definitions are neglected in order to meet public demands for concrete results (ibid.: 623). Luhmann interprets this as a symptom that a system defends itself against influences from the outside. Against this background, exaggerated claims can be interpreted as one form of reaction of scientists to public interest, i.e., as an undesirable side-effect of the medialization of science. As shown, the general interest journals play a relevant part in the medialization process, but they also have to cope with the inherent conflict resulting from divergent expectations on scientific robustness and newsworthiness.

I do not want to argue that papers published by *Science* and *Nature* are careless or flawed. Quite the contrary, both maintain a very high standard of scientific rigor. That is where their reputation originates. However, the encouraged visibility inside and outside of science is accompanied by a higher risk to overstretch results. In times when publication patterns become most relevant for scientific career paths worldwide, the bid for the top journals does challenge the entire system. While submissions in high-impact journals have steadily increased in the past decade, the acceptance rates remain stable at a very low level of under 10 percent or have even decreased (see McCook 2006). If the mission statement of a highly selective journal explicitly states that findings must be of general interest, the authors need to produce new results that will merit recognition by the scientific community and beyond. Bearing in mind the triaged decision-making process, manuscripts must attain editor's attention in the first place. It is the reviewers at the top journals who started to complain about the unsubstantiated bold claims authors increasingly made; they feel obliged to tone them down. In the end, there is no significant difference to papers published in lower impact journals. "There is a danger that authors are rewarded for a fundamentally dishonest approach," a *Nature* reviewer alerts (Brookfield 2003).

Due to their orientation on broad impact, both *Science* and *Nature* attract papers that bear the risk of overreaching scientific standards in order to gain attention of the editors. The *Science Committee,* which investigated the editorial handling in the Hwang case, points to the unintended side effects of the "published in top journals or perish" dynamics:

> *Science* (and *Nature*) have reached a special status. Publication in *Science* has a significance that goes beyond that of 'normal' publication. Consequently, the value to some authors of publishing in *Science*, including enhanced reputation, visibility, position or cash rewards, is sufficiently high *that some may not adhere to the usual scientific standards in order to achieve publication* (Science Committee Report 2006, my emphasis).

The transgression from making bold claims to trimming and cooking of data is fuzzy. Regular events of misconduct illustrate that intentional misrepresentations of a material fact are not easy to detect by the peer review system. Even if a reviewer raises concerns over the technical rigor of a study, the ultimate decision what to publish rests with the journal editors who need to balance (opposing) reviewers' votes. If published findings turn out to be wrong or even fraudulent, journal editors are forced to legitimate their publication decisions and to secure that the self-steering mechanism is working.

In the aftermath of the highly-publicized fraud case concerning Hwang, *Science* reacted with the implementation of a new policy on high-risk papers that warrant special scrutiny, i.e., papers that are likely to be especially visible or influential, for example on climate, energy, human health, etc. (Kennedy 2006; Science Committee Report 2006). *Science's* counterpart *Nature* decided to solicit independent verification for strong claims ahead of publication on a case-by-case basis (*Nature* 2007). This special requirement for highly visible papers deals with the structural discrepancy between the production and presentation of knowledge in scientific articles (as discussed above in Section 17.2). As trust in communications forms the basis of science it remains open whether a procedure to control the data production ahead of print is actually feasible in terms of time and financial outlay.[25]

Stem cell science is admittedly an exceptional case because of its ongoing politicization that explains both media attention and media orientation. Climate research might be another example that achieves similar public attention and scrutiny (see Chapter 14). It needs to be tested by further studies if the ambivalent role of the high-impact journals in framing scientific and public perception can be likewise reconstructed. As the mass media as well as the scientific publication system are shifting from print to the online era, future studies are needed to reflect this change in regard to the science-media coupling.

Acknowledgements I would like to thank Sascha Dickel, Fran Osrecki and the reviewers for their comments on a draft of this chapter.

---

[25] I only know of one example in which this special treatment for a cloning paper was used (see Byrne et al. 2007).

References

AAAS (n.d.). Short, 'intriguing' research articles focus of new feature in *Science*. http://www.aaas.org/news/releases/2002/0428brevia.shtml (last accessed on July 9, 2009).
Abbott, A. (2006a). Wissenschaft bei einer internationalen Fachzeitschrift I: Between peer review and a science journalism generator. In H. Wormer (ed.), *Die Wissensmacher. Profile und Arbeitsfelder von Wissenschaftsredaktionen in Deutschland*. Wiesbaden: VS Verlag für Sozialwissenschaften, pp. 299–313.
Abbott, A. (2006b). 'Ethical' stem-cell paper under attack. *Nature*, 443, 12.
Adam, D. (2002). Citation analysis: The counting house. *Nature*, 415, 726–729.
Aldhous, P. and E. S. Reich (2007). Fresh questions on stem cell findings. *New Scientist*, March 21, 2007.
Aldhous, P. and E. S. Reich (2008). Stem-cell researcher guilty of falsifying images. *New Scientist*, October 7, 2008.
Alexander von Humboldt Stiftung (2008). Publikationsverhalten in unterschiedlichen wissenschaftlichen Disziplinen. Beiträge zur Beurteilung von Forschungsleistungen. *Diskussionspapier Nr. 12*.
Atlas, M. C. (2004). Retraction policies of high-impact biomedical journals. *Journal of Medical Libraries Association*, 92, 242–250.
Bartlett, C., J. Sterne, and M. Egger (2002). What is newsworthy? Longitudinal study of the reporting of medical research in two British newspapers. *BMJ*, 325, 81–84.
Bazerman, C. (1988). *Shaping written knowledge. The genre and activity of the experimental article in science*. Madison, WI: The University of Wisconsin Press.
Bentley, R. A. (2007). Letter to the editors: Why do team-authored papers get cited more? *Science*, 317, 1496.
Blöbaum, B., A. Görke, and K. Wied (2004). *Quellen der Wissenschaftsberichterstattung*. Münster: Institut für Kommunikationswissenschaft.
Brookfield, J. (2003). The system rewards a dishonest approach. *Nature*, 423, 480.
Byrne, J. A., D. A. Pedersen, L. L. Clepper, et al. (2007). Producing primate embryonic stem cells by somatic cell nuclear transfer. *Nature*, 450, 497–502.
Camargo, A. and M. Winterhager (2006). Internationale Kooperation in der Stammzellforschung. Eine bibliometrische Analyse am Beispiel der Länder Deutschland, Kanada, Niederlande, Schweden, USA und Vereinigtes Königreich. In R. Wink (ed.), *Deutsche Stammzellpolitik im Zeitalter der Transnationalisierung*. Baden-Baden: Nomos Verlagsgesellschaft, pp. 45–71.
Chew, M., E. V. Villanueva, and M. B. Van Der Weyden (2007). Life and times of the impact factor: Retrospective analysis of trends for seven medical journals (1994–2005) and their editors' views. *Journal of the Royal Society of Medicine*, 100, 142–150.
Cole, J. and S. Cole (1971). Measuring the quality of sociological research: Problems in the use of the Science Citation Index. *The American Sociologist*, 6, 23–29.
COPE (2005). *Annual Report*. http://publicationethics.org/files/u2/2005_report.pdf (last accessed on August 29, 2010).
Corsi, G. (2005). Medienkonflikt in der modernen Wissenschaft? *Soziale Systeme*, 11, 176–188.
Couzin, J. (2006). ...And how the problems eluded peer reviewers and editors. *Science*, 311, 23–24.
Cronin, B. (1998). Metatheorizing citation. *Scientometrics*, 43(1), 45–55.
Cyranoski, D. (2004). Crunch time for Korea's cloners. *Nature*, 429, 12–14.
De Semir, V., C. Ribas, and G. Revuelta (1998). Press releases of science journal articles and subsequent newspaper stories on the same topic. *JAMA*, 280, 294–295.
Deutsche Forschungsgemeinschaft (DFG) (2005). *Publikationsstrategien im Wandel? Ergebnisse einer Umfrage zum Publikations- und Rezeptionsverhalten unter besonderer Berücksichtigung von Open Access*. Weinheim: Wiley-VCH Verlag.
Entwistle, V. (1995). Reporting research in medical journals and newspapers. *British Medical Journal*, 310, 920–923.

Esposito, E. (2005). Die Darstellung der Wahrheit und ihre Probleme. *Soziale Systeme*, 11, 166–175.
Franzen, M. (2009). Torwächter der Wissenschaft oder Einfallstor für die Massenmedien? Zur Rolle von *Science* und *Nature* an der Schnittstelle zwischen Wissenschaft und medialer Öffentlichkeit. In S. Stöckel et al. (eds.), *Das Medium Wissenschaftszeitschrift seit dem 19. Jahrhundert. Verwissenschaftlichung der Gesellschaft – Vergesellschaftung von Wissenschaft*. Stuttgart: Franz Steiner Verlag, pp. 229–252.
Franzen, M. (2011). *Breaking news: Wissenschaftliche Zeitschriften im Kampf um Aufmerksamkeit*. Baden-Baden: Nomos.
Fraser, V. and J. Martin (2009). Marketing data: Has the rise of impact factor led to the fall of objective language in scientific articles? *Respiratory Research*, 10. doi:10.1186/1465-9921-10-35.
Funtowicz, S. O. and J. R. Ravetz (1993). Science for the post-normal age. *Futures*, 25, 739–755.
Garfield, E. (1987). The 170 surviving journals that CC would have covered 100 years ago. *Current Contents*, 26, 164–173.
Garfield, E. (1996). What is the primordial reference for the phrase 'publish or perish'? *The Scientist*, 10, 11.
Garfield, E. (1998). Letters to the editor: The impact factor and using it correctly. *Der Unfallchirurg*, 48, 413.
Giles, J. (2006). The trouble with replication. *Nature*, 442, 344–347.
Göbel, A. (2006). Der "Heilige Geist des Systems"? Gesellschaftstheoretische Bemerkungen zum System der Massenmedien. In A. Ziemann (ed.), *Medien der Gesellschaft – Gesellschaft der Medien*. Konstanz: UVK Verlagsgesellschaft mbH, pp. 111–139.
Goffman, W. (1981). Journals. In K. S. Warren (ed.), *The ecology of the biomedical literature and information retrieval*. New York: Praeger, pp. 31–46.
Guhr, A., A. Kurtz, K. Friedgen, et al. (2006). Current state of human embryonic stem cell research: An overview of cell lines and their usage in experimental work. *Stem Cells*, 24, 2187–2191.
Hauskeller, C. (2002). *Humane Stammzellen – therapeutische Optionen, ökonomische Perspektiven, mediale Vermittlung*. Lengerich: Pabst Science Publishers.
Hilgartner, S. (1990). The dominant view of popularization: Conceptual problems, political uses. *Social Studies of Science*, 20, 519–539.
Hornbostel, S. (1997). *Wissenschaftsindikatoren. Bewertungen in der Wissenschaft*. Opladen: Westdeutscher Verlag.
Hwang, W. S., Y. J. Ryu, J. H. Park, et al. (2004). Evidence of a pluripotent human embryonic stem cell line derived from a cloned blastocyst. *Science*, 303, 1669–1674.
IMCJE (International Committee of Medical Journal Editors) (2007). *Uniform requirements for manuscripts submitted to biomedical journals: Writing and editing for biomedical publication*. Update from October 2007. http://www.icmje.org/2007_urm.pdf (last accessed on August 27, 2009).
Jasienski, M. (2006). It's incredible how often we're surprised by findings. *Nature*, 440, 1112.
Jiang, Y., B. N. Jahagirdar, R. L. Reinhardt, et al. (2002). Pluripotency of mesenchymal stem cells derived from adult marrow. *Nature*, 418, 41–49.
Kennedy, D. (2002). Next steps in the Schön affair. *Science*, 298, 495.
Kennedy, D. (2006). Responding to fraud. *Science*, 314, 1353.
Kiernan, V. (1997). Ingelfinger, embargoes, and other controls on the dissemination of science news. *Science Communication*, 18, 297–319.
Kim, K., K. Ng, and P. J. Rugg-Gunn (2007). Recombination signatures distinguish embryonic stem cells derived by parthenogenesis and somatic cell nuclear transfer. *Cell Stem Cell*, 1, 346–352.
Klimanskaya, I., Y. Chung, S. Becker, et al. (2006). Human embryonic stem cell lines derived from single blastomeres. *Nature*, 444, 481–485.
Knorr-Cetina, K. (1981). *The manufacture of knowledge. An essay on the constructivist and contextual nature of science*. Oxford: Pergamon Press.

Larivière, V. and Y. Gingras (2010). The impact factor's Matthew Effect: A natural experiment in bibliometrics. *Journal of the American Society for Information Science and Technology*, 61, 424–427.

Latour, B. and S. Woolgar (1986 [1979]). *Laboratory life*. Princeton, NJ: Princeton University Press.

Lawrence, P. A. (2003). The politics of publication. *Nature*, 22, 259–261.

Luhmann, N. (1990). *Die Wissenschaft der Gesellschaft*. Frankfurt/M.: Suhrkamp.

Luhmann, N. (1996). *Die Realität der Massenmedien*. Opladen: Westdeutscher Verlag.

Luhmann, N. (1997). *Die Gesellschaft der Gesellschaft*. 2 volumes. Frankfurt/M.: Suhrkamp.

Macilwain, C. (2010). Calling science to account. *Nature*, 463, 875.

Maddox, J. (1969). Journals and the literature explosion. *Nature*, 221, 128–130.

McCook, A. (2006). Is peer review broken? *The Scientist*, 20, 26.

Nath, S. B., S. C. Marcus, and B. G. Druss (2006). Retractions in the research literature: Misconduct or mistakes? *Medical Journal of Australia*, 185, 152–154.

Nature (2006). Three cheers for peers. *Nature*, 439, 118.

Nature (2007). Replicator review. *Nature*, 450, 457–458.

Nisbet, M. C., D. Brossard, and A. Kroepsch (2003). Framing science. The stem cell controversy in an age of press/politics. *Press/Politics*, 8, 26–70.

Nowotny, H., P. Scott, and M. Gibbons (2001). *Re-thinking science. Knowledge and the public in an age of uncertainty*. Cambridge, UK: Cambridge University Press.

Owen-Smith, J. and J. McCormick (2006). An international gap in human ES cell research. *Nature Biotechnology*, 24, 391–392.

Pahl, C. (1998). Die Bedeutung von Wissenschaftsjournalen für die Themenauswahl in den Wissenschaftsressorts deutscher Zeitungen am Beispiel medizinischer Themen. *Rundfunk und Fernsehen*, 46, 243–253.

Peters, H. P., H. Heinrichs, A. Jung, et al. (2008). Medialization of science as a prerequisite of its legitimization and political relevance. In D. Cheng et al. (eds.), *Communicating science in social contexts: New models, new practices*. Dordrecht: Springer, pp. 71–92.

Phillips, D. P., E. J. Kanter, B. Bednarczyk, et al. (1991). Importance of the lay press in the transmission of medical knowledge to the scientific community. *The New England Journal of Medicine*, 325, 1180–1183.

Rosenbaum, J. L. (2008). High-profile journals not worth the trouble. *Science*, 321, 1039.

Schäfer, M. S. (2009). From public understanding to public engagement: A comparison of mass media coverage on different science issues. *Science Communication*, 30, 475–505.

Schnabel, U. (2004). Wachhund oder Störenfried? – Zur Rolle der Presse im Umgang mit wissenschaftlichem Fehlverhalten. In Deutsche Forschungsgemeinschaft (ed.), *Wissenschaftliches Fehlverhalten – Erfahrungen von Ombudsgremien. Tagungsbericht Standpunkte*. Weinheim: Wiley-VCH, pp. 50–54.

Schwartz, L. M., S. Woloshin, and L. Baczek (2002). Media coverage of scientific meetings: Too much, too soon? *JAMA*, 287, 2859–2863.

Science (2007). General information for authors at http://www.sciencemag.org/about/authors/prep/gen_info.dtl (last accessed on December 11, 2007).

Science Committee Report (2006). Letter and report from committee examining *Science*'s peer review process for Hwang et al. papers [*Science*, 303, 1669 (2004) and *Science*, 308, 1777 (2005)] In D. Kennedy: Supporting online material for responding to fraud. *Science*, 314, 1353.

Siegfried, T. (2006). Reporting from science journals. In D. Blum et al. (eds.), *A field guide for science writers*. Oxford, New York: Oxford University Press, 2nd edition, pp. 11–17.

Snyder, E. Y. and J. F. Loring (2006). Beyond fraud – Stem-cell research continues. *NEJM*, 354, 321–324.

Stichweh, R. (1987). Die Autopoiesis der Wissenschaft. In D. Baecker et al. (eds.), *Theorie als Passion*. Frankfurt/M.: Suhrkamp, pp. 447–481.

Stollorz, V. (2008). Ist der Platz zwischen allen Stühlen der richtige Ort? Essay über die Frage, was Wissenschaftsjournalismus heute soll. In H. Hettwer et al. (eds.), *WissensWelten:*

Wissenschaftsjournalismus in Theorie und Praxis. Gütersloh: Bertelsmann Stiftung, pp. 566–582.
Testa, J. (2008). Playing the system puts self-citation's impact under review. *Nature*, 455, 729.
The PLoS Medicine Editors (2006). The impact factor game. *PLoS Medicine*, 3(6), e291.
Vogel, G. (2006). Wissenschaft bei einer internationalen Fachzeitschrift II: Journalism at a magazine-within-a-magazine. In H. Wormer (ed.), *Die Wissensmacher. Profile und Arbeitsfelder von Wissenschaftsredaktionen in Deutschland*. Wiesbaden: VS Verlag für Sozialwissenschaften, pp. 315–329.
Wade, N. (2002). Scientists make 2 stem cell advances. *New York Times*, June 21, 2002.
Wade, N. (2006). Journal clarifies stem cell report. *New York Times*, November 22, 2006.
Wade, N. and S. G. Stolberg (2002). Scientists herald a versatile adult. *New York Times*, January 25, 2002.
Weigold, M. F. (2001). Communicating science. A review of the literature. *Science Communication*, 23(2), 164–193.
Weingart, P. (2001). *Die Stunde der Wahrheit? Zum Verhältnis der Wissenschaft zu Politik, Wirtschaft und Medien in der Wissensgesellschaft*. Weilerswist: Velbrück.
Weingart, P. (2003). Growth, differentiation, expansion and change of identity – the future of science. In B. Joerges et al. (eds.), *Social studies of science and technology: Looking back ahead*. Dordrecht et al.: Kluwer, pp. 183–200.
Weingart, P. (2004). Öffentlichkeit der Wissenschaft – Betrug in der Wissenschaft. In Deutsche Forschungsgemeinschaft (ed.), *Wissenschaftliches Fehlverhalten – Erfahrungen von Ombudsgremien. Tagungsbericht Standpunkte*. Weinheim: Wiley-VCH, pp. 41–49.
Weingart, P., S. Salzmann, and S. Wörmann (2008). The social embedding of biomedicine: An analysis of German media debates 1995–2004. *Public Understanding of Science*, 17, 381–396.
Weiss, R. (2006). Stem cells created with no harm to human embryos; but concerns are raised about the technique. *Washington Post*, August 24, 2006.
Whitley, R. (1985). Knowledge producer and knowledge acquirers. In T. Shinn et al. (eds.), *Expository science: Forms and functions of popularisation*. Sociology of the Sciences Yearbook IX. Dordrecht: Reidel, pp. 3–28.
Williams, A. and S. Clifford (2009). *Mapping the field: A political economic account of specialist science news journalism in the UK national media*. Report funded by the Department for Business, Innovation and Skills and commissioned by the Expert Group on Science and the Media.
Wormer, H. (2008). "Wie seriös ist Dr. Boisselier?" – Quellen und Recherchestrategien für Themen aus Wissenschaft und Medizin. In H. Hettwer et al. (eds.), *WissensWelten: Wissenschaftsjournalismus in Theorie und Praxis*. Gütersloh: Bertelsmann Stiftung, pp. 345–362.
Zuckerman, H. and R. K. Merton (1971). Patterns of evaluation in science: Institutionalization, structure, and function of the referee system. *Minerva*, 9, 66–100.

Chapter 18
Practitioner's Perspective: Science as a Public Resource: Rules of Engagement

Liza Gross

18.1 Opening Access to Scientific Information

> Most of the fundamental ideas of science are essentially simple, and may, as a rule, be expressed in a language comprehensible to everyone (Albert Einstein).

The "scientific community" is conventionally defined in the most restrictive sense, describing only experts trained in the scientific method to obtain new knowledge within a specific field. In this sense, the scientific community includes experts who produce this new knowledge and write up their results for publication in journals as well as those who read others' work in specific fields. However, the technical nature of these reports necessarily excludes the general public, which lacks the specialized knowledge required to understand the jargon, data, and implications of the results presented. When the Public Library of Science made the transition from grassroots advocacy organization to publisher by launching *PLoS Biology* in 2003, it rejected this narrow definition. In keeping with its primary objective of making science a public resource by ensuring free and unrestricted access to the ever-expanding corpus of scientific and medical literature[1] – 800,000 new articles appeared in PubMed in 2008 alone (Neylon and Wu 2009) – PLoS sought to redefine the scientific community, particularly regarding the readers and users of the content, to include teachers, students, physicians and the general public (Bernstein et al. 2003). The decision to expand the notion of scientific community flows naturally from the PLoS founders' rationale for becoming a publisher: not because the world needs more biomedical journals, but because U.S. taxpayers, who underwrite tens of billions of dollars of research each year, have a right to access – and the chance to understand – the results of that research (Joseph 2008). Open-access publishers present content online that is available for anyone, anywhere to read, use,

[1] PLoS allows free, unrestricted access and use of all content published in PLoS journals by using the Creative Commons Attribution License, http://creativecommons.org/licenses/by/2.5/.

L. Gross (✉)
PLoS Biology
e-mail: lgross@plos.org

S. Rödder et al. (eds.), *The Sciences' Media Connection – Public Communication and its Repercussions*, Sociology of the Sciences Yearbook 28,
DOI 10.1007/978-94-007-2085-5_18, © Springer Science+Business Media B.V. 2012

or distribute in any way they choose (provided that the original source is cited). Making science a true public resource requires removing intellectual, as well as physical, barriers to scientific research. PLoS believes that the general public should have access to the scientific literature, tools to understand that literature, and even the opportunity to engage in scientific debates (see Section 18.3).

As long as subscription and pay-to-view barriers remain in place, thwarting efforts to find hidden connections among disparate ideas within and across scientific disciplines, the digital revolution's potential to accelerate discovery and expand the frontier of human knowledge will be squandered. But if the immediate benefits of unfettered access to the world's primary literature redound only to the narrowly defined scientific community, is that enough? Making the concepts and methods of science accessible to the broader public – finding ways to ensure that intellectual access proceeds hand-in-hand with physical access – is an integral part of PLoS's mission to make science a public resource. This commitment is codified as one of the organization's core principles:

> Our mission of building a public library of science includes not only providing unrestricted access to scientific research ideas and discoveries, but developing tools and materials to engage the interest and imagination of the public and helping non-scientists to understand and enjoy scientific discoveries and the scientific process.[2]

Making the basic research findings published in a science journal accessible to an interested lay audience has not been a priority for most scientific journals. *PLoS Biology* editors sought to move beyond that tradition by creating article types not typically found in other science journals, including "synopses" and "primers." Both article types serve as mini science education modules and are designed to enhance public understanding of a key area of biology.

Whereas most articles in scholarly journals are written by scientists for scientists (aside from those journals that include news sections), synopses are written for a general audience by professional science writers to introduce non-scientists to fundamental concepts and methods in biology and, importantly, to the nature of scientific progress. Synopses were designed as a counterpoint to the type of mainstream media stories that suggest science progresses in great leaps rather than incrementally, an approach that ultimately confuses and frustrates readers who can't understand, for example, how a drug or food can be reported as beneficial one year and unhealthy the next (Rowe 2000).

Synopses (which originally accompanied each research article but now supplement a select subset) describe the advances reported in a paper in a way that reveals science as an evolutionary rather than revolutionary process, with knowledge shifting to accommodate new evidence and observations. Toward this end, synopses help non-scientists understand a study's results by explaining why the research question is important, what evidence supports the hypothesis tested, how the researchers tested their hypothesis, what evidence the experiments produced,

---

[2] PLoS Core Principles, http://www.plos.org/about/what-is-plos/core-principles/.

why the interpretation of the evidence appears sound, how the findings change current thinking, what the findings can't answer, and how they suggest new avenues of investigation.

PLoS Biology also complements a subset of research articles with "primers" written by leading scientists to present a concise introduction to a field, including enough context and background for naïve readers to appreciate the significance of the advance reported in the paper. Rather than focus primarily on the paper, as synopses do, primers function more as tutorials on an important aspect of biology, describing recent advances in the field, how the reported findings move the field forward, the next set of questions raised by the study, and the types of experiments that could explore these questions. Authors are asked to avoid or explain technical jargon to make the big questions in the field, experimental approach, relevant primary literature, and principal findings of the paper accessible to a broad audience. This may involve explaining a new technique or approach that facilitates the study of new biological questions; describing an emerging discipline, which might be at the interface of existing disciplines; setting the record straight on a frequently misunderstood biological phenomenon; or introducing a significant biological question or phenomenon that has recently attracted significant research focus or investment.

18.2 Run-Away Media

Like most other scientific journals, PLoS journals distribute notices to the mass media to publicize a subset of published papers. Press releases are issued for papers with potential interest to a cross-section of researchers, educators, policymakers, and members of the general public. From the beginning, editors took pains to ensure that press releases communicated the principal findings of published results without sensationalizing the implications of the work, hoping to prevent the sensationalized news stories that distort not only the significance of the findings at hand but the very nature of scientific progress. The commitment to providing straightforward, "unembellished" press releases complements PLoS's mission to enhance public understanding of science by featuring educational materials and resources designed to make the concepts, methods, and process of scientific research accessible to a broad audience (with synopses and primers, for example, as described above).

Though PLoS staff typically work with public information officers at authors' institutions to craft evenhanded accounts of the papers published, which are released through the PLoS press list and external media outlets, authors are free to issue their own press releases. In such cases, PLoS has no control over what is written and authors' interests in promoting their work could theoretically lead to overblown claims about its significance. In at least one case, that's just what happened.

In May 2009, PLoS ONE published a paper that described the discovery of a 47-million-year-old primate fossil, *Darwinius masillae,* nicknamed Ida, after the daughter of one of the authors (Franzen et al. 2009). The publication of the paper coincided with a “multi-platform,” multi-city media campaign orchestrated by Atlantic Productions, which calls itself a “factual production” company, on behalf of the authors.[3] The multimedia publicity blitz, which included a book, TV documentary, and Web site,[4] launched with the unveiling of the fossil at New York City’s American Museum of Natural History, with a high-profile press conference hosted by Mayor Michael Bloomberg. The mayor, wearing a T-shirt emblazoned with the title of the History Channel documentary of the discovery, “The Link: This Changes Everything,” stayed on message by calling the remarkably well-preserved fossil a “missing link.” At the press conference, the scientists who described the fossil in the PLoS ONE paper echoed similar statements. Jorn Hurum, paleontologist at the Natural History Museum in Oslo, Norway, which owns the fossil, said, “This fossil rewrites our understanding of the evolution of primates. . . . It will probably be pictured in all the textbooks for the next 100 years.” Coauthor Jens Franzen of the Senckenberg Research Institute of Frankfurt said: “When our results are published, it will be just like an asteroid hitting the earth. [. . .] She is the eighth wonder of the world” (Leake and Harlow 2009). Predictably, a barrage of media reports parroted the press materials, calling Ida a “missing link” and “the missing link that proves Darwinism.”[5]

In the paper, the authors’ claims were more measured. “Note that *Darwinius masillae,* and adapoids contemporary with early tarsioids, could represent a stem group from which later anthropoid primates evolved, but we are not advocating this here, nor do we consider either *Darwinius* or adapoids to be anthropoids” (Franzen et al. 2009). In this passage, the authors note that they are not claiming that the fossil belongs to the class of primates that gave rise to anthropoids – higher primates such as monkeys, apes, and humans – that is, they are not suggesting that *Darwinius* represents an ancestral species in the human lineage. They also state:

> Defining characters of *Darwinius* ally it with early haplorhines [higher primates, including monkeys, apes, and humans] rather than strepsirrhines [including lemurs and lorises]. We do not interpret *Darwinius* as anthropoid, but the adapoid primates it represents deserve more careful comparison with higher primates than they have received in the past (Franzen et al. 2009).

Though the language is somewhat ambiguous, neither passage portrays the fossil as a missing link in human evolution.

Almost immediately after the high-profile media events, scientists and science journalists began to criticize the claim that Ida is a missing link, as it was reported in nearly every mainstream news story about the discovery. Most critics said that while the fossil is an amazing specimen, its characteristics place it among the fossil primates related to lemurs, not apes or humans. This massive publicity operation

---

[3] http://www.atlanticproductions.co.uk/.

[4] http://www.revealingthelink.com/.

[5] http://www.thisislondon.co.uk/standard/article-23695169-scientists-reveal-the-missing-link-that-proves-darwinism.do.

turned what was an exciting discovery of a remarkably preserved ancient fossil – the most complete fossil of such an early primate – into a case study of how not to communicate science. "Many lines of evidence indicate that *Darwinius* has nothing at all to do with human evolution," said Chris Kirk, associate professor of anthropology at the University of Texas at Austin. "Every year, scientists describe new fossils that contribute to our understanding of primate evolution. What's amazing about *Darwinius* is, despite the fact that it's nearly complete, it tells us very little that we didn't already know from fossils of closely related species."[6]

Beyond misrepresenting the significance of the fossil under study, the media hype reinforced misperceptions about how evolution works. The notion of a "missing link" presents the false impression that life evolves along a straight line leading to humans as the crown of creation, rather than as a messy web that branches in all directions, sometimes splitting off into evolutionary dead ends. Even worse, saying that one fossil proves Darwinian evolution is true suggests that mountains of evidence, from the fossil record to molecular and genetic studies, hadn't already established beyond doubt that evolution is scientific fact, the unifying theme of all biology.

In a talk at Oslo's Natural History Museum, Atlantic Productions CEO Anthony Geffen argued that the media attention surrounding Ida was good for science because it raised the public's interest in science. In just a few months, he said, the campaign had managed to get people around the world talking about Ida. "It's an icon for millions and millions and millions of kids that can tell you right now around the world of a way of them thinking about science and thinking about evolution."[7] But many scientists worry that this type of media hype will undermine the public's trust in science and scientists by misrepresenting science and the way scientists gather and evaluate evidence and knowledge. In the case of Ida, the television documentary was scheduled before the paper was written. The press campaign was developed before the paper was submitted. The book was written before the scientific paper was peer reviewed.[8] Now that the media hoopla has passed, will most people have a better grasp of primate history or evolutionary theory? Or will they think that until Ida, scientists had been searching for the missing link in human evolution or, even worse, looking for proof that evolution is true?

18.3 Inviting Public Participation

On a positive note, because the paper was published in PLoS ONE – an open access journal – anyone can read and examine the evidence put forth in the paper. And thanks to advances in Web-based technology, readers can add their own analysis of the results by posting comments and notes that appear online with the original paper.

[6] http://www.utexas.edu/cola/depts/anthropology/news/O14102.

[7] http://www.nhm.uio.no/besokende/faste-utstillinger/ida/seminar/geffen/.

[8] http://www.guardian.co.uk/commentisfree/2009/may/20/fossil-ida-evolution-darwinius.

Anyone interested in learning more about what *Darwinius* does – and doesn't – add to the scientific record, can do so by reading the paper itself along with readers' comments on the paper. As of September 2011, there were nearly 98,000 page views and 13 comment threads, including one about the media hype.[9]

With these Web-based tools, open access online publishing can invite anyone to engage in scientific discourse, allowing the public to participate in a way that was not possible with print journals. Traditionally, print journals invite readers to participate in scholarly discourse by submitting letters to the editor. *PLoS Biology* (along with the other PLoS journals) experimented with an electronic version of letters to the editor with "reader responses," in which comments were submitted to the journal's Web site and staff editors decided whether they constituted valid contributions to the scientific record and should be posted as a response to an article. Although reader responses were published online and anyone could submit a response, responses were handled like traditional letters or correspondence, and required a level of scientific expertise. The new commenting tool, described above, removes the requirement of scientific credentials, allowing anyone to contribute to the discussion, provided the comments are not inflammatory or otherwise inappropriate.

This commenting tool, which invites active participation from readers, was introduced in March 2009 as part of PLoS's "article level metrics," or ALM, program. ALMs include a group of Web-based tools designed to provide more meaningful measures of an individual article's value than the traditional measure, in which articles are judged based on the impact factor of the journal in which it appears.[10] The flaws of judging an individual article's worth based on the impact factor – a measure of the average number of citations for a journal – have been covered elsewhere (Neylon and Wu 2009), but a single statistic shows how little a journal's impact factor reveals about a specific article: *Nature* editor-in-chief Phil Campbell reported that 89% of *Nature's* 2004 impact factor was generated by just 25% of their papers. Article level metrics attempt to address this flawed evaluation approach by developing indicators, collected after an article has been published, that are tied more directly to individual articles.

The tools currently available on all PLoS journal articles include expert ratings, notes within the article, comments, downloads and page views, citations (such as Scopus and PubMedCentral), blog and media coverage, and social bookmarks (for example, CiteULike and Connotea). By providing the means to capture data on every article published in the journal, these tools provide the means to reveal an individual article's utility, value, and impact on multiple levels. The ALM program is still in the early phase of development, and PLoS is still using these tools to acquire data and present it in the form of dynamic cumulative statistics (such as reader views, downloads, ratings, citations). This data is also available for anyone to access, so that third parties can verify and replicate the data, or even better, develop new uses for it.

---

[9] http://www.plosone.org/article/comments/info%3Adoi%2F10.1371%2Fjournal.pone.0005723.

[10] Research in Context, http://article-level-metrics.plos.org/.

The primary goal of article level metrics is to provide the scientific community with more appropriate indicators to determine the value, quality, and impact of individual studies, but thanks to the tools that collect these metrics, their potential applications are far wider. *PLoS Biology* is developing an education series designed to help teachers bring cutting-edge research into undergraduate classrooms, by using research articles published in open-access journals as the basis for exploring fundamental questions in biology. The Web-based tools developed for PLoS's article level metrics program offer educators a unique opportunity to annotate these articles with a range of teaching materials – from online tutorials, archived videotaped seminars and lectures to demonstrations, animations, and interactive media – that provide context, explain complicated concepts, and explore the implications of the research. Such efforts will make the lessons and insights of dedicated teachers available to anyone with an interest in the topic.

As these educational resources increase the accessibility of more and more of the primary research literature, the data collected through the article level metrics program may well lay the foundation for a novel social experiment.[11] Online open access journals can reach anyone with access to a computer. And anyone – regardless of expertise, scientific standing, or institutional affiliation – can use tools developed for the article level metrics program to add notes, comments, and ratings or write blogs that "trackback" to any scientific paper.[12] By allowing individuals outside the traditional confines of the scientific community to eavesdrop on and even join scientists' conversations, online annotating and commenting tools invite public participation in an arena that has traditionally inhabited a world apart.

Whether this (unintentional) democratization of scholarly communication can influence scientific discourse or change the way scientists think remains to be seen. But social scientists have a new opportunity to find out: PLoS makes the entire dataset for all article level metrics across all journals available as a summary Excel file, which is updated periodically.[13] Because these data are openly provided and publicly accessible, social scientists can begin to study the nature of this novel form of public participation in scholarly communication and ask new questions about how or if the entrée of the "extra scientific" public into scientific discussions affects scientific practice.

When the PLoS founders – Harold Varmus, Patrick Brown, and Michael Eisen – announced their foray into publishing "to catalyze a revolution in scientific publishing," they did so hoping that one day everyone with access to a computer and an Internet connection would benefit from the vast wealth of scientific and medical knowledge. "This online public library of science will form a valuable resource

[11] Since PLoS launched Article-Level Metrics in March 2009, several different groups and individuals have independently analyzed various aspects of the dataset. Here is a sample of these efforts: http://article-level-metrics.plos.org/.

[12] Guidelines for Notes, Comments, and Corrections, http://www.plosbiology.org/static/comment Guidelines. action.

[13] Future releases of the platform may automate this function and/or provide data via an application program interface.

for science education, lead to more informed healthcare decisions by doctors and patients, level the playing field for scientists in smaller or less wealthy institutions, and ensure that no one will be unable to read an important paper," the founders wrote in the inaugural issue of *PLoS Biology* (Brown et al. 2003). As more research articles acquire supplemental materials designed to increase the intellectual accessibility of the concepts, methods, and practice of science, the ever-expanding online library of scientific literature – and the conversations among an expansive scientific community – may well begin to realize its full potential to accelerate discovery and push the boundaries of human knowledge.

References

Bernstein, P., B. Cohen, C. MacCallum, et al. (2003). *PLoS Biology* – We're open. *PLoS Biology*, 1(1), e34. doi:10.1371/journal.pbio.000003.pbio.0000034.

Brown, P. O., M. B. Eisen, and H. E. Varmus (2003). Why PLoS became a publisher. *PLoS Biology*, 1(1), e36. doi:10.1371/journal.pbio.0000036.

Franzen, J. L., P. D. Gingerich, J. Habersetzer, et al. (2009). Complete primate skeleton from the middle Eocene of Messel in Germany: Morphology and paleobiology. *PLoS One*, 4(5), e5723. doi:10.1371/journal.pone.0005723.

Joseph, H. D. (2008). Testimony on behalf of the Scholarly Publishing and Academic Resources Coalition (SPARC) alliance for taxpayer access association of research libraries before the subcommittee on courts, the internet, and intellectual property committee on the judiciary regarding H.R 6845, The fair copyright in research works act. http://judiciary.house.gov/hearings/pdf/Joseph080911.pdf (last accessed on August 19, 2010).

Leake, J. and J. Harlow (2009). Origin of the specious – Times Online. London: *The Sunday Times*. http://www.timesonline.co.uk/tol/news/uk/science/article6350095.ece (last accessed on August 19, 2010).

Neylon, C. and S. Wu (2009). Article-level metrics and the evolution of scientific impact. *PLoS Biology*, 7(11), e1000242. doi:10.1371/journal.pbio.1000242.

Rowe, S. B. (2000). Food for thought: An in-depth look at how the media report. nutrition, food safety, and health. *AMWA Journal*, 15(4), 5–9.

Part VII
Conclusions

Chapter 19
Dimensions of Medialization. Concluding Remarks

Peter Weingart, Simone Rödder, and Martina Franzen

19.1 Theoretical Considerations

The studies assembled in this volume focus on the same phenomenon: the relation of scientists, scientific organizations and scholarly communication to the mass media. This amounts to an extension of the public to which communication from science is directed – from the public of disciplinary specialists to a mostly unspecified public addressed by the mass media. As a starting point for empirical investigation, all authors agree on the relevance of the mass media as the major arena for the representation and framing of public opinion in modern society, and thus as conduit to different publics. This agreement exists irrespective of the varying interpretations that are presented in the individual chapters.

The extension of publics appears to be a general characteristic of societies labelled "media societies" because of the pervasive influence of the modern mass media and the effects not only on science but also on all other spheres. However, medialization assumes very different forms in each of them.

Shinn and Whitley (1985) have argued convincingly that the demarcation between the internal and external publics to which scientific communication is addressed is not as sharp as especially the scientists make it to be. Following Fleck, they argue that any communication between disciplines and even between specialized research areas pre-supposes a certain "translation" (or popularization) of one specialized language into another. Thus, they suggest replacing the traditional communication model with its sharp division between scientific and popularizing communication by a continuum ranging from purely internal to popular communication. In other words, the thrust of their argument was to question the specificity and inaccessibility of scientific communication which has also served to legitimize a lack of accountability. This has paved the way for carrying the analysis further and to respond to more recent developments because some conditions have changed

P. Weingart (✉)
Bielefeld University
e-mail: weingart@uni-bielefeld.de

S. Rödder et al. (eds.), *The Sciences' Media Connection – Public Communication and its Repercussions*, Sociology of the Sciences Yearbook 28,
DOI 10.1007/978-94-007-2085-5_19, © Springer Science+Business Media B.V. 2012

since then. While Shinn and Whitley were concerned with tearing down conceptual barriers in the study of communication within science and between disciplines, we have focused on the effects of communication by science to external audiences on science itself.

It may now be claimed that the extension of the publics of science is a necessity considering the precarious position of science in mass democracies. The opening to the public is necessary because the pressure on the institution increases: Its costs are rising steadily (regardless of whether the burden is borne primarily by state governments or by industry) and, perhaps even more important, the immediate relevance of its product for society is a matter of contention. Beneficial effects are recognized by the public but so are risks created by new knowledge.

Science thus has a multifaceted problem of legitimacy. But the solution is obviously not simply to "open" the institution to public participation (e.g., voting). Functional differentiation and specialization in the mastering of knowledge prevent an easy translation of public interests into research problems, let alone how to do the research. The often heard call for the "democratization" of science is an expression of this problem of legitimacy, and communicating with the public is seen as a means to secure that legitimacy.

Some argue that public engagement with science and technology will and should make scientific knowledge "socially robust" (Nowotny et al. 2001) and the research enterprise accountable (Jasanoff 2003). A variety of participatory formats like consensus conferences or citizen juries bear witness to a participatory turn in science and technology policies. Participatory procedures have been institutionalized in science policy since the 1980s to assess the risks and benefits of new technologies by those who will be affected (Joss and Durant 1995). While the efficiency of standardized procedures is contested, improvements for processes and methods have been proposed to achieve the aim of more accountability and credible assessments of science due to wider public engagement (Jasanoff 2003). Against the background of uncertainties in knowledge production and the evolving risks of new technologies it has been argued that public participation is urgently required (e.g., the concept of "extended peer review" as promulgated by Funtowicz and Ravetz 1993). Nowotny et al. (2001) claim that science's cognitive authority has been eroded (mode-2 science) and that the quality and utility of research are today negotiated in the "agora", a problem-generating and problem-solving environment where scientific knowledge is contextualized by multiple actors. Together with the hope for a socially robust knowledge production this contextualization, in which the mass media play a relevant part, is simultaneously regarded as an end to the autonomy of science as a separate institution. This implies a dilution of criteria in the assessment of science, which some see as a danger to the quality and reliability of scientific knowledge (e.g., Weingart 2001) while others argue that the quality of "post-academic" or "post-normal" science cannot be assured anymore by the peer review system alone (e.g., Funtowicz and Ravetz 1993).

It is an empirical question whether science's autonomy in defining the research agenda and assessing scientific quality by epistemic criteria actually erodes "when the context speaks back" (Nowotny et al. 2001). Both, hopes and concerns, point to changes in the processes of knowledge production welcomed by some as a democratization of science that is based on the involvement of the lay public in knowledge production or seen as a threat to the credibility of science as a source of reliable knowledge. It is premature to interpret anecdotal evidence or organizational transformations as a reversal of the functional differentiation of society. But it is plausible to assume that the anticipation of the mass media's news value criteria backfires on science on different levels. We have proposed the concept of "resonance" to capture adaptations as well as resistance to medialization and thus the processing of differences between systems (see Chapter 2). As argued above (see Chapter 1), medialization as an analytical concept points to the dilemma in which scientists and scientific organizations find themselves because having to respond to expectations to communicate with the general public has become part of their legitimating exercises. This is also indicated by a terminological shift from the paternalistic "popularization" to the more neutral "science communication". However, differences exist over the reach and implications of these demands.

In the chapters of this book, resonances of the medialization of science are described and interpreted with different concepts and within different theoretical frameworks. In addition to the perspective of systems theory, the distinctions between "front stage" and "backstage" (1), and between "talk" and "action" (2), are applied to enhance our understanding of the sciences' media connection.

(1) The assumption underlying the first, theatrical metaphor is that the front stage is the place where actors perform in front of and addressing a broad public, the understanding on both sides being that what is happening on the front stage is a presentation, to some degree fictitious (Goffman 1959). It is, in any case, not to be mistaken with what happens backstage, invisible to the audience. The construct of a presentation on the front stage distinguished from the "reality" of the backstage implies, as an analytical problem, the effects that the activity on the front stage has on the backstage activity. The distinction would not be very interesting by itself if it were not for the repercussions of "mere" presentation on the "actual" activity as such.
(2) Similarly, the distinction between "talk" and "action" suggests that the legitimation and rationalization of motives, decisions and behaviour does not correspond with the action to which it refers (Brunsson 1989). Again, "talk" in this context pertains to "presentation" to a public, akin to public relations, propaganda, but also, less pejoratively, to appeal, impress, convince and to persuade. It shields "action", i.e., what is actually done and happens, from direct scrutiny. One may discard mission statements of universities as "just" talk directed to policymakers and the public but the question is if this talk will remain without any resonance on the actual activities of the organization. The communication of press and public relations offices of universities, research institutes or scientific journals to the media could be denigrated as "mere" propaganda. But can this propaganda be completely insulated from what it is supposed to represent?

19.2 Empirical Findings

The chapters of this volume all point to the same question: What happens to science as a social institution responsible for the production of reliable knowledge under the influence of the mass media? With the duality of hopes and concerns in mind, we now look systematically at empirical evidence. What do the data tell us, first, with regard to hopes for a democratizing impact of medialization?

To begin with, we may take the quantity of media coverage as a symptom of medialization conducive to provide transparency and public insight into what is going on in science. Case studies of science news reporting on a single scientific field or a particular debate usually demonstrate high levels of mass media attention (for reviews, see Lewenstein 1995; Schäfer 2010). But contrary to impressions from individual cases, a longitudinal analysis of science news in periodicals, newspapers and books covering the years between 1820 and 2006 reveals that coverage does not rise continuously but rather follows contingent developments and shifting discourses in society. *Martin Bauer*'s picture of long-term fluctuation in the public attention to science, which he gains from combining the quantitative evidence from several studies, surely needs to be interpreted with caution and tested in other contexts. Nonetheless, the quantitative picture suggests "that the most recent period of public attention to science, which began in the 1980s, is indeed historical. It is longer than any other wave of expansion in the past 200 years, and the amplitude is without precedent" (Chapter 3, p. 53).

With regard to hopes for a democratizing impact of high media attention, however, other chapters gather rather disillusioning evidence. *Richard Elliot* shows for UK news coverage of regenerative medicine that more intense coverage does not imply that participatory claims and their resonance in science also increase. On the contrary, the framing in terms of "scientific revolutions" and "medical miracles" suggests that journalists do not challenge scientific authority but tend to reinforce and amplify the optimistic predictions of scientific and clinical experts. Neither does more intense coverage necessarily lead to a loss of credibility of scientific knowledge, as one might expect as a paradoxical effect of a closer science-media coupling. Rather, a re-stabilization of the normative and cognitive expectations placed on science can be observed in the news reporting. *Arlena Jung* shows by way of an in-depth analysis of the frames in news coverage of the contested field of stem cell research in a comparison with epidemiology in the German press that the epistemological status of scientific knowledge is not called into question; the general public does not change its normative expectations of scientific knowledge as reliable and trustworthy in general. Instead the credibility of science is called into question in concrete cases.

Likewise, *Joan Haran* focuses her interest on the issue of public trust in science. By referring to a unique organization at the science/media interface, the UK Science Media Centre (see Chapter 13), she critically assesses how the public of science is constructed and how strategic communication tasks are defined in the particular cases of cloning and stem cell science. In terms of democratic governance of science, she identifies a tension in the Science Media Centre's communication strategy that "erases the public as stakeholders through its narrow focus on policymakers and communications practitioners". Haran concludes: "Whether informing and engaging the public about science could ever be affected through media relations is an open question" (p. 255).

Hopes for open debates and more public participation are specifically voiced with regard to new media formats. But emancipatory claims that are linked to the internet as such, and in particular to web 2.0 formats such as blogs, remain an illusion as the account by *Brian Trench* shows. Trench takes his analysis of "climate-gate", the hacking of emails at a Climatic Research Unit in the UK, as evidence of the sensitivity and uncontrollability of the new media. Disseminated through blogs as well as traditional mass media, the leak of emails triggered widespread reactions from scientists and laymen alike and a discussion on the purported manipulation of climate data by the Intergovernmental Panel on Climate Change (IPCC). New media formats thus enable public scrutiny of scientific communication, but the singular example may be a precursor of a rather questionable development. Instead of the rise of an "extended peer review" (Funtowicz and Ravetz 1993), the new media are abused for conversations that "easily degenerate into name-calling or focus on trivial aspects" (Chapter 14, p. 286). This may well "deter scientists from engaging in an online medium that is inherently predisposed to personal, affective communication" (Chapter 14, p. 287) and will eventually prevent scientists from making use of these formats.

Furthermore, novel forms of scientific publishing have emerged with the internet. Open access is the buzzword – making scientific publications accessible online without any subscription fees. One of the successful endeavours of open-access-journals is the *Public Library of Science* (PLoS). This non-profit organization started with the biomedical journal *PLoS Biology* in 2003 and today comprises seven journals. From her perspective as a senior editor with *PLoS Biology*, *Liza Gross* presents its ambitious mission to provide not only physical but also intellectual access to the journal's issues for a broader public. In what ways the internet public reacts to what PLoS has to offer remains to be seen, and studied. In this regard, the data that the journal makes available on every article published are a valuable resource for social scientists. But while online annotating and commenting tools invite public participation and data on every article's utility, value and impact are collected on multiple levels. A quote from the organization's core principles phrases the aim of the game in the familiar PUS sound:

> Our mission of building a public library of science includes not only providing unrestricted access to scientific research ideas and discoveries, but developing tools and materials to engage the interest and imagination of the public and helping non-scientists to understand and enjoy scientific discoveries and the scientific process (Chapter 18, p. 354).

A deficit-model approach is also still the dominant model underlying scientist's attitudes towards science communication, as another chapter suggests on the basis of international survey data. This attitude is accompanied, however, by a widespread acceptance and anticipation of media criteria among scientists and press officers in research institutions. *Hans Peter Peters* interprets this and other findings from a wealth of data as a "media-based governance mechanism" for science which "might even be considered to be participatory in nature, involving the 'public' in an indirect way in the governance of science" (Chapter 11, p. 238).

But how does this relate to findings that, for most periods in science news coverage, the scientific agenda determines the agenda of the mass media – and not the other way round – (Rödder and Schäfer 2010), despite the selectivity of news values? How the mass media report on science is a concern from the perspective of news value research that has so far been hardly studied. In their chapter, *Franziska Badenschier* and *Holger Wormer* combine a practical, a theoretical and an empirical approach to determine the applicability of the general news values concept, which was originally developed from content analyses of political journalism, for science coverage. Their analyses point to the need to distinguish between science news outside specific science pages and science journalism that is triggered by genuinely scientific events. While the common catalogue of news values is appropriate to describe the selection of science news in the first case, i.e., motivated by the daily news agenda, more specific factors describe the selection processes for news from science. Examples are "surprise" and "considerations of scientific relevance".

To sum up, the hopes for democratization that are often connected with mass media communication about science and especially the emancipatory claims attached to new media formats and developments, such as open access, remain for the most part wishful thinking. Even though a high level of traditional media attention (at least for certain fields) and new media formats extend the communication opportunities between science and the public, this does not necessarily imply that the underlying communication models shift from the deficit-model of the Public Understanding of Science movement to more participatory approaches to science communication.

What can we say, then, with regard to concerns that the front stage interferes with the backstage, that PR takes over or that talk determines action? Conflicting expectations that scientists and science communicators in institutional PR offices encounter in their interactions with journalists are discussed and evaluated in a number of chapters; on the basis of quantitative, qualitative and anecdotal evidence from several countries.

The fossil IDA that we refer to in the introduction as a case that raised widespread worries among scientists, science editors and science journalists alike is discussed in detail by *Liza Gross* as an instance of "how not to communicate science" (Chapter 18, p. 357). The IDA case is a prime example of a media event that involved strategic communication by scientists and where one could observe the aggrandisement of claims in the media spotlight, claims that were not backed up by the scientists' own journal publication. But this case apart, is there evidence for a trend that media visibility is by now incorporated in the professional role of scientists?

On the basis of a survey of stem cell researchers and epidemiologists, *Hans Peter Peters* infers that media contacts by researchers are by now institutionalized, implying that their organizational context, i.e., their affiliation with a university etc., becomes more important than the norms of scientific communities. Peters' findings show, however, that peer responses to media visibility remain context-sensitive, despite a catalytic effect of institutional PR. These ambiguities reappear in the contribution by *Paul Nolte*, himself a social scientist with a high profile in German media. His personal account suggests that colleagues' allegations "that media contact corrupts scholarship and damages one's academic integrity" (Chapter 9, p. 186) do not avert media visibility per se but induce justification strategies. Reasons such as "being asked by the media" surface in Nolte's account, when he justifies his public role at the same time as describing it.

This is in line with evidence from interviews with researchers of the publicly and privately funded Human Genome Projects in France, Germany, the UK and the USA that reveal a deep ambivalence of scientists towards highly visible colleagues. Visibility is in need of justification because publics other than scientific peers are not the audience to judge scientific claims and thus not the addressees of a scientist's primary role as researcher. *Simone Rödder* concludes from her analysis of the genome case that media visibility is not an option for the average researcher, and it is policy talk rather than empirical fact that it is by now built into the role expectation of every professional scientist, including being eligible for reputation.

Looking at the same case – the high profile competition between a public consortium and a private firm to finish the first draft of the human genome – *Stephen Hilgartner* applies a dramaturgical perspective to reveal strategic action with respect to science communication: "Conceiving of media orientation as a form of theatrical self-consciousness casts a spotlight on questions about precisely what actors seek to make visible to whom and when" (Chapter 10, p. 213). Hilgartner's study of different facets of interaction between scientists and the media in the Genome Project focuses on the theatrical metaphor of the actor and distinguishes four roles of media orientation: actors shaping media coverage, actors as self-conscious members of the audience, actors as commentators of media coverage and actors as builders of media relations infrastructure.

A differentiation of social roles, this time based on a differentiation theory perspective, also allows probing a little deeper with regard to the difficulty to draw a line between scientific and popularized knowledge. In his chapter, *Fran Osrecki* distinguishes three roles: the media expert, the public intellectual and the hybrid role of the "diagnostician of the present state of society". The corresponding types of texts that scientists produce in these roles are categorized into the genres "general (social) theories" and "diagnoses of the present". The latter is characterized by an argumentative structure that serves the mass media's need for surprise and novelty with regard to the research object itself – the new "XY-society". Such diagnoses may influence theories and the academic discourse in general and are thus examples of a resonance to media expectations in the social sciences.

The aspects of strategic communication, legitimation and dissemination of information certainly underlie the establishment of science PR on the organizational level. A transformation in the identity of PR staff – from a scientific self-understanding via a journalistic perception of public relations to a professional understanding of the PR job – is evident in the use of marketing rhetoric such as "branding" (Chapter 11). Several types of PR activities at the interface of science and the mass media can be differentiated. One is the public relations departments of universities and research institutes. A second type is the press offices of scholarly journals. A third type includes the establishment of a national public relations agency such as the Science Media Centre (SMC) in the UK, working to promote the voices and views of the scientific community to the national news media "when science hits the headlines" (Chapter 13, p. 257).

From her practitioner's perspective as director of the SMC, *Fiona Fox* gives insights into the history, philosophy and operation modes of the centre, which is often considered as an international model for media relations activity in science. The institutionally independent SMC supports scientists to engage with public debates and learn the rules of the media game. The SMC is pro-active in anticipating science news stories by regular press briefings and comments to embargoed papers. It also provides an expert service of "media-friendly scientists" (Chapter 13, p. 261). According to Fox, the underlying task is to improve the quality of science reporting by facilitating a journalistic balance in offering conflicting views from eminent scientists on a range of scientific issues. In the aftermath of the UK GM food debate, the SMC wants

> to ensure that never again will the UK have a national media debate about an area of science without the best scientists taking up their rightful part in that debate. If at the end of discussions the UK still says no to GM then that is simply a healthy democracy (Chapter 13, p. 260–261).

However, whether the SMC's "pro-science" approach can really help to restore public trust in science is contested by Haran (Chapter 12).

Ulrike Felt and *Maximilian Fochler* widen the scope of medialization beyond the classic media to a plurality of contexts and formats that are governed by "similar logics" (Chapter 7, p. 134). They observe a strategic use of the media by academic scientists to secure resources and to attract young people to science careers. Their analysis of the Austrian life science landscape reveals resonances such as impediments to the development of research fields that have been critically discussed in the public sphere, e.g., green biotechnology. Promotional presentation activities led to a disillusionment of researchers once they realise that glossy brochures and other medialized accounts only convey a front stage image, which is quite distinct from the backstage reality. They conclude from a range of empirical data that

> the feedback of media representations into science, as well as the very practices researchers engage in to medialize their research, influence and change research cultures and practises, along with how the researchers understand what living in and doing research means to them (Chapter 7, p. 151).

The resonance of mass media logics in science is furthermore addressed by looking at the editorial practice and PR activities of scientific journals. *Martina Franzen* argues that the high-impact journals play a decisive role in framing the discourses in scientific fields because scientific and public attention to new findings is structured by the decision-making of different journal types and their promotional activities. The two multidisciplinary journals *Science* and *Nature* are well-known for their PR efforts and media visibility of their publications. *Franzen's* study of the role of these journals in the stem cell debate comes to the conclusion that a conflict between the expectations of scientific rigour, on the one hand, and general interest, respectively newsworthiness, on the other, becomes virulent in these journals and can disturb the communication within science. "If the mission statement of a highly selective journal explicitly states that findings must be of general interest, the authors need to produce new results that will merit recognition by the scientific community and beyond" (Chapter 17, p. 347). However, overstretching the benefits of research results comes as the undesirable side effect of anticipating editorial programs favoring broad impact. This, in turn, may have negative effects on science's public image.

A displacement of scientific communication to newspaper pages is another issue that raises concerns about what happens to scholarly communication. *Susann Wagenknecht* demonstrates in her case study that a recent debate between archaeologists and historians on the status and contemporary significance of the ancient city of Troy was confined to the "culture sections" (Feuilleton) of leading German newspapers with no counterpart in scholarly journals. This outlet provided a much greater visibility than would have been the case with any scientific publication. The relevance of the case, which, although rare, is by no means unique, lies in the fact that the public debate led the discussants to be much harsher in the attacks on their opponents. Wagenknecht observes that this sound bite style prevailed when the debate returned to the academic setting of a conference.

19.3 Conclusions

The accounts presented in this volume reveal a multifaceted picture of the science-media relationship. The chapters demonstrate that one-dimensional indicators are not sufficient to grasp the complexity of the sciences' media connection. As *Hilgartner* (Chapter 10) remarks the public relations efforts of organizations often work to *prevent* media attention. Thus, the pure quantity of media coverage is not indicative of structural changes in science.

The framework of differentiation theory and the concept of "resonance" allow for (functional) analysis on different levels. As assumed, different system levels appear to be susceptible to irritation and subsequent resonance to different degrees. There is plentiful evidence for the establishment of PR on the organizational level that administer the front stage performance or the talk of scientific organizations. Besides organizational adaptations to media interest, the anticipation of the mass media's selection criteria in the core of scientific communication itself is not easy to measure and to interpret. Here, only case-by-case analyses can be applied to distinguish between distinct modes of communication in science and the mass media.

We learn furthermore that there are considerable differences between disciplines. Those fields whose results can be readily commercialized and those which are highly policy-relevant and absorb large amounts of resources are evidently more intensely involved in communicating to the public than those to which these criteria do not apply. But also the humanities and social sciences can attract so much attention from the mass media that scholars are tempted into waging their disputes on newspaper pages and in the argumentative style of mass media texts. With regard to disciplinary differences, it is worth considering Osrecki's objection to the applicability of the medialization concept to the social sciences and the humanities:

> It is mainly a model for analyzing 'hard' sciences, because it presupposes an autonomous core of knowledge production. [...] In social sciences, it is far more difficult to locate such an autonomous sphere and, consequently, to work with the medialization concept. Not because there exist no clear and rigorous social scientific procedures of validation and certification. Social sciences can indeed distinguish between true and false. Yet, unlike in the case of 'hard' sciences, social scientists address the publics of mass media in largely *overlapping* roles (Chapter 16, p. 311).

The chapters come to diverging conclusions in how far media visibility is by now part of the professional role of scientists and whether promotional "talk" is set apart from knowledge production processes (Chapters 7, 8, 11 and 16). More towards the "action" side – with an eye to public impact – is the strategic choice of a model organism such as the colourful butterfly in genetic research (Chapter 11), generating diverging interpretations by scientists, science communicators and by observers of medialization processes. The replacement of the well established model organism *Drosophila melanogaster* by a colourful butterfly poses the question whether the continuity of research is assured. Even if the quality of research on butterflies is not called into question, the suitability may well be. The same can be questioned in stem cell science where the bulk of research from 1998 to 2002 focused on the transdifferentiation of adult stem cells – a politically highly desired method to circumvent the ethical dilemma around the derivation of embryonic stem cells. Experimental articles on this issue were predominantly published in high-impact journals that promise a maximum exposure of the work and the publication titles were uncommonly catchy. Titles such as "Turning Brain into Blood" or "Generalized Potential of Adult Neural Stem Cells" were easily adopted by the mass media and stimulated the debate on stem cell regulation policies worldwide. It later became apparent that the transdifferentiation of adult stem cells did not fulfil what had been promised. Further studies showed that the experiments lacked the appropriate technical methods to prove the argument. The stem cell case illustrates that high visibility can set the research agenda in scientific fields, and it is arguable whether this was a fruitful research experience in order to improve the technical standards or just a blind alley (see Franzen 2011). Of course, the choices of one model organism and of one research method are hardly enough proof of a widespread phenomenon.

Thus, the principal conclusion based on the range of analyses at hand is a twofold one: On the one hand, the hopes attached to science communication having a "democratizing effect" are disappointed. On the other hand, fears that orientation to the media and their logic of selection and representation will be detrimental to the production of reliable knowledge also prove to be exaggerated. Incidents where such effects can be observed are rare and hard to come by.

References

Brunsson, N. (1989). *The organization of hypocrisy. Talk, decisions and actions in organizations*. Chichester: Wiley.

Franzen, M. (2011). *Breaking news: Wissenschaftliche Zeitschriften im Kampf um Aufmerksamkeit. Reihe Wissenschafts- und Technikforschung*. Baden-Baden: Nomos.

Funtowicz, S. O. and J. R. Ravetz (1993). The emergence of post-normal science. In R. V. Schomberg (ed.), *Science, politics and morality: Scientific uncertainty and decision making*. Dordrecht: Kluwer, pp. 85–123.

Goffman, E. (1959). *The presentation of self in everyday life*. New York: Doubleday Anchor.

Jasanoff, S. (2003). Technologies of humility: Citizen participation in governing science. *Minerva*, 41, 223–244.

Joss, S. and J. Durant (eds.) (1995). *Public participation in science. Consensus conferences in Europe*. London: Science Museum.

Lewenstein, B. (1995) Science and the media. In S. Jasanoff et al. (eds.), *Handbook of science and technology studies*. Thousand Oaks: Sage, pp. 343–360.

Nowotny, H., P. Scott, and M. Gibbons (2001). *Re-thinking science. Knowledge and the public in an age of uncertainty*. Cambridge, UK: Cambridge University Press.

Rödder, S. and M. S. Schäfer (2010). Repercussion and resistance: An empirical study on the interrelation between science and mass media. *Communications*, 35, 249–267.

Schäfer, M. S. (2010). Taking stock: A meta-analysis of studies on the media's coverage of science. *Public Understanding of Science*. Published online before print December 1, 2010, doi: 10.1177/0963662510387559.

Shinn, T. and R. Whitley (1985). *Expository science: Forms and functions of popularisation*. Sociology of the Sciences Yearbook IX. Dordrecht: Reidel.

Weingart, P. (2001). *Die Stunde der Wahrheit? Zum Verhältnis der Wissenschaft zu Politik, Wirtschaft und den Medien in der Wissensgesellschaft*. Weilerswist: Velbrück Wissenschaft.

References

Brunsson, N. (1989). *The organization of hypocrisy: Talk, decisions and actions in organizations*. Chichester: Wiley.

[illegible]

Biographical Notes

Franziska Badenschier studied science journalism at the Universities of Dortmund (Germany) and Strasbourg (France) and was a trainee with *Spiegel Online*. In 2010, she finished her master in science journalism studies. Currently she is working as freelance science journalist for print, online, and radio media. In the journalistic practice she mainly covers medicine, biology, chemistry, life sciences, and food sciences. Her fields of interest in journalism studies are the development of science journalism and the quality of science journalism especially in terms of research sources for science journalists, style of writing and narrative style. Her first scientific paper dealt with the boom of science coverage in German newspapers (in: J&MC Quarterly 2008; with C. Elmer and H. Wormer).

Martin W. Bauer read Psychology and Economic History (Bern, Zurich and London) and is Professor of Social Psychology and Research Methodology at the London School of Economics (LSE). A former Research Fellow of the Science Museum, he currently directs the MSc Social and Public Communication, edits the Journal *Public Understanding of Science*, and is a regular visitor to Brazil. He investigates science in common sense through theory building, attitude surveys, media monitoring and qualitative enquires. Publications include "Genomics and Society" (Earthscan 2006, with G. Gaskell); "Journalism, Science and Society" (Routledge 2007, with M. Bucchi); and forthcoming "Atoms, Bytes and Genes – Public Resistance and TechnoScientific Responses" (Routledge) with papers in *Nature*, *Science*, *Nature Biotechnology*, *PUS*, *Genetics and Society*, SSS, IJPOR, *Science Communication* and DIOGENE.

Richard Elliott is currently completing his PhD at the Institute for Science and Society, University of Nottingham. His thesis considers public and media perceptions of regenerative medicine and builds on his background studying molecular biology and science communication at the universities of Edinburgh and Bath. He has also worked for Nature Publishing Group and the British Science Association and interned with the American Association for the Advancement of Science. His research interests include the sociology of expectations and the public understanding of science and he recently co-edited the book: "Communicating Biological Sciences – Ethical and Metaphorical Dimensions" (Ashgate 2009).

S. Rödder et al. (eds.), *The Sciences' Media Connection – Public Communication and its Repercussions*, Sociology of the Sciences Yearbook 28,
DOI 10.1007/978-94-007-2085-5, © Springer Science+Business Media B.V. 2012

Ulrike Felt has studied physics and mathematics finishing her PhD in 1983. She then moved to the field of science and technology studies. Since 1999 she is professor for social studies of science and head of the department for social studies of science at the University of Vienna. Her research interest is mainly focused on changing knowledge cultures and their institutional (especially academic) dimensions; science communication and engagement; science, democracy and governance; Ethical, Legal, and Social Aspects of Sciences and Technologies (ELSA) research. Her work is often comparative between national context and technological or scientific fields (especially life sciences, biomedicine and nanotechnologies). From 2002 to 2007, she has been the editor of the Journal *Science, Technology, and Human Values*.

Maximilian Fochler was trained as a sociologist and finished his dissertation in the social studies of science in 2007. He currently is recipient of an Apart fellowship of the Austrian Academy of Sciences at the Department of Social Studies of Science of the University of Vienna. His research focuses on three aspects of techno-scientific innovation in modern knowledge societies: the shifting cultures and practices of research in academia and in hybrid spaces between academia and business; the governance of innovation; and the relations between technosciences and their publics, especially in mechanisms of public engagement with emerging technosciences. He develops these lines of research in projects dealing with the life sciences and their ethical and social implications.

Fiona Fox has a degree in journalism and 24 years of experience in working in media relations for high profile national organisations. Her career includes stints working for the Equal Opportunities Committee, National Council for One Parent Families, and CAFOD (a leading aid agency). Despite having no background in science, Fiona managed to persuade a distinguished panel of eminent scientists to take a risk and appoint her to become the founding Director of the Science Media Centre which opened in April 2002. The main remit of the Centre is to help restore public trust in science by persuading more scientists to engage more effectively with the big controversial science stories that hit the headlines.

Martina Franzen studied sociology, philosophy and German literature and language studies in Kiel and Bremen. In 2010 she finished her dissertation in sociology on the role of scientific journals at the interface of science and the media at Bielefeld University. Currently she is a researcher at the Institute for Science and Technology Studies (IWT) at the University of Bielefeld and co-heads a research project on medialization as a specific governance structure of science as part of the programme "New governance of science," commissioned by the German Federal Ministry of Education and Research. Her fields of interest cover the sociology of science and the media, peer review and scientific misbehavior. Recent publications include "Breaking News: Wissenschaftliche Zeitschriften im Kampf um Aufmerksamkeit" (Baden-Baden: Nomos 2011).

Liza Gross edits the "magazine" section of the biomedical journal *PLoS Biology*, where she recently launched a new series featuring public engagement in science initiatives, edited by Claire Marris and Nikolas Rose. Before joining *PLoS*, she experimented with novel approaches to public understanding of science as lead writer for a new collection of biology exhibits at the Exploratorium, a hands-on science museum. She continues to find inspiration in the vision of Exploratorium founding director Frank Oppenheimer, who thought that giving people the tools to understand science would not only help them understand the world around them but also inspire them to work to make it a better place. Gross also works as a freelance science journalist.

Joan Haran is a Research Fellow in Media, Culture and Genomics at Cesagen, Cardiff University. She is co-author of "Human Cloning in the Media: From Science Fiction to Science Practice," and is currently working on a monograph provisionally entitled "Genetic Fictions: Science, Gender and Genre." Her research is concerned with gender and genre boundaries in the representation of technoscience, including the ways in which "experts" mobilise claims about the fictiveness of the claims and/or the irrational subjectivity of those perceived to threaten the legitimacy of their expertise. She has a long-standing interest in the sociology of human reproduction and – what are now known as – assisted reproductive technologies.

Stephen Hilgartner is Associate Professor and Chair of the Department of Science and Technology Studies, Cornell University. His research examines the social dimensions and politics of contemporary and emerging science and technology. His book *Science on Stage: Expert Advice as Public Drama* – which examines how the authority of scientific advisory bodies is produced, contested, and maintained – won the Carson Prize from the Society for Social Studies of Science in 2002. His ethnographic research on genomics focuses on access, ownership, and control. Recent publications include "Intellectual Property and the Politics of Emerging Technology" (*Chicago-Kent Law Review* 2010), a chapter on expert knowledge about risk (in *Comunicar los riesgos*, Biblioteca Nueva 2009), a special issue in *Science and Public Policy* (October 2008) on anticipatory knowledge, and a forthcoming paper on "Selective Flows of Knowledge in Technoscientific Interaction" to appear in the *British Journal for History of Science*.

Arlena Jung finished her dissertation on communication processes between science and politics in the American bioethics council "The President's Council on Bioethics" in 2007 at the Institute for Science and Technology Studies (IWT) and the department of sociology at Bielefeld University. She is currently working in the research project "Scientization of society or socialisation of science?". Her main fields of research are the interface between science, mass media and politics, theories of the public and mass media, systems theory and phenomenology. Current publications include "Identität und Differenz. Sinnprobleme der differenzlogischen Systemtheorie" (Bielefeld: transcript 2009).

Paul Nolte is a professor of modern and contemporary history at the Free University Berlin. Previously, he has held positions at Bielefeld University, where he also finished his Ph.D. in 1993 and his Habilitation in 1999, and at Jacobs University, Bremen. He has been a fellow at the Center for European Studies, Harvard

University, and at the Center for Advanced Study (Wissenschaftskolleg) Berlin. Most recently, he was a Visiting Professor at the University of North Carolina, Chapel Hill, in 2010–11. Nolte's interests focus on the postwar social and political history of Germany, with current projects on the development of the German middle classes, and on the 20th century transformation of democracy. His most recent book publication is "Religion und Bürgergesellschaft" (Religion and Civil Society), published with Berlin University Press in 2009. He is also the chief editor of *Geschichte und Gesellschaft. Zeitschrift für Historische Sozialwissenschaft*, a leading peer-reviewed journal.

Fran Osrecki studied sociology and science and technology studies at the University of Vienna. In 2010 he finished his dissertation in sociology at the University of Bielefeld. His fields of interest cover the sociology of science, sociological theory, and the sociology of organizations. During the last years his main focus of research was the relationship of social sciences and mass media. Recent publications include "Die Diagnosegesellschaft: Zeitdiagnostik zwischen Soziologie und medialer Popularität" (Bielefeld: transcript 2011).

Hans Peter Peters studied social science and physics at the universities of Cologne and Bochum. In 1984 he finished his dissertation in social science. Currently he is a researcher at the Institute of Neuroscience and Medicine: Ethics in the Neurosciences (INM-8), Research Center Jülich, Germany, and Adjunct Professor of Science Journalism at the Institute for Media and Communication Studies, Free University of Berlin. His research deals with public communication of science and technology, in particular with the science-media interface, and with social contexts of the neurosciences. Recent publications include "Science Communication: Interactions with the Mass Media" (*Science*, 321, 2008) and "Scientists as Public Experts" (in: Handbook of Public Communication of Science and Technology, ed. by M. Bucchi and B. Trench, New York: Routledge 2008).

Simone Rödder works at the Institute for Science and Technology Studies (IWT), Bielefeld University, where she currently co-heads a project on the presentation and production of knowledge in mathematics, molecular biology and recent history as part of the programme "New governance of science," commissioned by the German Federal Ministry for Education and Research (BMBF). Simone's research is located at the intersection of the sociology of science and the sociology of the media and includes theory and practice of science journalism and science communication. Simone has an academic background in biology, sociology and science communication and holds an interdisciplinary Ph.D. from the University of Bielefeld. She has been trained as a journalist and worked for several newspapers and magazines.

Brian Trench studied languages at Trinity College Dublin and cultural sociology at Université de Bordeaux and University of Birmingham. He worked for 20 years as a journalist in magazines, newspapers and broadcasting, specialising in later years in science and technology. He joined Dublin City University as a journalism lecturer in 1992 and helped set up a Masters in Science Communication there in 1996. He has lectured and published widely on science communication education and training, on Internet journalism and Internet communication in science and on communication models in science communication. He retired from Dublin City University in 2010 but remains active in EU-funded projects on public engagement with research and scientific culture. He is co-editor (with M. Bucchi) of "Handbook of Public Communication of Science and Technology" (Routledge 2008).

Susann Wagenknecht graduated in history, philosophy and sociology of science from Bielefeld University in 2010. She is currently enrolled at Aarhus University Graduate School of Science and is working on a dissertation about "Scientific Practice as Collective Endeavour" (working title) in the Department of Science Studies.

Peter Weingart is Professor Emeritus of Sociology and Science Policy at the University of Bielefeld, Germany and former director of the Institute for Science and Technology Studies (IWT 1993–2009) as well as of the Center for Interdisciplinary Research (ZiF 1989–1994). He is Visiting Professor University of Stellenbosch (since 1994), a fellow of STIAS, member of the Berlin-Brandenburg Academy of Sciences as well as the Academy of Engineering Sciences (acatech). Current research interests are science advice to politics, science-media relation, and science communication. He assumed the editorship of MINERVA in 2007. Recent publications: with J. Lentsch (eds.): "Wissen – Beraten – Entscheiden. Form und Funktion wissenschaftlicher Politikberatung in Deutschland" (Weilerswist: Velbrück Wissenschaft 2008); "Die Stunde der Wahrheit" (Weilerswist: Velbrück Wissenschaft 2005); with B. Hüppauf (eds.) "Science Images and Popular Images of the Sciences." Routledge Studies in Science, Technology and Society (London: Routledge 2007).

Holger Wormer studied chemistry and philosophy in Heidelberg, Ulm and Lyon. Between 1996 and 2004 he was science editor at the German nationwide newspaper *Süddeutsche Zeitung*. Since 2004, he is an ordinary Professor of Science Journalism at TU Dortmund University where he has developed the only complete (bachelor and master) course in science journalism at a German University. Until today he has taught several hundreds of journalists and scientists in professional trainings for science journalism and communication. He is member of several juries, such as for the science communication prize of the German Science Foundation (DFG). His research projects focus mainly on the quality of science reporting (latest project: www.medien-doktor.de). For his journalistic work he got two journalism awards.

MIX
Papier aus verantwortungsvollen Quellen
Paper from responsible sources
FSC® C105338

If you have any concerns about our products,
you can contact us on
ProductSafety@springernature.com

In case Publisher is established outside the EU,
the EU authorized representative is:
Springer Nature Customer Service Center GmbH
Europaplatz 3, 69115 Heidelberg, Germany

Printed by Libri Plureos GmbH
in Hamburg, Germany